D0986848

The Compton Effect

Stephen Deutch photo / Courtesy of AIP Niels Bohr Library

THE Compton Effect

TURNING POINT IN PHYSICS

ROGER H. STUEWER
UNIVERSITY OF MINNESOTA

SCIENCE HISTORY PUBLICATIONS·NEW YORK

First published in the United States by
Science History Publications
a division of
Neale Watson Academic Publications, Inc.
156 Fifth Avenue, New York, N.Y. 10010

Sole world distributor
excluding the United States, its possessions, and Canada
McGraw-Hill International Book Company

Library of Congress Cataloging in Publication Data:

Stuewer, Roger H
The Compton effect.

Includes bibliographical references.
1. Compton effect. 2. Electromagnetic theory—History. 3. Compton, Arthur Holly, 1892-1962.
I. Title.
QC794.6.S3S88 539.7′54 74-5486
ISBN 0-88202-012-9

Designed and manufactured in the U.S.A.

Table of Contents

PREFACE

Arthur Holly Compton, in late 1922, calculated that a quantum of radiation undergoes a discrete change in wavelength when it experiences a billiard-ball collision with an electron at rest in an atom, and his X-ray scattering experiments confirmed this change in wavelength. This phenomenon soon became known as the Compton effect, and for his discovery Compton received the Nobel Prize of 1927, sharing it with C.T.R. Wilson.

Now, since Compton's derivation postulated that the incident radiation consists of quanta, and since it was Einstein who introduced the light quantum hypothesis into physics in 1905, it is natural to assume that there was a direct historical or genetic connection between Compton's derivation and Einstein's hypothesis; in other words, that Compton in late 1922 somehow learned of Einstein's hypothesis, immediately applied it to the preceding collision problem, and calculated the change in wavelength. As I became more and more interested in the history of physics and began reading some of Compton's original papers, however, I soon realized that this direct connection simply did not exist. Instead, I became convinced that far from stemming from an inspired application of Einstein's hypothesis, the actual route traversed by Compton in making his discovery had been very long and difficult. From Martin J. Klein's papers, I also gained a great deal of detailed knowledge, as well as a sense of perspective, on the origin, evolution and influence of Einstein's insights into the nature of radiation.

My general impression from all of my reading was that Compton's early ideas were not obviously connected *either* with Einstein's 1905 light quantum hypothesis, *or*—even more surprisingly—with Compton's own 1922 discovery. It seemed to me, therefore, that it would be unusually interesting to trace and analyze in detail the theoretical and experimental work in radiation physics bearing on the background, discovery, and immediate impact of the Compton effect. This program, moreover, seemed particularly practicable and inviting at the time (1966–1967), because Compton's original research notebooks had recently become available for study. I therefore had every hope that by integrating the information gleaned from these notebooks with that obtained from other primary and secondary source materials, I would be able to produce an accurate and balanced historical account of the origins and immediate influence of Compton's discovery. From the outset, I conceived my study as focusing primarily on conceptual issues rather than, for example, on social or cultural issues. Furthermore, I did not view my study as constituting primarily a personal or scientific biography of Arthur H. Compton, since I knew that I would treat only the earliest phase of his extraordinarily full and productive life. Wherever appropriate, of course, I decided to include sufficient biographical data on Compton and other physicists to lend continuity to the personal aspects of my account.

The results of my attempt constitute the present book. Its organization is simply stated: Chapters 1 and 2 deal primarily with developments in radiation theory, particularly concerning the nature of X-rays, prior to Compton's entry into the field. They contain an attempt to summarize the most relevant aspects of Compton's intellectual heritage. In Chapters 3 to 6, I concentrate primarily on the development of Compton's thought between 1913 and 1924. There is an attempt, first, to isolate the reasons why only certain of the concepts and experiments previously discussed appeared initially to be most significant to Compton. Second, there is an attempt to trace how Compton's own theoretical insights and experiments, as well as those of other physicists, gradually led him to modify his views and to make his famous discovery. Finally, there is an attempt to discuss some of the immediate reactions to Compton's discovery. Chapter 7 concludes with an attempt to display the impact and significance of Compton's discovery in somewhat broader perspective, to show some of the respects in which it represented a turning point in physics. I have tried to use a consistent symbolic notation throughout the entire discussion. I have also assumed that the reader has a basic technical vocabulary and is familiar with at least the principal characteristics of X-ray absorption and emission, radioactivity, Thomson scattering, and Compton scattering. Appendix I was provided for those readers who would like to review these concepts before reading the body of the book.

The primary source documents on which this history was based are of three types: (1) Published journal articles, reports, and the like. A list of abbreviations of the works cited is given in Appendix II. (2) The Arthur Holly Compton Research Notebooks, the originals of which are deposited in the Archives of Washington University, St. Louis, and copies of which are deposited in the Center for History of Physics, American Institute of Physics, New York. With the help of Mr. Richard H. Lytle, former Washington University Archivist, and with the support of a grant from the National Science Foundation administered by the Center for History of Physics under the directorship of Dr. Charles Weiner, I prepared a comprehensive description and tables of contents of these thirty-two notebooks in 1967. (3) Relevant correspondence, lecture notes, tape transcripts and other documents. I studied these deposited in the Center for History of Physics, in the Einstein Archives in Princeton, and in the Archive for History of Quantum Physics located in the library of the American Philosophical Society in Philadelphia, in the library of the University of California at Berkeley, and in the library of the Universitets Institut for Teoretisk Fysik in Copenhagen.

I would like to extend my sincere thanks at this point to Mrs. S. K. Allison, Mrs. Peter Debye, and Frau Franca Pauli for permission to quote from their husbands' writings; to Professors E. C. Kemble, John C. Slater, and John A. Wheeler for permission to quote from their writings; to Professor Aage Bohr and Sir Lawrence Bragg (now deceased) for permission to quote from their fathers' correspondence; to Professor R. E. Norberg for sending me and permitting me to reproduce the pictures of Compton's apparatus and the 1922–1923 Washington University physics staff; to Professor A. O. C. Nier for showing me and permitting me to reproduce the picture of the 1916-1917 University of Minnesota physics staff; to Professor A. B. Pippard for sending and permitting me to reproduce the picture of the 1920 Cavendish Laboratory physics research staff; to Dr. Charles Weiner and Mrs. Joan N. Warnow for their gracious hospitality and assistance in using the materials in the Center for History of Physics; to Mr. Murphy Smith for his help in using the materials in the Archive for History of Quantum Physics; to Dr. Otto Nathan for permission to quote from Einstein's correspondence; to Miss Helen Dukas for her cheerful and informed help with Einstein's papers; and to Drs. M. Rooseboom and P. van der Star for their help with the Lorentz correspondence. I would most especially like to thank Mrs. Arthur Holly Compton, who not only granted me permission to quote from her husband's notebooks, papers, and correspondence, but who also repeatedly encouraged me throughout the course of my research and writing.

I carried out much of my early work on this book at the University of Wisconsin during 1966–1967 while holding a National Defense Education Act Fellowship and working on my doctoral thesis under the direction of Professor Erwin N. Hiebert (now on the faculty of Harvard University). Professor Hiebert first stimulated my professional interest in the history of science, and since then he has offered me so much formal and informal guidance, as well as intellectual and personal companionship, that a simple expression of thanks at this point seems hopelessly inadequate. Similarly, after I joined the faculty of the University of Minnesota Center for Philosophy of Science and School of Physics and Astronomy in 1967, Professors Herbert Feigl, Morton Hamermesh, and Grover Maxwell, as well as my many other colleagues at Minnesota, created such a congenial intellectual and personal atmosphere for me in which to further my studies that it is impossible to thank them adequately. Financial support which I received both from the Center through its grant from the Carnegie Corporation, from the University of Minnesota Graduate School, and from the National Science Foundation, is also gratefully acknowledged.

During the later phases of my work the advice and criticism of certain other scholars has been particularly helpful. These include Professor Max Jammer, who visited the Minnesota Center for Philosophy of Science for several weeks in 1968; Professor Martin J. Klein, who gave me concrete help in several instances, and whose publications have been a constant guide and inspiration to me; the late Professor Imre Lakatos, who was a frequent visitor at the Minnesota Center, and who repeatedly provided me with intellectual challenge and sustained encouragement; and Professor Robert S. Shankland, who read the entire manuscript, offering me most insightful and detailed help, advice, and criticism, and who has also been a constant source of encouragement to me through his long and gracious correspondence. Professor Shankland's introduction to the *Scientific Papers of Arthur Holly Compton: X-Ray and Other Studies* (Chicago: University of Chicago Press, 1973) provides a valuable complement to my own work. I also benefited from conversations or correspondence with H. H. Barschall, Joan Bromberg, Jon Dorling, John Earman, Paul Forman, M. W. Friedlander, Stanley Goldberg, John L. Heilbron, Armin Hermann, Gerald Holton, A. L. Hughes, Marjorie Johnston, Thomas S. Kuhn, E. E. Miller, John S. Rigden, Arthur E. Ruark, Richard Schlegel, Alan E. Shapiro, Katherine J. Sopka, Howard Stein, and J. H. Van Vleck.

Finally, I would like to thank Neale Watson for his editorial assistance, and my wife Helga, as well as Barbara McLaughlin and Maurine Bielawski, for their help in typing and preparing the manuscript.

ROGER H. STUEWER

Minneapolis, Minnesota
January 1974

CHAPTER 1

X-Rays: Pulses, Particles, Waves or Quanta?

A. Introduction

The nature of X-rays was a subject of discussion, debate, and, at times, intense controversy among physicists for roughly three decades, from the time of Röntgen's discovery (1895) until shortly after Compton's scattering experiments (1922). During this period their nature often seemed to most contemporary physicists to be unproblematic: earlier and unsettling theories and experiments, for one reason or another, lost their sense of urgency. As a consequence, the actual route followed by Compton in arriving at his quantum theory of scattering was arduous and complex. He spent, he recalled in 1961, "five years in an unsuccessful attempt to reconcile certain experiments on the intensity and distribution of scattered x rays with the electron theory of the phenomenon that had been developed by Sir J. J. Thomson."[1] Why, however, did Compton approach the scattering problem from the point of view of "the electron theory of the phenomenon that had been developed by Sir J. J. Thomson"? Why at the outset of his investigations did he not simply apply Einstein's light quantum hypothesis to the problem?

To answer these questions it will be necessary to sketch relevant developments in physics between 1895 and 1916, when Compton began his first postdissertation studies. Only then will we be in a position to understand and explore the long development of Compton's own thought, and to appreciate the astonishment with which his conclusions were greeted. Historically, those conclusions were of the utmost importance: they forced physicists to accept the most profound change in their conception of the nature of radiation since the days of Young and Fresnel. At the same time, they were a key element in the complex structure of theory and experiment that led to the creation of modern quantum theory, itself a thorough revolution in the foundations of physics. Few subjects, therefore, are as worthy of an in-depth historical analysis as the background, discovery, and impact of the Compton effect.

B. The Pulse Theory of X-Rays and Its Reception (1897–1911)

1. Thomson's Classical Theory of Scattering and Barkla's Confirmation of It

Röntgen, immediately after he discovered X-rays, established most of their puzzling properties—for example, their apparent inability to be reflected, refracted, or polarized. These properties appeared to set them completely apart from the usual electromagnetic radiations, infrared, visible, and ultraviolet light. Röntgen therefore proposed a tentative, alternate hypothesis. Noting that physicists "have known for a long time that there can be in the ether longitudinal vibrations besides the transverse light-vibrations," he asked: "Ought not, therefore, the new rays to be ascribed to longitudinal vibrations in the ether?"[2] This suggestion, which was completely heretical from the point of view of Maxwell's theory, was soon challenged. New hypotheses were soon advanced. The search for consistent and fruitful models had begun.

The first new hypothesis that gained widespread acceptance was proposed independently in 1897 by G. G. Stokes (1819–1903), Professor of Mathematics at Cambridge University, and Emil Wiechert (1861–1928), then a student at the University of Königsberg. Both Stokes and Wiechert were less impressed by the baffling properties of X-rays than by the means by which they were ordinarily produced—by bombarding a thin metal plate with a beam of charged particles. Such a succession of rapidly decelerating particles should generate a stream of independently moving, transverse, electromagnetic pulses, "analogous," in Stokes' words, "to the 'hedge-fire' of a regiment of soldiers."[3]

In 1903 this pulse hypothesis was modified and endowed with new significance by Stokes' colleague, the Cavendish Professor of Physics and discoverer of the electron, J. J. Thomson (1856–1940).[4] Thomson recognized, as he pointed out in his Silliman Lectures at Yale University,[5] that the relatively small ionization produced when X-rays traverse gases indicated that this radiation possessed a "structure." Characteristically, he translated his ideas into a concrete physical model. X-rays, he suggested, should be viewed as "tremors in tightly stretched Faraday tubes" which extend through the ether and give it a "fibrous structure." An advancing wavefront, then, "instead of being, as it were, uniformly illuminated, will be represented by a series of bright specks on a dark ground, the bright specks corresponding to the places where the Faraday tubes cut the wave front."[6] When such a traveling tremor struck a gas molecule, it should ionize it, a relatively improbable event.

Thomson's modified and picturesque pulse model of X-rays offered a plausible interpretation of ionization phenomena, but it ultimately became firmly established as a consequence of another problem that Thomson attacked. In one of his deepest insights, Thomson realized that if transverse electromagnetic pulses pass through matter, they should be *scattered* by the electrons in it; that is, they should be able to force the electrons into oscillation, causing them to reradiate secondary pulses in all directions. This was an extremely bold conjecture, since no scattering experiments of any kind had been carried out as yet. Thomson published his solution to this problem in the first (1903) edition of his book, *The Conduction of Electricity Through Gases,*[7] the famous volume that served as a textbook and experimental guide at the Cavendish Laboratory.[8] It developed, however, that this solution was off by a factor of exactly one-half, a point that Thomson corrected in the second edition of 1906.[9]

The basic steps in Thomson's calculation are easily outlined. First, he derived the Larmor formula,

$$P = \frac{2}{3}\frac{e^2a^2}{c^3} \tag{1.1}$$

for the rate P at which energy is radiated into free space at velocity c by an electron (charge e, mass m) being accelerated at the rate a. He then introduced his key assumption, that the electron's acceleration originated in its response to an incident "square-wave" electromagnetic pulse of amplitude A and "thickness" d. This meant that it underwent a constant acceleration eA/m for a time d/c, so that the total radiated energy E is given by

$$E = \frac{2}{3}\frac{e^2}{c^3}\int_0^{d/c} a^2dt = \frac{2}{3}\frac{e^2}{c^3}\int_0^{d/c}\left(\frac{eA}{m}\right)^2 dt = \frac{2}{3}\frac{e^4A^2d}{m^2c^4}. \tag{1.2}$$

Finally, dividing this result by the total radiant energy $(1/4\pi)A^2d$ incident on the electron, and multiplying it by a factor N, the number of electrons per unit volume in the scatterer, Thomson obtained what is now known as the "Thomson scattering coefficient" (or "Thomson cross section"),

$$\sigma_0 = \frac{8\pi}{3}\frac{Ne^4}{m^2c^4}. \tag{1.3}$$

For our purposes the most noteworthy feature of Thomson's result is that although he explicitly assumed that the incident electromagnetic radiation consists of pulses, no indication of this assumption appears in the above expression—which is the reason, of course, that it is given exactly as above in modern textbooks.

Thomson's calculation attracted attention almost immediately after it first appeared in 1903. C. G. Barkla, who had been an 1851 Exhibition Scholar under Thomson at the Cavendish from 1899–1902, but who had then left for the University of Liverpool where he received his D.Sc. degree in 1904,[10] saw that by measuring the total amount of radiation scattered by a substance he could determine N, the number of electrons per unit volume.[11] In this way he obtained the first solid evidence indicating that for light elements the number of electrons in an atom is approximately numerically equal to half its atomic weight (a result that Thomson himself built upon in work that later became of great importance for the development of Rutherford's nuclear atom[12]).

These experiments, however, were only the beginning for Barkla: in 1905–1906 he carried out further scattering experiments of key significance for Thomson's theory. Stimulated by a suggestion of Professor Wilberforce, Barkla saw that if Thomson's theory were correct he could devise a test for the polarization of X-rays.[13] In an early step in his derivation of the Larmor formula, Thomson used the fact that the electromagnetic amplitude reradiated by an accelerated electron is given by (ae/rc) sin θ, where θ is the angle between the acceleration vector $\mathbf{a}$ and the vector $\mathbf{r}$ from the electron to the point at which the scattered radiation is observed. No scattered radiation whatsoever should therefore appear in, or opposite to, the direction of motion of the electron ($\theta = 0, \pi$), whereas a maximum amount should appear at right angles thereto ($\theta = \pm\pi/2$). In 1905 Barkla tested this prediction by scattering X-rays from a single block of carbon (a "single-scattering" experiment) and found indications of its correctness, whereupon in 1906 he carried out a more elegant "double-scattering" experiment.[14] This latter experiment (which Barkla had actually intended to carry out before the former one, but had judged infeasible owing to intensity considerations) yielded conclusive evidence that X-rays may be polarized.

Barkla had therefore succeeded where Röntgen and every other experimentalist since Röntgen had failed. His experiments were recognized as a triumph for Thomson's theory of scattering, and consequently also as a triumph for the Stokes-Wiechert-Thomson electromagnetic pulse theory of X-rays. Only later did it become clear that Barkla's experiments required only the *transversality* of the electromagnetic vibrations. Since Thomson had analyzed the X-ray-electron interaction by explicitly assuming that the incident X-rays were transverse electromagnetic *pulses,* this was the theory that was taken to be substantiated at the time.

No one was more impressed with Barkla's work than Thomson. In 1907 he restated his conviction that X-rays were nothing but pulses or tremors traveling along stretched Faraday tubes, producing an advancing wavefront resembling bright specks on a dark field. "In fact," he wrote, "from this point of view the distribution of energy is very [much] like that contemplated on the old emission theory, according to which the energy was located on moving particles sparsely disseminated throughout space. The energy is as it were done up into bundles and the energy in any particular bundle does not change as the bundle travels along the line of force."[15] He added that this picture also appeared to account, for example, for the photoelectric effect.

2. *Early Work on Gamma Rays*

The pulse theory of X-rays, while strikingly substantiated by Barkla's polarization experiments, was soon challenged. This challenge arose out of the quest to determine the nature of another baffling radiation, later known as gamma (or γ) radiation, which was discovered in 1900 by the French physicist Paul Villard.[16] Villard observed γ-rays emitted by radioactive elements and established two of their fundamental properties, namely, their high penetrating power and their inability to be deflected by a magnetic field.

Both properties, argued Friedrich Paschen (1865–1947) of the University of Tübingen, could be accounted for if γ-rays were very rapidly moving charged particles. This contention, however, was soon opposed by A. S. Eve

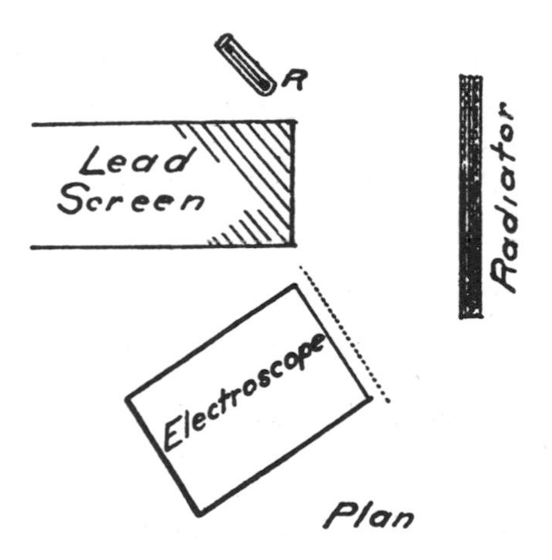

FIG. 1. Eve's 1904 experimental setup.

of McGill University in 1904.[17] Eve based his opposition on the results of scattering experiments (Fig. 1[18]) in which he inserted various absorbers into either the direct (primary) beam of γ-rays from a radium source *R*, or into the scattered (secondary) beam in front of a detecting electroscope. He was able to show that for a number of different scatterers (radiators) the secondary radiation was "much less penetrating" than the primary radiation, and from this and other observations he concluded that ". . . γ rays either consist of particles practically devoid of electric charge, or are of the type of Röntgen rays, or have a special character of their own hitherto unknown."[19] In the event that they were similar to Röntgen rays, he suggested that any "dissimilarity between Röntgen and γ rays . . . is probably due to the fact that the Röntgen pulses are more broad than those which constitute the γ rays."[20]

Three years after Eve reached that conclusion, J. J. Thomson, in a paper which we have already mentioned,[21] extended the analogy between the two radiations by comparing their spatial extensions. He estimated from energy considerations that while X-ray "bundles of energy" are surrounded by about 1 liter of space—a very "coarse" structure—γ-ray "bundles of energy" occupy a very much smaller volume. Furthermore, Thomson concluded, since these widely spaced, highly energetic "units possess momentum as well as energy they will have all the properties of material particles, except that they cannot move at any other speed than that of light. Thus we can readily understand why many of the properties of the γ rays resemble those of uncharged particles moving with high velocities."[22] The momentum Thomson mentioned, of course, was the electromagnetic momentum which had been theoretically predicted by Bartoli, Boltzmann, and Maxwell,[23] and experimentally detected at the turn of the century by Lebedev,[24] and, more quantitatively, by E. F. Nichols and G. F. Hull.[25] In essence, therefore, Thomson held that γ-rays could readily be brought within the compass of the electromagnetic pulse theory.

3. The First Phase of the Bragg-Barkla Controversy

Thomson's position was challenged to the core by William H. Bragg (1862–1942).[26] While born in England and educated at Cambridge University, Bragg at the moment was isolated both physically and intellectually from Thomson and Barkla, since, in 1884 (with Thomson's support) he had become the Professor of Mathematics and Physics at the University of Adelaide in South Australia. There, until he was over 40 years of age, he had led "a pleasant and useful life as a popular teacher and good friend in the Adelaide community," but "produced nothing that could be called research."[27] Almost unexpectedly, after finding "something that seemed to him to ask for experiment,"[28] Bragg entered the laboratory in 1904 and carried out his first important researches, which dealt with the range of α-particles in matter.[29] These experiments were actually designed to test certain predictions of two current atomic models, J. J. Thomson's (a sphere of positive charge containing widely spaced electrons), and Philipp Lenard's (widely spaced positive and negative pairs or "dynamids" dispersed throughout the atomic volume). For our purposes, however, their most important consequence was that three years later, in 1907, Bragg saw in Lenard's "dynamids" the basis for a corpuscular theory of γ-rays.[30] His single-minded advocacy of his views generated and sustained the first particle-wave—or rather, particle-pulse— controversy in this century.

Bragg felt that the most revealing characteristic of γ-rays was that their emission was always accompanied by the emission of α- and β-particles. Was it not reasonable to suppose, therefore, that a γ-ray was simply an α-particle and a β-particle electrically bound together to form (as Bragg assumed) an uncharged or "neutral pair"? "Röntgen himself proposed in the third of his memoirs a theory of this nature,"[31] Bragg noted, and Lenard's "dynamids" represented a similar suggestion. Such a pair would have only a "local action" and therefore readily penetrate metals and other substances—even though its electric moment, and hence its penetrating power, might be changed in the process. Since it was electrically neutral as a whole, it would not be deflected by electric or magnetic fields; nor would it undergo any appreciable reflection or refraction. If in passing through matter it were broken apart or "resolved," it would give rise to a rapidly moving β-particle, the velocity of which would be independent of the *intensity* of the primary pairs. Moreover, the number of pairs resolved per unit length would be proportional to the total number present, yielding the exponential law of absorption.

Gamma rays were known to exhibit all of these properties, and, Bragg asserted, since they were also "amongst the properties" of X-rays, "an hypothesis which will suit one form of radiation will also so far suit the other."[32] Bragg felt that all of the evidence in favor of the pulse theory—for example, Barkla's polarization experiments and Haga and Wind's recent diffraction experiments[33]—was "indirect," "a little over-rated," and could be accounted for on his neutral pair theory by making certain assumptions on how such pairs interact with matter. If the pulse theory accounted more readily for these experiments, this advantage was offset by the grave difficulties it experienced in trying to explain, for example, either ionization phenomena or the properties of β-rays ejected by X-rays from substances. Even Marx's experiments,[34] which seemed to prove that X-rays travel with the velocity of light, were to Bragg compatible with the assumption that X-rays consist only *partially* of pulses, that they consist "mainly of neutral pairs" traveling with a velocity "yet undetermined."[35]

Bragg's ideas were hardly in print before they were challenged by Barkla in a note in *Nature*.[36] Barkla had meanwhile extended his 1905–1906 scattering experiments by pursuing a further consequence of Thomson's theory of scattering, namely, that the angular distribution $I(\theta)$ of the scattered radiation should vary as $(1 + \cos^2\theta)$, where θ is the scattering angle. Thus the ratio of the maximum intensity (at $\theta = 0, \pi$) to the minimum (at $\theta = \pm \pi/2$) should be 2, and Barkla had found reasonably close agreement with this figure in experiments in which he had scattered X-rays from carbon. He therefore regarded these experiments as a "strong confirmation" of the pulse theory—a conclusion which directly conflicted with Bragg's views. Barkla considered his conclusion to be particularly justified because he had calculated on Bragg's theory, assuming that the direction of emission of secondary neutral pairs was independent of the direction of the incident radiation, that ordinarily the value should be very much larger.

Three months later, Bragg responded to Barkla.[37] In the meantime, he had enlisted the aid of his Adelaide colleague, J. P. V. Madsen, Lecturer on Electrical Engineering, and had proven experimentally that when γ-rays traverse thin layers of carbon and lead, considerably more secondary radiation is scattered forward than backward.[38] (A. S. Mackensie of Dalhousie University, Halifax, Nova Scotia, had reached a similar conclusion one year earlier.[39]) To Bragg, this observation not only invalidated Barkla's neutral pair calculation (see above), it also directly contradicted the prediction, which followed from Thomson's $(1 + \cos^2\theta)$ variation, that the angular distribution should be completely symmetric in the forward and backward directions. The "most remarkable want of symmetry" that Bragg and Madsen had found was therefore "fatal to the ether pulse theory of the γ rays. Moreover, all our experiments so far show that, on the whole, the [secondary] kathode radiations . . . possess momentum in the original direction of motion of the rays, and this shows that the rays are material."[40] Bragg therefore had found "a decisive experiment" in favor of his neutral pair theory.

Two weeks later, Barkla replied: ". . . [recent X-ray experiments] supply what appears to me to be absolutely conclusive evidence in favor of the ether pulse theory. For . . . I have found that the intensity of [backward scattered X-radiation] . . . is within 5 per cent of that calculated on the ether pulse theory. . . . If Prof. Bragg can suggest a distribution of ejected pairs that will produce such close agreement between the calculated and experimentally determined intensities, it will be time to consider the theory further."[41] Specifically, Barkla had found[42] that, depending upon whether the primary X-rays were "soft" or "hard" (i.e., of low or of high penetrating power), the ratio of the maximum to the minimum intensity scattered by carbon was between 1.85 and 1.45—close enough to the predicted value of 2 to satisfy Barkla. Bragg in turn responded by pointing out that his γ-ray experiments—and even Barkla's own "hard" X-ray experiments—yielded far poorer results, which he stated, was "no compliment to the ether pulse theory."[43]

Almost simultaneously, J. J. Thomson added the weight of his authority to the controversy. Remarking that there was "much the same difference" between the ether pulse and neutral pair theories as "between the undulatory and emission theory of light,"[44] Thomson concluded that the former possessed two outstanding advantages over the latter: it could easily account for the polarization of X-rays and for the fact that they apparently travel with the velocity of light. Sensing the need for new experiments, however, Thomson encouraged R. D. Kleeman, an 1851 Exhibition Scholar from Bragg's University of Adelaide, "to investigate the properties of the secondary γ rays from different substances exposed to the γ rays of radium, and to compare their properties with those of the primary rays."[45]

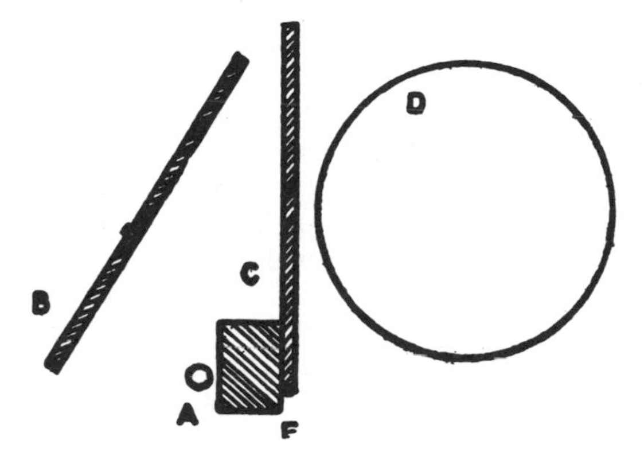

FIG. 2. Kleeman's 1908 experimental setup.

Using an apparatus (Fig. 2) similar to Eve's, Kleeman allowed γ-rays from a source, *A*, to first strike various scatterers, *B*, and then pass through various absorbers, *C*, before entering an ionization chamber, *D*. He found, with Eve, that "on the whole the coefficients of absorption of the secondary rays are much greater than those of the primary rays."[46] He interpreted these results as "probably due in part to a difference in the breadth of the pulses, the pulses in a group of secondary rays being probably broader than those in the corresponding group of primary rays."[47]

4. Later Stages in the Development of Bragg's Thought

Much of the succeeding controversy between Bragg and Barkla involved variations on the themes sketched above. Barkla's experiments, in which X-rays were scattered from carbon, appeared to agree with the polarization and angular distribution predicted by Thomson's theory of scattering, while Bragg's experiments, in which γ-rays were scattered from metal plates, appeared to disagree with the angular distribution predicted by Thomson's theory. The pulse theory could apparently account for the difference in hardness between the primary and secondary rays, while the neutral pair theory apparently experienced difficulties, but not insurmountable ones, on this score. In general, Barkla took his experiments to confirm the pulse theory, while Bragg took his to confirm the neutral pair theory. Neither Barkla nor Bragg was able to view the other's experiments in a detached manner; each tended to interpret the other's experiments in the light of his own theoretical commitments.

This was true even when entirely new evidence was uncovered later, for example, in 1908 when Barkla and his colleague C. A. Sadler found[48] that secondary X-rays in general consist of two distinct types: (1) the ordinary Thomson-scattered X-rays, and (2) entirely new, homogeneous X-rays, whose hardness depended only on the element emitting them. Not surprisingly, Barkla felt that his discovery of these *characteristic* X-rays, for which he eventually won the Nobel Prize, "verified the ether pulse theory in a more striking way than I ever anticipated."[49] Bragg of course disagreed. He felt that at the very least Barkla's discovery necessitated a "radical alteration" in the pulse theory. In any case, he felt that Barkla's characteristic X-rays could be accounted for on the neutral pair theory by assuming that the incident pairs undergo "some transformation" in the target element.[50]

The strength of the conceptual wall separating the two physicists was further revealed when Bragg discussed some of his own later experiments. Guided by his neutral pair theory, Bragg suspected that a primary γ-ray pair, if "resolved" in a scatterer, should generate β-rays whose velocity should be independent not only of the intensity of the primary γ-rays but also of the nature of the scatterer. Careful experiments confirmed his suspicions: γ-rays incident on a number of *different* scatterers gave rise to β-rays of the *same* velocity, and further experiments proved that their velocity depended only on the hardness of the primary γ-rays.[51]

Bragg realized that these results, which years later became of key importance, were capable of different interpretations which he would have to examine—and discount—if he were to maintain his neutral pair theory. The first derived from early ideas of P. Lenard and A. S. Mackensie, and held that the incident γ-ray "is a pulse which merely pulls the trigger" of an atomic gun. If this picture were correct, however, Bragg asked: "How is it that the pulses always find the guns pointing in the direction in which they are traveling themselves, so that the motion of the shot is a continuation of their own line of flight?"[52]—a most implausible situation. A second possible interpretation derived from Thomson's suggestion that X-rays consist "of 'bundles of energy' occupying only a very small portion of the wave-front, the rest of the front being blank." Even if this idea led "at once" to natural explanations of gaseous ionization, of the "absence of a relation between the velocity of the secondary electron . . . and . . . the intensity of the radiation and the nature of the atom," and of the "concentration of momentum," it nevertheless appeared to Bragg to be extremely implausible. It entailed, in the first place, a "very special and complicated structure of the aether." It meant, in the second place, "that the energies of the primary electron, the bundle, and the secondary electron are all equal," that the "only links" between the primary and secondary electrons would be "little bundles of energy moving with the speed of light. . . ." This picture, Bragg felt, was unreasonable: "The difficulties of this theory are exactly those which would naturally arise in the attempt to transfer the properties of a material particle to an immaterial disturbance."[53]

"Replace the bundle of energy by a neutral pair, and the whole affair seems simple enough," concluded Bragg. The preponderance of secondary β-radiation in the forward direction is then "simply" explained by assuming that as the γ-ray pair is resolved, "the negative flies on and the positive becomes ineffective." That the speed and penetration of the secondary β-rays increase with the penetrating power of the primary γ-rays "is also an obvious consequence of the hypothesis. The faster the γ particle is moving, the greater the initial speed of the negative." That the speed of the β-ray does not depend upon the "nature or condition" of the scattering atom is also "readily explainable: the electric field of the atom merely dissolves the bonds that connect the pair. It is not able to affect the speed of the negative set free."[54]

These arguments and others applied with equal force to X-rays and to γ-rays, as recent experiments emphasized. Thus, C. D. Cooksey of Yale University had found,[55] qualitatively speaking, that X-rays and γ-rays produce the same type of asymmetrically emitted β-rays. To Bragg this was "the last experiment required to show that all the properties . . . are true for X-rays as well as for γ rays, *mutatis mutandis*."[56] Bragg himself, however, soon recorded one additional experiment[57] when he found that X-rays, like Eve's and Kleeman's γ-rays, are softened when scattered.

Bragg's tantalizing critique of Thomson's "bundle of energy" hypothesis was heightened even further through the interpretation which Bragg's Adelaide colleague, J. P. V. Madsen, placed on concurrent experiments of his own.[58] The primary object of Madsen's experiments was to *quantitatively* determine the ratio of the total amount of secondary γ-radiation "forward scattered" by lead and zinc plates to the total amount "backward scattered." However, it was when he decided to broaden his search and look for "the existence of modified or softened γ rays" in the secondary radiation that he was led to his most interesting speculations. Eventually, while using aluminum and zinc scatterers, he found "a considerable difference in the quality" between the forward-scattered and back-scattered γ-rays. He suggested that to understand this softening we consider what might befall a *homogeneous* bundle of "hard," corpuscular, neutral pair γ-rays when passing through matter. These neutral pairs, he wrote, would undoubtedly "suffer collisions; the effect of such collision is to change the direction of motion of the incident primary ray—in other words, to scatter it; at the same time the scattered ray loses a certain amount of energy—it has become softened; this softening may be due either to a change in its speed or to a change in [the electric] moment of the γ pair, or it may be both."[59]

Madsen envisioned, therefore, a "billiard-ball" scattering process. Unfortunately, however, he did not believe that this process actually accounted for the softening he had just observed. Believing γ-rays to be ordinarily *inhomogeneous,* he postulated instead a selective scattering effect; namely, that more soft γ-rays are back-scattered than hard γ-rays, making the back-scattered radiation appear softer on the whole than the forward-scattered radiation.[60]

What, one might ask, prevented the two Adelaide colleagues, Bragg and Madsen, from combining their very suggestive ideas on scattering and arriving at an entirely new, even revolutionary, theory? The answer seems to be rather straightforward, for there was one point on which they were in complete agreement: both believed that γ-rays were particles, pure and simple. Perhaps future discussions would have enabled them to break through this conceptual barrier, but they were not to be given the chance to engage in them. By 1909 Bragg's researches had established his reputation, and he received a call to the University of Leeds as Cavendish Professor of Physics. At age 47, after twenty-four years in Australia, he returned home to England.

It must have been shortly after Bragg's arrival in Leeds when he learned that two Continental physicists, Johannes Stark and Arnold Sommerfeld, were also currently engaged in a dispute over the correct interpretation of certain X-ray and γ-ray phenomena, particularly the asymmetry in the intensity of the scattered radiation. For the moment it suffices to point out that Stark was advocating Einstein's light quantum hypothesis, while Sommerfeld was advocating the Stokes-Wiechert pulse theory. Bragg naturally saw the relevance of this dispute to his own work, and hence was prompted to exchange several letters with Sommerfeld between February 7, 1910, and July 7, 1911.[61] He wrote Sommerfeld that he agreed with him that "Dr. Stark's quantum" had nothing to do with the problem, but at the same time he maintained that his neutral pair theory explained the asymmetry and other observations. Sommerfeld was not convinced: secondary β-rays moving at relativistic velocities, if suddenly stopped, would generate pulses of electromagnetic radiation confined to the forward direction.[62] Nevertheless, replied Bragg, in a short time they would spread out into space, which conflicted with their well-known particulate behavior, as strikingly reaffirmed by C. T. R. Wilson's recent and beautiful cloud chamber photographs.[63] Once again, enough evidence could be mustered on either side of the question, and neither physicist convinced the other of his point of view.

The dominant theme in Bragg's letters to Sommerfeld was Bragg's insistence on a model of X-rays and γ-rays that reflected their concentrated, quantum-like behavior. He emphasized repeatedly that *one* X-ray or γ-ray is "concerned in the making" of *one* cathode ray or β-ray. Recognizing, however, that the inverse transformations also occur, he was soon led to espouse a "double transformation" hypothesis (X-ray to cathode ray, cathode ray to X-ray).[64] This hypothesis, he argued, could even serve as a basis for replacing Barkla's explanation of his characteristic X-rays; namely, that a pulse "[shakes] an atom in passing and [makes] it give out its own characteristic quivers." Moreover, it could be generalized to include all of the "radiant entities," the α, β, γ, X and cathode rays: "The form of the entity may change, γ into β, X into cathode ray, and so on; but there is so little change in anything but form that practically we may assume continuity of existence."[65] If the entity changed its direction, but not its form, it would be simply scattered, like Geiger and Marsden's α-particles.

Bragg remained convinced, therefore, that the simplest "working model" capable of describing X-ray and γ-ray transformations was the neutral pair model. It correlated the evidence and suggested new experiments—as Bragg himself proved once again when he demonstrated, in very important experiments with H. L. Porter,[66] that X-rays ionize substances only indirectly, through the action of the β-rays they produce. Gradually, however, Bragg was forced to modify his original model as evidence accumulated that the presumed positive component, the α-particle, was doubly charged. As early as 1908 he had suggested that this component might be a particle that "has hitherto been received with little favor," a particle having the same mass and charge as the negative electron, namely, the "positive electron."[67] By 1910 he had replaced this inspired guess with a nameless "quantity of positive electricity" which neutralizes the charge of the electron "but adds little to its mass."[68] That such a positive particle had not yet been isolated experimentally did not impress Bragg as a major difficulty—current technology was probably inadequate for its detection.[69]

In spite of his convictions, however, Bragg (as he explicitly wrote Sommerfeld) had "no wish to dogmatise"; nor was he at all "adverse to a reconcilement of a corpuscular and a wave theory." Nevertheless, he was greatly—

perhaps overly—impressed with the difficulties of the wave (or pulse) theory, and just as Sommerfeld tended to minimize the "spreading difficulty" of the pulse theory, so Bragg tended to minimize, for example, the "polarization difficulty" of his neutral pair theory. At one point he remarked that Barkla's experiments only demonstrated that more X-rays may be deflected in one direction than in another—and when striking an obstacle, "a billiard ball with side on does as much, or more exactly still, a spinning tennis ball."[70] He even discounted such experiments as Whiddington's,[71] which proved that a certain minimum amount of energy was necessary for X-rays to excite the characteristic X-radiations of a substance: "The requisite energy is that of the characteristic X corpuscle of that substance."

In sum, by the autumn of 1911 Bragg still held that: "The 'corpuscular theory' of X-rays is really an induction from experimental facts. The 'neutral-pair' theory which I have described at various times is at any rate a simple working model. But the 'ether-pulse theory' is at present little more than an aspiration."[72] The reporter from *Nature* who heard Bragg express these views at a meeting of the British Association could not resist a pun: he wrote, "Pulses of genuine delight ran through the meeting while Prof. Bragg expounded his views, and it was clear that many were impressed by the cogency of his arguments."[73] Evidently Bragg would not relinquish his corpuscular theory until entirely new and striking evidence would force him to do so.

C. Experimental and Theoretical Work Stimulated by the Bragg-Barkla Controversy (1908–1914)

1. On the Asymmetric Emission of Secondary Beta Rays

As a rule, scientific controversies as extended, open, stimulating, and significant as the Bragg-Barkla controversy attract a great deal of attention. But two physicists at two different locations may well regard different aspects of the controversy as worthy of further research, pursue those avenues, and thereby uncover new experimental facts or achieve unanticipated theoretical insights. Thematically speaking, the Bragg-Barkla controversy stimulated experimental research on three important subjects between 1908 and 1914: the forward-backward asymmetry in the secondary β-ray distribution, the forward-backward asymmetry in the secondary X-ray distribution, and the difference in hardness or penetrability between secondary and primary X-rays or γ-rays. The exploration of each of these avenues, which we shall now discuss, answered questions of direct relevance to the controversy—or raised others of far broader significance.

Bragg had regarded the first point, the preponderance of secondary β-rays ejected in the forward direction from a metal plate by γ-rays, as telling evidence in favor of his neutral pair theory. The first outsider to speak concerning this question was C. D. Cooksey of Yale University (Sheffield Scientific School), in 1908.[74] Following "closely Prof. Bragg's method of procedure," Cooksey attempted to determine whether or not X-rays also produce the same asymmetry. He found that, depending upon the substance irradiated, 50 to 90 percent more β-radiation was produced in the forward than in the backward direction, which at least agreed qualitatively with Bragg's γ-ray experiments. Nevertheless, Cooksey categorically rejected Bragg's interpretation: "I cannot agree with Prof. Bragg that the evidence is conclusive that X-rays and γ rays must consist of some type of radiation other than electromagnetic pulses." Thus, "an electromagnetic pulse possesses momentum also in the direction of propagation. Though little is known of the mechanism of the production of secondary kathode rays by ether pulses, it is not unreasonable to suppose that an ether pulse could contribute some of its momentum to the secondary kathode particles, causing them to go more in the direction of propagation of the primary than in any other."[75]

Two years later (1910), similar but even more germane experiments were carried out in O. W. Richardson's laboratory at Princeton University by Otto Stuhlmann, Jr.[76] Stuhlmann coated ("sputtered") a thin film of platinum onto a thin quartz plate and allowed *ultraviolet light* to be incident on either the film side or the quartz side of the plate. He, too, found a very noticeable forward preponderance in the secondary β-radiation. Moreover, independently and virtually simultaneously, R. D. Kleeman (now D.Sc. and Mackinnon Student of the Royal Society at Emmanuel College, Cambridge) reported an identical conclusion: "an electron liberated by ultra-violet light has a component of motion in the direction of propagation of the exciting light."[77] Somewhat later, C. D. Cooksey, taking account of subsequent work of his own, summarized the experimental situation as follows: "If the mechanism of production of cathode rays is the same with ultra-violet light as it is with X-rays, the [asymmetry] . . . does not seem to vary much over a range of [wavelengths] . . . corresponding to the [characteristic] X-rays from tin [$\approx$ .5Å] up to that corresponding to ultra-violet light [$\approx$ 3000Å]."[78]

Now if these experiments "stood alone," as Stuhlmann remarked, they would constitute a "strong argument" in favor of a corpuscular theory of radiation. But he realized that they did not stand alone, and hence that their interpretation was ambiguous, as both Kleeman and Cooksey also emphasized. In fact, most physicists on either side of the controversy must have realized that Stuhlmann's, Kleeman's, and Cooksey's experiments could easily be interpreted as establishing a bridge between γ-rays and X-rays, on the one

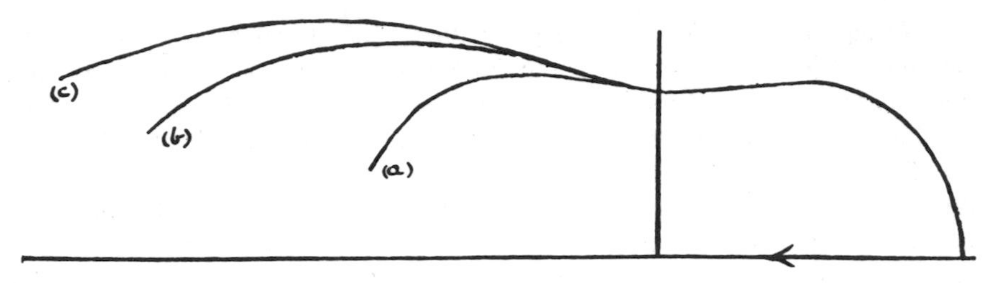

FIG. 3. Owen's sketch showing the increase in "excess scattering" as the incident X-rays became softer and softer (curve *a* to *b* to *c*). The X-rays are incident from the left.

hand, and ultraviolet light, on the other—and almost every contemporary physicist believed that ultraviolet light consisted of electromagnetic waves. Bragg's corpuscular theory therefore undoubtedly suffered a serious setback as a consequence of these experiments.

2. *On the Asymmetric Emission of Secondary X-Rays (Excess Scattering)*

The second avenue explored experimentally in more detail was the angular distribution $I(\theta)$ of the secondary X-radiation. Bragg, using γ-rays, had recorded very definite departures from the perfectly symmetrical $(1 + \cos^2\theta)$ variation predicted by Thomson's theory, but, as Barkla's experiments indicated, the situation was much less certain for X-rays. Some of the first new evidence on the question was provided in 1911 by J. A. Crowther,[79] who was working under J. J. Thomson at the Cavendish Laboratory, and who found a definite forward-backward asymmetry when X-rays were scattered by a thin aluminum plate. A similar observation was made at roughly the same time by C. G. Barkla and T. Ayres for X-rays scattered by carbon,[80] and, not surprisingly, Barkla and Ayres did not see their results as constituting a threat to the pulse theory. They argued that the deviations from Thomson's theory were no doubt "capable of special explanation."[81] Barkla soon refuted an alternate suggestion of Crowther's.[82]

Crowther, however, was not discouraged by Barkla's criticism: he was spurred on to greater efforts. First, he found a systematic increase in the forward-backward asymmetry as the atomic weight of the scatterer increased.[83] Secondly, he varied another experimental parameter—the quality or hardness of the incident X-rays—and found a systematic increase in the asymmetry as they became softer and softer.[84] This variation was clearly displayed in a diagram (Fig. 3) published almost simultaneously by E. A. Owens,[85] an 1851 Exhibition Scholar and Crowther's colleague at the Cavendish. Owen's diagram shows that, as the incident X-rays (note arrow for direction) become softer and softer (curve *a* to *b* to *c*), the secondary radiation in the forward direction becomes more and more intense.

This phenomenon was appropriately termed "excess scattering," and for several years there was considerable confusion over its interpretation.[86] One early explanation was suggested in 1914 by H. A. Wilson, who traced its origin to the coexistence of two different effects: (1) a symmetrical scattering from the "amorphous portion" of the target, and (2) Bragg-type, presumably asymmetrical reflection from "internal crystals" in the target.[87] Another, more fruitful, line of reasoning began with D. L. Webster in 1913.[88] Webster argued that the electrons in a heavy atom are spaced so closely that they do not scatter independently but cooperate to produce constructive interference in the forward direction.[89] Analyzing the problem in terms of the pulse theory, he found indications that his approach was correct, and his arguments were subsequently extended, first by C. G. Darwin (1914),[90] then by Peter Debye (1915)[91] and finally by J. J. Thomson (1916).[92] Debye found, for example, that the angular distribution $I(\theta)$ for radiation scattered coherently by Z electrons should vary as Z^2, in contrast to Thomson's linear variation—a conclusion that was also reached by Darwin and C. G. Barkla[93] from a general qualitative argument.

In sum, we see that while "excess scattering" eventually found a ready interpretation based fundamentally on an extension of Thomson's theory of scattering, the appropriateness and validity of this interpretation was by no means immediately evident. The route to its explanation illustrates the difficulty with which the ultimately relevant discrepancies from Thomson's theory were sifted out from the mass of observations.

3. *On the Difference in Hardness between Primary and Secondary X-Rays and Gamma Rays*

The third and final avenue of experimental research opened up by the Bragg-Barkla controversy, the one centering on the difference in hardness between the primary and secondary X-rays or γ-rays, ultimately became the most important of all. Here Bragg's and Madsen's work provided the direct stimulus for entirely new and highly significant experiments carried out in 1910 by D. C. H. Florance[94] in Rutherford's laboratory at the University of Manchester. Using the simple (and self-explanatory) apparatus shown in Fig. 4, Florance first studied the angular distribution $I(\theta)$ for radium γ-rays scattered by an iron plate (the radiator) and confirmed that the distribution departs markedly from Thomson's $(1 + \cos^2\theta)$ variation. At that point he hit upon a highly original idea: to measure quantitatively the difference in

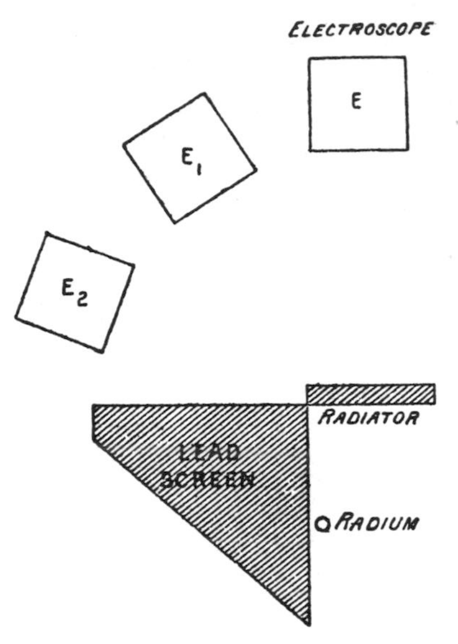

FIG. 4. Florance's 1910 scattering apparatus.

hardness between the primary and secondary γ-rays *as a function of the scattering angle θ.*

The only measure of the hardness (or quality, or penetrating power) of radiation known to physicists in 1910 was the linear-absorption coefficient μ, defined by the equation $I(x) = I(0)e^{-\mu x}$. (The closely related *mass*-absorption coefficient μ/ρ, defined by the equation $I(x) = I(0)e^{-(\mu/\rho)\rho x}$, where ρ is the density of the absorber, was also frequently employed instead of the *linear*-absorption coefficient μ.) To determine μ (or μ/ρ), one simply measures (for example, with an electroscope) the intensities $I(0)$ and $I(x)$ of the radiation before and after it has passed through a given *absorber* of thickness x. By performing *both* measurements, first with the absorber in the primary beam and then with the absorber in the secondary (or scattered) beam, one may calculate the corresponding absorption coefficients, and thereby measure the difference in hardness between the primary and secondary radiation. In practice, the intensities $I(0)$ and $I(x)$ are generally corrected in each measurement for "background radiation" by taking identical measurements with the *scatterer* removed.[95]

This in general was the method Florance used to determine the difference in quality between primary and secondary γ-rays scattered by plates of carbon, iron, and lead through scattering angles of 25° and 55°. The corresponding absorption coefficients (lead absorber) which he calculated are summarized in the following table:[96]

[Scatterer or] Radiator	Absorption coefficient of		
	primary radiation	secondary radiation	
		$\theta = 25°$	$\theta = 55°$
5 cm carbon	= .68	= 1.20	= 1.77
10 cm carbon	= .65	= 1.18	= 1.70
2.2 cm iron	= .65	= 1.17	= 1.68
5 cm iron	= .59	= 1.05	= 1.55
.416 cm lead	= .625	= 1.11	= 1.65

These measurements—the first of their kind—and others clearly demonstrated, in Florance's words, that "for all radiators the secondary γ radiation gradually becomes softer as the electroscope is moved further away from the normal position. . . ." Furthermore, they proved that the *ratio* of the absorption coefficients for any two radiators remains constant as the scattering angle is changed. (Compare, for example, "10 cm carbon" with "2.2 cm iron"; the ratio of the absorption coefficients is very nearly unity at both scattering angles.) "This points to the conclusion," Florance wrote, "that the secondary radiation is the primary radiation scattered."

If the secondary radiation were characteristic X-radiation "it would be expected that the quality would depend on the material." "It is important to notice that Bragg and Madsen . . . have shown that the character of the β radiation caused by γ-rays is independent of the atom in which it arises, and depends solely on the nature of the γ-rays to which it is due. The present investigation shows that this is also true for the secondary γ radiation." In sum: "The quality of the secondary γ radiation shows no sudden change from that of the primary. There is simply a gradual softening the more the secondary radiation is deflected from its original direction. The gradual softening is the same for every radiator. Other investigators have shown that β-rays are scattered in their passage through matter. The scattering of γ rays appears to be analogous to the scattering of β-rays."[97]

The significance of Florance's remarkable conclusions must be judged in the light of his belief concerning the precise relationship between the softening of the radiation and the nature of the scattering process. It was in fact identical to Madsen's: "The primary γ-rays possess a wide range of penetrating power. The softening of the secondary radiation that has been observed is the result of this heterogeneity of the primary rays. The softer radiation is more scattered than the harder radiation. . . ."[98] The extent to which Florance accepted Bragg's neutral pair theory is uncertain, but he must have realized that his acceptance of this "selective scattering" explanation of the softening conflicted with Thomson's theory of scattering, which excludes any change in quality of the radiation in the scattering process.

In 1912, two years after Florance had carried out these remarkable experiments, some closely related observations were made at the University of Liverpool by Charles A. Sadler, the physicist who had shared in Barkla's 1908 discovery of the characteristic X-rays, and Paul Mesham, Lecturer and Demonstrator in Physics.[99] (Barkla himself had left the University of Liverpool for the University of London in 1909.) Exploiting techniques they had learned from Barkla, Sadler and Mesham were able to overcome formidable experimental difficulties, and, instead of using the intrinsically heterogeneous, direct beam from an X-ray tube as a primary source of X-rays, they found they could use the very homogeneous, characteristic X-rays emitted, for example, by iron. Their experiments therefore were entirely novel, and, completely contrary to their expectations, they found that even these characteristic X-rays when scattered by carbon became "distinctly *less penetrating* than the primary exciting beam."[100]

Three possible hypotheses occurred to Sadler and Mesham to account for their unanticipated observation: (1) that an impurity was present in the carbon which emitted the softer secondary radiation; (2) that the primary beam was actually inhomogeneous and a selective scattering effect took place; and (3) that the carbon itself emits a feeble, soft radiation. By means of a series of often ingenious arguments and experiments, Sadler and Mesham proved that *none* of these hypotheses was tenable. They had reached an impasse. They had exhausted all of their hypotheses and had failed "to provide a complete solution of this peculiarly elusive problem." One might perhaps think that more than one unknown characteristic radiation had been excited by the primary rays. "The authors incline more to the belief," they concluded, "that an actual modification of the X ray occurs in its passage through matter, a general softening taking place. . . . One thing appears to be firmly established, viz., the harder the rays, the more profoundly are they modified in their passage through matter."[101]

Sadler and Mesham's experiments (1912), as well as those of J. P. V. Madsen (1909) and D. C. H. Florance (1910), stimulated further experimental work by J. A. Gray of McGill University in 1913. "Florance," wrote Gray, "ascribed the secondary γ rays to scattering of the primary rays, as the quality of the secondary rays appeared to be independent of the nature of the radiator. He obtained the important result that the secondary or scattered rays became less penetrating as the angle of scattering increased. . . ." What troubled Gray, however, was his hunch that Madsen's and Florance's selective scattering explanation of the softening, the one "usually given," was "probably not sufficient" to account for the facts. Gray therefore felt it was necessary to re-examine the "question of the scattering of γ rays. . . ."[102]

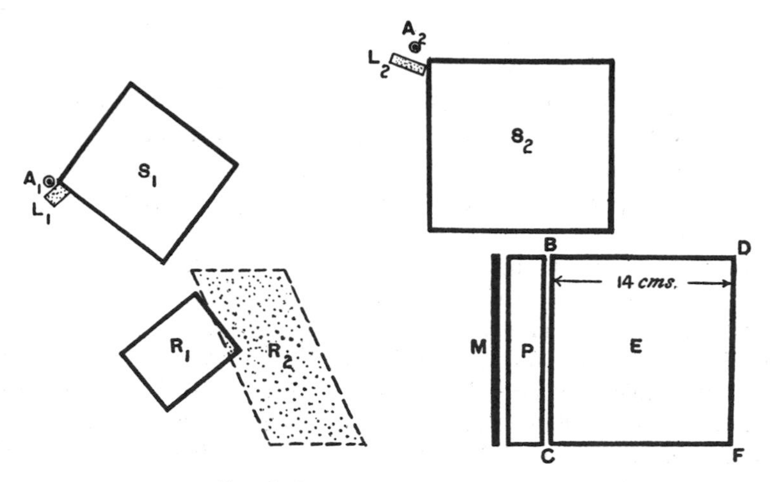

FIG. 5. Gray's 1913 scattering setup.

Gray's apparatus is shown in Fig. 5[103]—the picture is actually a composite of two distinct setups labeled by the subscripts 1 and 2, respectively. Consider the former arrangement, in which primary γ-rays from the source A_1 pass through a lead absorber L_1, are scattered by the radiator R_1 (carbon, iron, or lead), and finally, after passing through a lead absorber M of variable thickness and a lead filter P of constant thickness, enter the detecting electroscope E. In this arrangement, the scattering angle θ is about 50°; in the analogous $A_2 - L_2 - R_2 - E$ arrangement, it is about 110°. Now for a given primary absorber L_1 (or L_2) and a given radiator R_1 (or R_2), Gray gradually increased the thickness of the secondary absorber M to determine the quality of the scattered radiation. By changing the thickness of the primary absorber L_1 (or L_2), he was also able to determine the quality of the primary radiation.

Some of Gray's numerous measurements were particularly suggestive. First, he found that, for a given secondary absorber thickness M (say, 3.0 mm), the recorded intensity (in relative units) was nearly the same for γ-rays scattered by carbon (2.39) and by iron (2.35). When scattered by lead, it was lower—which was to be expected, since lead absorbs much more radiation internally than either carbon or iron. Gray concluded from these and similar measurements that the "quantity of radiation scattered per unit mass is approximately independent of the nature of the radiator." In other words, the intensity of the scattered radiation is approximately proportional to either the atomic weight or the atomic number of the scatterer.

Second, Gray observed that if he changed the scattering angle from 50° to 110° (while keeping M constant at 3.0 mm), the intensity of the radiation scattered by carbon decreased from 2.39 to 0.37, which was *almost exactly the same decrease* as that observed when scattered by iron (2.35 to 0.38). Gray concluded: "The quality of the radiation scattered depends on the angle of scattering, and not on the nature of the radiator"—he had therefore clearly confirmed "Florance's result that the greater the angle of scattering the softer the radiation."

Finally, Gray found that for a given scattering angle (say, 110°) and for a given scatterer (say, carbon), the recorded intensity was reduced by approximately the same amount (to 12% of its original value) *either* when a very thin absorber M (3.0 mm) was inserted into the secondary beam *or* when a very thick absorber L_2 (2.94 cm) was inserted into the primary beam. Now, if the radiation underwent no change in quality in the scatterer, the radiation entering it and leaving it would be identical, and therefore *identical* absorber thicknesses M and L_2 should have been required. Gray concluded: "It is evident . . . that some change must take place in the quality of the γ rays when they penetrate matter, either direct or by scattering. The possible production of characteristic radiations is not considered, as the quality of the secondary rays is very nearly independent of the nature of the radiator."[104]

Two possible explanations for the change in quality occurred to Gray: either the primary rays were somehow continuously transformed into softer rays solely as a result of passing through increasing thicknesses of matter; or, when "homogeneous γ rays are scattered there is a change of quality, the scattered rays being softer the greater the angle of scattering." The first possibility could be ruled out, since it was known that homogeneous X-rays passing through increasing thicknesses of matter were only diminished in intensity, and not softened. Thus, Gray concluded, we are

> forced to the second explanation, and this is quite sufficient. The change in quality is probably very small, when the angle of scattering is small, as experiment shows that in this case the scattered rays do not differ much in penetrating power from the primary. A certain similarity to the scattering of α rays may here be noted. When an α ray is scattered through a large angle there is little doubt that it loses velocity, the loss being greater the greater the angle of scattering. . . .
>
> Sadler and Mesham found that X rays are softened by scattering. The change in quality appeared to be greater the greater the penetrating power of the X-rays, and this point is supported by the present experiments. . . . It seems quite possible that the change in quality is small for very soft X rays. It will be seen that the rays scattered through 50° are still much softer than the primary rays even after passing through a centimetre of lead, so the softening must happen to every type of γ ray scattered.[105]

D. C. H. Florance, the physicist who inaugurated this striking series of experiments, had the last word on the subject in 1914—before the outbreak of

World War I disrupted research in laboratories everywhere. While Florance reported new experiments,[106] they were in essence closely related to his earlier ones. What was entirely new was Florance's interpretation of his experiments. He now wholeheartedly endorsed Sadler and Mesham's, and Gray's, conclusions:

> If the secondary γ radiation we have been examining is a result of scattering, then the scattering of γ rays must involve some modification in the primary γ rays. The early view that scattering consisted in the sifting out of the various components of the original primary beam is no longer tenable. This is obvious from the fact that after the primary beam has passed through 1 cm. of lead and then through a lead radiator 1 cm. thick, there is still present a soft secondary radiation which must have been completely absorbed had it been an untransformed contituent of the primary beam. . . .
>
> We seem, therefore, to be reduced to two conclusions, that either these secondary γ rays are the primary rays scattered and in the course of scattering the rays have lost energy and have become altered in type, or a complete transformation has taken place and these radiations are true secondary [i.e., characteristic] radiations. This latter hypothesis is not supported by experimental evidence. The secondary radiation is not in any way characteristic of the material from which it emerges. . . .
>
> It is necessary to adopt the first view, that the primary γ rays during the process of scattering lose energy and are in consequence modified in type.[107]

D. Einstein's Light Quantum Hypothesis and Its Reception (1905–1911)

1. Introduction

The theories and experiments discussed so far are perhaps not widely known today. By contrast, everyone today has heard of Einstein's 1905 light quantum hypothesis, and therefore it may have seemed strange that although we have discussed researches of a relatively late date, virtually no mention has been made of Einstein's work. This obvious omission, rather than being arbitrary, has been in keeping with the history of the period: neither Thomson, nor Barkla, nor Bragg, nor most of the other physicists we have mentioned, was in any fundamental way influenced by Einstein's 1905 light quantum hypothesis.[108] By the time Thomson, Barkla, and Bragg, working in England and Australia, became aware of Einstein's hypothesis, their own research programs had been established firmly enough to guide their investigations. They derived little insight from Einstein's unfamiliar arguments and bold hypothesis.

Bragg, whose views were superficially at least most akin to Einstein's, was explicit on this point. Remarking that "it is easy to miss a single reference when one is in a very isolated laboratory. . . ,"[109] he stated: "When I first put forward the neutral pair theory I was ignorant of the work of Einstein and was guided only by the results of experimental investigation on the behavior of the new [γ] rays. I did not think of carrying over the idea to the theory of light; on the contrary, I had hopes of proving that no connection existed between the two kinds of radiation."[110] By 1910–1911, as we have seen, he still saw no close connection between his hypothesis and "Dr. Stark's quantum." Ironically, Bragg did not realize how very closely he was echoing words of Einstein himself when in 1912 he vigorously urged the "search for a possible scheme of greater comprehensiveness, under which the light wave and the corpuscular X ray may appear as the extreme presentments of some general effect."[111]

Initially at least, the lack of influence was mutual: Einstein did not once refer to X-rays in his 1905 paper. Two reasons come immediately to mind: First, the nature of X-rays, as we have seen, was very problematic in 1905; second, and far more fundamental, Einstein's approach and motivation were entirely different from all other contemporary physicists. His basic concern was not the nature of X-rays or γ-rays; his concern was the nature of radiation. His light quantum hypothesis, which he himself termed "very revolutionary,"[112] challenged perhaps the major triumph of nineteenth century physics, the electromagnetic wave theory of light. It inaugurated a largely independent chain of theoretical and experimental developments, which we shall now sketch, building upon the work of Martin J. Klein.

2. *The Origin of Einstein's Hypothesis*

Einstein gave two penetrating arguments for light quanta, one negative, the other positive. His negative argument was that classical radiation theory is fundamentally inadequate: it cannot account for the observed spectral distribution of black-body radiation, and it leads inexorably to the blatantly false conclusion that the total radiant energy in a cavity is infinite—the "ultraviolet catastrophe," as Ehrenfest later termed it.[113]

Einstein developed his positive argument in two stages. First, he derived an expression for the entropy of black-body radiation in the Wien's law spectral region (the high-frequency, low-temperature region) and used this expression to evaluate the difference in entropy $S - S_o$ associated with a change in volume $v - v_o$ of the radiation at constant energy E: His result was

$$S - S_o = \frac{E}{\beta\nu} \ln\left(\frac{v}{v_o}\right), \tag{1.4}$$

where ν is the frequency of the radiation and β is the constant in the exponential of Wien's law. Secondly, he considered the case of n particles moving about freely and independently inside a given volume v_o and, by using the statistical version of the second law of thermodynamics in conjunction with his own strongly physical interpretation of the probability, he determined the change in entropy $S - S_o$ of the system if all n particles were accidentally found in a given sub-volume v at a randomly chosen instant of time. His result was

$$S - S_o = R\left(\frac{n}{N_o}\right) \ln\left(\frac{v}{v_o}\right), \tag{1.5}$$

where R is the ideal gas constant and N_o is Avogadro's number. Now, from the striking formal similarity between these two expressions, which shows that $E \leftrightarrow n(R\beta/N_o)\nu$, Einstein came to what he considered to be an inevitable conclusion: "Monochromatic radiation of small energy density (within the validity range of the Wien radiation formula) behaves in thermodynamic theoretical relationships as though it consisted of distinct independent energy quanta of magnitude $R\beta\nu/N_o$."[114] The fact that Einstein consistently used the cumbersome combination of constants $R\beta/N_o$, instead of Planck's constant h, merely serves to emphasize that Einstein's arguments were entirely independent—even of Planck's.

Einstein pointed out that his light quantum hypothesis suggested natural interpretations of (1) Stokes' rule of fluorescence, according to which the frequency of the fluorescent radiation emitted by an element is always less than, or equal to, the frequency of the incident light; (2) the ionization of gases by ultraviolet light; and (3) the photoelectric effect, in which ultraviolet light incident on a metal plate causes the emission of electrons. Einstein concluded that the photoelectric effect, now the most famous of the three phenomena, should be governed by the equation

$$\Pi e = \frac{R\beta\nu}{N_o} - P, \tag{1.6}$$

where Π is the (positive) potential "large enough to prevent a discharge" of the plate, e is the charge of the electron, and P is the amount of energy required to just remove an electron from the plate. Einstein laconically remarked, "If the derived formula is correct, then Π must be a linear function of the frequency whose slope depends on the nature of the material being studied."[115] This was an extremely bold prediction, since the only extensive experiments that had been carried out to date, Lenard's "trail-blazing" *(bahnbrechende)* experiments of 1902,[116] indicated only that the energy of the emitted electrons increases with increasing frequency: no specific functional dependence had been definitely established—nor would be until 1915.

To arrive at equation (1.6), of course, Einstein assumed that the entire energy of the incident quantum is transferred to the electron in the metal. It is well worth noting that Einstein also did "not exclude the possibility that electrons can absorb only parts of the energy of light quanta."[117] In that case the above equation would become an inequality:

$$\Pi e \leqq \frac{R\beta\nu}{N_0} - P. \tag{1.7}$$

The routes taken in recognizing the full significance of this inequality are, of course, the subject of this book.

3. Einstein's Subsequent Insights and Their Reception

For several years after 1905, Einstein was, as he wrote his friend J. J. Laub in 1908, "ceaselessly occupied with the question of the constitution of radiation and [was] . . . in correspondence on this question with H. A. Lorentz and Planck." One year earlier he had been eminently successful in applying quantum ideas to the puzzling problem of the specific heats of solids,[118] but his thoughts constantly returned to the nature of radiation. He was convinced that "This quantum question is so incredibly important and difficult that everyone should busy himself on it." He added: "I have already succeeded in working out something which may be related to it, but I have serious reasons for still thinking that it is rubbish."[119] It appears probable that in this earthy self-criticism Einstein was referring to results he later published in 1909 in an article entitled "On the Present State of the Radiation Problem."[120] He set the tone of this article in his opening remarks: "I communicate the following in the opinion that it is to advantage if everyone who has reflected seriously on this matter share their views, even if they have not been able to bring them to a definite result."[121] Between 1905 and 1909, Einstein had not succeeded in finding a comprehensive theory of radiation, but he had discovered new arguments for taking his light quantum hypothesis seriously.

Einstein's new arguments dealt with what is known in statistical mechanics as fluctuation phenomena, a field of study whose fruitfulness Einstein himself had first demonstrated (notably in his work on Brownian motion[122]) but one which was, at the same time, very unfamiliar to most of his contemporaries. He now displayed its applicability to the energy and momentum fluctuations of black-body radiation.

Consider a cavity of volume V containing black-body radiation at constant temperature T, and let ϵ denote the instantaneous departure of the energy of the radiation from its equilibrium value. The first result that Einstein obtained, by an elegant (and now standard) analysis, was the following expression for the mean square energy fluctuation $\overline{\epsilon^2}$ of the radiation:

$$\overline{\epsilon^2} = \frac{R}{N_o} T^2 V d\nu \frac{d\rho}{dT}. \tag{1.8}$$

Sincc this expression depends upon the spectral distribution or energy density $\rho(\nu,T)$ of the radiation, Einstein had to specify this function to complete his calculation. But which of the various functions should be chosen? Four years earlier he had rested his argument for light quanta on Wien's law. Now Einstein saw that he could turn the entire argument around. Planck's law had been experimentally verified again and again. Why not accept it as a law of nature and see what might be deduced from the mean square energy fluctuation predicted from it? Einstein therefore substituted Planck's law into equation (1.8) and obtained a result equivalent to

$$\overline{\epsilon^2} = V d\nu [h\nu\rho + \frac{c^3}{8\pi\nu^2} \rho^2], \tag{1.9}$$

where c is the velocity of light. By a further, related argument, Einstein concluded that the mean square *momentum* fluctuation $\overline{\Delta^2}$ of black-body radiation, during a given time interval and at constant temperature, is given by:

$$\overline{\Delta^2} = K d\nu [h\nu\rho + \frac{c^3}{8\pi\nu^2} \rho^2], \tag{1.10}$$

where K is a constant (its exact value is not important for our present purposes).[123]

To Einstein the striking common feature of equations (1.9) and (1.10) was that both consisted of a sum of two terms, each of which could be simply interpreted. The second term, as Einstein argued on dimensional grounds, would be precisely the expression which would be obtained if the Rayleigh-Jeans law were universally valid; it would arise naturally from a consideration of the interference of waves. The first term, as Einstein concluded from statistical mechanical considerations, would be precisely the expression which would be obtained if Wien's law were universally valid; it would arise naturally "if the radiation consisted of independently moving point quanta of energy $h\nu$."[124] The *coexistence* of *both* terms therefore implied a coexistence of waves and quanta in black-body radiation, its wave characteristics dominating at low frequencies and its quantum characteristics dominating at high frequencies. This paper of Einstein's therefore heralded the birth of what came to be known as the "wave-particle duality," although Einstein himself did not emphasize this point at this time: his main purpose, apparently, was to demonstrate that Planck's law was in no way incompatible with his light quantum hypothesis.[125]

Einstein concluded his paper with a plea for the experimental testing of the quantum theory, "one of the most important tasks which contemporary physics has to solve."[126] He delineated three possibilities. The first concerned the implications of the quantum theory for the specific heats of solids. These experiments, when eventually carried out by Walther Nernst, were decisive in motivating Nernst to accept the quantum theory and spread its influence.[127] The second concerned the predictions of the quantum theory regarding the absorption of radiation, for example, by gases. These predictions, even though Einstein had already discussed them at length in his 1905 paper, had "unfortunately . . . remained unobserved until now." The third encompassed those elementary processes in which the energy of the final state was never greater than that of the initial state. Included in this group were investigations on Stokes' rule of fluorescence, on the velocities of cathode rays produced by Röntgen rays, and on "the interesting application which Herr Stark has made of the quantum theory to explain the spectral energy distribution in the lines emitted by canal rays."[128]

Einstein was here referring to Johannes Stark, then *Privatdozent* at the Technische Hochschule in Hanover, and to Stark's explanation of the origin of a companion spectral line emitted by hydrogen "canal rays" (ionized hydrogen molecules). In very general terms, Stark had assumed that in collisions with stationary gas molecules the canal rays can undergo changes of form (and hence energy) only if the energy available is greater than, or equal to, $h\nu$.[129] Curiously, Stark's willingness to accept and develop quantum ideas—which was very much against the trend of the times—may have been motivated to some degree by their very unorthodoxy. As Armin Hermann has written, "Opposition to commonly taught opinions *(Lehrmeinung)* always seemed to Stark a necessity of life."[130] In any event, although Stark did not mention Einstein in a 1907 paper, his earliest, in which he discussed the photoelectric effect, the two physicists had previously corresponded, and by late 1907 both recognized their mutual interest. On February 22, 1908, Einstein expressed the hope of meeting Stark soon to discuss with him personally *"physikalische Dinge."*[131]

This hope was realized at the 81st assembly of the Gesellschaft Deutscher Naturforscher und Ärtze held in Salzburg in September 1909, which was the first conference of German scientists in which Einstein participated. It also took place, incidentally, shortly after both Einstein and Stark had accepted new positions, the former having exchanged his carefree life as a patent officer in Bern for an *ausserordentliche* professorship at the University of Zürich, the latter having exchanged his *Dozentur* at Hanover for a professorship of experimental physics in the Technische Hochschule in Aachen, a move strongly supported by Sommerfeld, who had left Aachen for Munich in 1906. It was at Aachen that Stark eventually achieved fame for his discovery of the "Stark effect" (1913), but for our history it is more important to note that by the time Stark had taken up residence there, he had become firmly convinced, as he later wrote, that from "an epistemological standpoint, the light quantum hypothesis is simpler than the ether-wave hypothesis."[132] No doubt it was this conviction that made Stark extremely receptive to the ideas which Einstein expressed at Salzburg.

In his Salzburg paper, "On the Development of our Views on the Nature and Constitution of Radiation,"[133] Einstein argued that there was undeniably an "extensive group of data"—on photochemical reactions, on the photoelectric effect, and on the emission of radiation by atoms and molecules —"which show that light has certain fundamental properties that can be understood much more readily from the standpoint of the Newtonian emission theory than from the standpoint of the wave theory." He intended, therefore, to provide a foundation for his opinion "that the next phase of the development of theoretical physics will bring a theory of light that can be interpreted as a kind of fusion of the wave and emission theories. . . ." This meant that "a profound change in our views of the nature and constitution of light is indispensable."[134] He pointed out that unfortunately while his own "so-called relativity theory" proved that the transmission of radiant energy is equivalent to a transference of mass, it changed nothing regarding "our conception of the structure of radiation, especially the distribution of energy in irradiated space. . . ."[135]

Einstein saw as the root of the wave theory's difficulties the circumstance that all elementary processes consistent with Maxwell's equations do not have inverse processes which are actually found in nature. A spreading spherical wave never contracts to its point of origin—even though Maxwell's equations are perfectly consistent with such a process. Furthermore, radiant energy apparently remains localized while traveling in space, as indicated for example by the observation that, if cathode rays fall on a metal plate P_1 and produce Röntgen rays, the Röntgen rays in turn can eject secondary cathode rays from another metal plate P_2 some distance away from P_1. Most significantly, the velocity of the secondary cathode rays was observed to be of the same order of magnitude as that of the primary cathode rays. It depends "as far as we know today, neither on the separation of the plates P_1 and P_2, nor on the intensity of the primary cathode rays, but exclusively on the velocity of the primary cathode rays. . . ." These facts suggested that Röntgen rays were not spreading spherical waves; they were far more consistent with the old "emission theory of Newton." In sum, Einstein told his Salzburg audience that the "emission of radiation appears to be a directed elementary process. Furthermore, the impression is gained that the process of the production of Röntgen rays in P_1 and the production of secondary cathode rays in P_2 are essentially inverse processes."[136]

But to recognize the wave theory's inadequacies was unfortunately not equivalent to resolving them. Important clues, however, could be gathered from Planck's radiation formula, in particular by asking "whether from it something may be deduced concerning the constitution of radiation." This stance, which was identical to the one he had adopted earlier in the year, once again led Einstein to rederive and discuss the expression, equation (1.10), which he had obtained earlier for the mean square momentum fluctuations of black-body radiation. He now, however, emphasized its full significance: "It is as if two independent different causes were present which produced a fluctuation in the radiation pressure." But the major difficulty still remained. This result, while striking, did not offer much of a clue to the comprehensive theory of light Einstein was seeking. Suppose that the phenomena of interference and diffraction were still unknown, but that one had derived equation (1.10), where ν in the *wave term* was regarded as a "parameter of unknown meaning" determining the color. "Who," Einstein asked, "would have enough imagination to construct the wave theory of light on this foundation?"[137]

While a comprehensive constructive theory of light therefore still eluded Einstein, he had nevertheless formulated certain ideas on where to begin in order to develop such a theory. In his earlier paper, he had suggested that the dimensional equivalence of e^2/c and h was significant.[138] This might well be the clue, he then felt, to the modification required in the general theories of molecular mechanics and electrodynamics which would account for light quanta. By the time of the Salzburg meeting he was no closer to finding this modification, but he had formed a more concrete mental picture of what he termed the "most natural conception" of light. He suggested that the "behavior of the electromagnetic fields of light is just as bound up with singular points as is the behavior of electrostatic fields according to the theory of electrons." Consider one of these electromagnetic singular points. "I imagine, roughly, such a singular point to be surrounded by a field of force which essentially possesses the character of a plane wave and whose amplitude decreases with distance from the singular point."[139] Many such overlapping fields of force could perhaps produce the electromagnetic wave field —a picture which might prove capable of reconciling the particle and wave manifestations of light. With characteristic self-criticism, however, Einstein added that until this idea led to an exact theory, no value should be attached to it.

In the extensive discussions that followed Einstein's paper at Salzburg, the first person to respond was Planck. As Martin Klein has remarked, it might perhaps be expected that of all the physicists present, Planck would be the one most likely to appreciate Einstein's arguments. Planck, however, in spite of Einstein's cautionary remarks, saw the light quantum hypothesis as being completely opposed to the usual electrodynamics. He thought that Maxwell's equations would have to be given up—and that, he said, "appears to be a step which in my opinion is not yet called for."[140]

The man who answered Planck, and apparently the only physicist who supported Einstein,[141] was Johannes Stark.

Did Planck believe that there were no reasons for regarding light as concentrated in quanta? How else could one understand the fact that electromagnetic radiation leaving a Röntgen tube and traveling through "large distances, up to 10 meters, is still concentrated enough to act on a single electron?" Planck answered: "With Röntgen rays it is another matter; I should not like to maintain much concerning them. Stark has introduced something in favor of the quantum theory; I would like to introduce something against it. This is the occurrence of interference at the colossal path difference of hundreds of thousands of wavelengths. If a quantum interferes with itself, it has to have a [spatial] extension of hundreds of thousands of wavelengths. That is certainly . . . a difficulty."[142]

Stark agreed, but expressed the hope that it was not an insuperable one. Einstein, who had the last word in the discussion, was of the same opinion: "The interference phenomena may perhaps not be so troublesome as imagined, for the following reasons: one dare not assume that radiation consists of quanta which do not interact with each other. On that view it would be impossible to explain interference phenomena." He repeated his idea that a light quantum might be a "singularity surrounded by a large vector field," and he suggested, as before, that the superposition of a large number of these fields might account for interference. His final words were completely unambiguous: "I do not see a fundamental difficulty in interference phenomena."[143]

4. *Stark Concludes Quanta Possess Linear Momentum*

Einstein's Salzburg audience included, besides Planck and Stark, Arnold Sommerfeld, Heinrich Rubens, Max Born, P. S. Epstein, Fritz Reiche, Lise Meitner, and many others, most of whom he met personally for the first time in Salzburg. Of those present, however, no one seems to have been influenced more deeply by Einstein's paper and the succeeding discussions than Johannes Stark. Immediately on his return to Aachen, Stark began writing a series of truly remarkable papers which were subsequently published in the *Physikalische Zeitschrift.* All were based upon the light quantum hypothesis, and all dealt primarily with the experimental problem discussed at length by Einstein at Salzburg: the production of Röntgen rays by cathode rays, and vice versa.

In his first paper,[144] Stark drew a far-reaching, very general conclusion, one which no one before him had drawn explicitly. He wrote, in the clearest possible terms:

According to the ether-wave hypothesis, the total electromagnetic momentum, even in an elementary emission process, is zero. . . . And if . . . an electron capable of emitting [radiant energy] collides with another material particle . . . , the sum of the mechanical momenta remains constant ($m_1\mathbf{v}_1 + m_2\mathbf{v}_2 = \text{const}$), and a transformation of mechanical into electromagnetic momentum, or vice versa, is impossible.

In this respect, the conclusions of the light quantum hypothesis are different. According to this hypothesis, the total [radiant] momentum emitted by an accelerated electron is different from zero and, indeed, its absolute value is given by $h\cdot n/c$, where h is Planck's quantum of action and n the frequency. Its direction is given by the direction of motion of the center of gravity of the quantum of radiation. . . . In a collision, if $m_1\mathbf{v}_1$ is the mechanical momentum of the electron, $m_2\mathbf{v}_2$ that of the struck particle before the collision, $m_1\mathbf{v}_1'$ and $m_2\mathbf{v}_2'$ these quantities after the collision, and if the electromagnetic momentum emitted during the collision is $(hn/c^2)\mathbf{c}$, where $\mathbf{c}$ is the velocity of propagation of its center of gravity, according to the law of conservation of momentum the following equation holds:

$$m_1\mathbf{v}_1 + m_2\mathbf{v}_2 = m_1\mathbf{v}_1' + m_2\mathbf{v}_2' + \frac{hn}{c^2}\mathbf{c}. \qquad [1.11]$$

If the elementary process takes place from left to right (in the sense of this equation), mechanical momentum will be transformed into electromagnetic momentum; if it takes place from right to left, electromagnetic momentum will be transformed into mechanical momentum.[145]

While the value hn/c (or $h\nu/c$) for the momentum of a light quantum was implicit in Einstein's equation (1.10), Stark was the first physicist to state

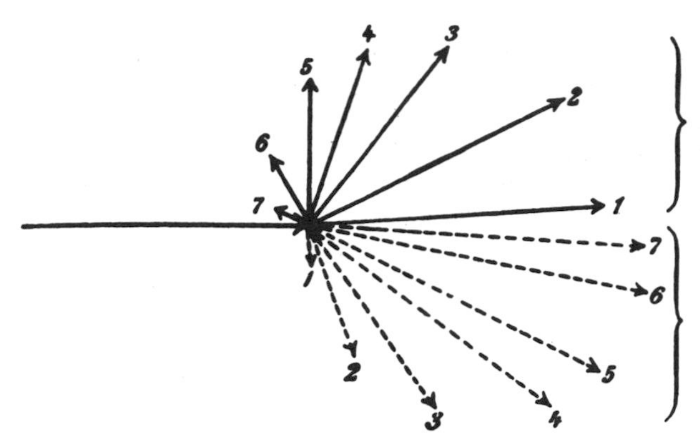

FIG. 6. Stark's sketch showing the production of X-rays (dashed lines) in a cathode ray–cathode ray collision.

this fact explicitly. And Stark's statement is the first clear statement in the literature of the law of conservation of momentum as applied to a process involving the interaction of radiation and matter. Moreover, Stark was fully aware of its significance. He immediately applied it to the problem Einstein had posed: the production of X-rays by cathode rays. He accompanied his analysis with the following diagrams:

[In Fig. 7, Stark explained,] "is shown . . . the vector parallelogram (after the impact) for the mechanical momentum [of the incident and struck electrons] and the electromagnetic momentum of the emitted quantum of radiation. For the sake of simplicity, we assume that the mechanical momentum [$m_2\mathbf{v}_2$] of the struck particle is, relative to $m_1\mathbf{v}_1$, zero before the impact. . . . For this special case, we have drawn in [Fig. 6] . . . the directions (broken lines) in which quanta of radiation are emitted as a result of the acceleration of the electrons on impact; the momentum of each quantum is proportional to the length of the arrow. [Note that by adding a given cathode ray vector (solid line) to its corresponding X-ray vector

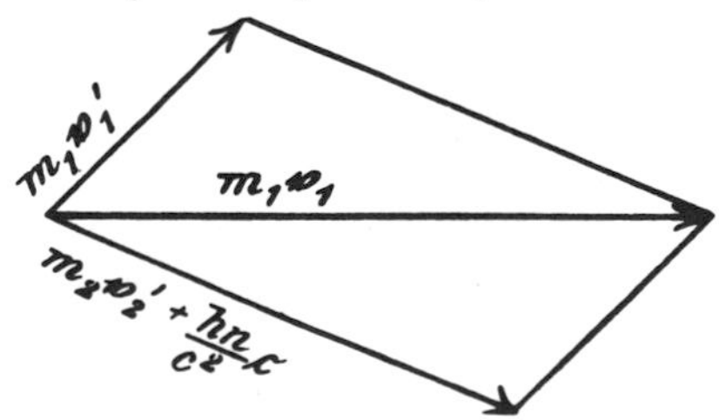

FIG. 7. Stark's vector diagram showing conservation of momentum in the collision process in Fig. 6.

(broken line) one always obtains the incident cathode ray vector.] The distribution of the radiation in space remains constant under rotation of the lower half-plane about the axis of the incident cathode rays. In [Fig. 6] . . . one sees at a glance that . . . the distribution of the emitted [Röntgen] rays, according to the light quantum hypothesis, is asymmetric with respect to the plane (normal plane) passing through the point of impact and normal to the direction of the incident bundle of cathode rays. It is probable that Röntgen radiation will be emitted in the direction of the incident cathode ray bundle, but not in the opposite direction.

Furthermore, on the basis of the light quantum hypothesis, we find the following special consequence for the ideal special case of Röntgen ray emission. For different (large) values of the accelerations [experienced by the incident electrons,] the absolute value hn/c of the emitted electromagnetic momentum is different. Since h and c are constant, we find Röntgen ray quanta of different frequencies in different directions. Thus, Röntgen rays of different absorbabilities will be emitted by the same material. The intensity J_n of Röntgen radiation of a certain frequency is, therefore, a function [1] of the angle α between the direction of emission and the direction of the incident cathode ray bundle, [2] of the frequency n, and [3] evidently also of the velocity v_1 of the incident bundle of cathode rays. Thus, we have to set $J_n = zf(\alpha,n,v_1)$.[146]

Written as they were in November 1909, these paragraphs clearly demonstrate Stark's keen physical insight and imagination at this time, even though his work in general was by no means free of error.[147] He concluded his paper with a plea for the experimental testing of his predictions. He specifically pointed out that, if the angular asymmetry in the intensity and the angular variation in the absorbability of the secondary X-rays existed, this would constitute experimental proof in favor of the light quantum theory. A further test was also possible. Thinking perhaps of Einstein's words in Salzburg, and of his own reply to Planck, Stark wrote that one should determine as precisely as possible if "Röntgen rays, even at great distances from their source, cause cathode rays to be emitted whose energy is of the order of magnitude of the energy of the individual primary cathode ray electrons. . . ."

He pursued this line of argument one step further by explicitly pointing out that the

. . . inverse of the phenomenon of the emission of Röntgen rays by cathode rays is the emission of secondary cathode rays by the absorption of Röntgen rays. It is to be suspected that if Röntgen rays are absorbed in a thin plate, more secondary cathode rays are emitted on the reverse side of the normal plane [i.e., in the direction of the incident beam] . . . than on the front side. Furthermore, the cathode rays emitted toward the front side [i.e., opposite to the direction of the incident beam] have a smaller velocity than those emitted toward the reverse side.

The β-rays of the radioactive elements are fast cathode rays; the γ-rays are probably high frequency Röntgen rays. Carrying out the above analogy, it may be concluded that β-rays on impact on the atoms of a solid will produce very few γ-rays in the direction opposite to their direction [of motion]. . . . Conversely, γ-rays from radioactive elements may emit, at the point of their absorption in solid bodies, more and faster secondary β-rays in their direction of motion than in the opposite direction.[148]

It is true that in his analysis Stark never explicitly set up the additional equation expressing conservation of energy, incorporating the correct relativistic mass variation into it—he did not solve the problem in detail. It is also true that Stark's attention was focussed on the production and absorption of X-rays rather than on their scattering. Nevertheless, there can be little doubt that Stark's logic led him a fair distance along the route to making a discovery of the first magnitude. At first sight, it is perhaps even somewhat surprising that Stark, who was well known for such tactics,[149] did not come forward with something approaching a priority claim when this discovery—the subject of this book—was actually made some fourteen years later.[150] One can only speculate on the reasons for Stark apparently not doing so. Perhaps they were linked to his later, blind antagonism toward Einstein's theory of relativity, as well as toward the "dogmatism" of the quantum theory. Perhaps this was another, later, and very ironical instance of Stark's "Opposition to commonly taught opinions."

But what was the immediate fate of Stark's theory? As it developed, Stark in a sense had directed the bulk of his discussion to the "wrong problem," to the problem of the production of X-rays by cathode rays, which was the inverse of the problem that later would become so important for the quantum theory of radiation. Stark's paper, in fact, was hardly in print before his theory was challenged—by Arnold Sommerfeld. (We recall that it was the subsequent Stark-Sommerfeld controversy that attracted W. H. Bragg's attention and prompted Bragg to initiate a correspondence with Sommerfeld.)

Sommerfeld's attitude toward Stark's quantum theory of X-ray production was as negative as his attitude toward Bragg's neutral pair theory of X-rays and γ-rays. As he wrote: "The observations recently published by J. Stark, which seem to point to an explicit asymmetry in the Röntgen radiation emitted by a carbon anticathode, motivate me to make known my ideas, ideas that have occupied me for a long time. I begin with the familiar Wiechert-Stokes view, which sees the origin of Röntgen radiation in the stopping of the cathode ray electron." In the production of the other type of Röntgen radiation, Barkla's characteristic (fluorescent) X-radiation, Sommerfeld pointed out, "an absorption and emission of energy takes place in the atom. It is quite possible that here Planck's quantum of action plays a role. . . . It does not seem to have anything to do with the first, 'stopping part', however. . . .

The properties of this part should be attempted to be understood from purely electromagnetic theoretical considerations."[151] And Sommerfeld then went on to develop his well known theory of *Bremsstrahlung.*

This theory proved that the pulse of radiation emitted by a rapidly stopped electron is confined to the region between two coaxial, narrow cones opening in the forward direction with apex at the electron. Or, as a later account put it, Sommerfeld proved that his theory "would give just the sort of space distribution of intensity that Stark had found, and also that the Doppler effect itself would give his changes of hardness with direction, all on the basis of Stokes' theory of electromagnetic pulses."[152] Sommerfeld therefore felt justified in asserting that the radiation had "absolutely the character of a projectile and in its energy localization is no longer appreciably different from a corpuscular radiation or from the hypothetical light quantum."[153]

Sommerfeld's challenge very quickly stimulated a correspondence with Stark. At first it was polite, but with touches of sarcasm. On December 4, 1909, Sommerfeld disputed Stark's contention that the asymmetric emission of secondary X-rays would offer proof of the quantum hypothesis, remarking that: "If such a justly famous researcher as yourself commits such a serious error, he should set it right immediately in the interests of the judicious readers [of the paper]."[154] Six days later, Sommerfeld wrote Stark that "Nothing could be further from my mind than to want to initiate a controversy with you. This [controversy] would be very unequal, since you are my superior in experimental ideas, and I am yours in theoretical clarity."[155]

Initially, Stark refused to bend to Sommerfeld's arguments. He responded to Sommerfeld in print, arguing that "the asymmetric emission of momentum, and the appearance of a radiation pressure, is right from the beginning an integral part of my views just as it is of the views of Mr. Sommerfeld. In order to place the experimental problems on a base of easy to understand considerations, I have combined this theoretical idea with the Einstein light quantum hypothesis. It is true that this hypothesis is independent of Maxwell's differential equations of the electromagnetic field. However, at least until now, Mr. Sommerfeld has not published a proof that this hypothesis contradicts Maxwell's equations."[156] It also had the great advantage of readily explaining the apparent localization of radiant energy, which was its "main content."[157]

In spite of Stark's display of bravado, however, it soon became obvious that he was shaken by Sommerfeld's theory and criticism. Their personal relationship deteriorated very rapidly.[158] Moreover, in a final paper on the subject, Stark shifted his entire point of view. He focussed his attention not on *microscopic* elementary interactions, but on a *macroscopic* picture, developing equations to describe, for instance, the "asymmetric emission of negative electrons in the photoelectric effect."[159] Stark's later canal ray impact experiments, which were related to his theoretical ideas, were unsuccessful.[160]

There exists a brief but very interesting contemporary impression of Stark's work. In September 1910, the Japanese physicist Hantaro Nagaoka, who since 1906 had been Professor of Theoretical Physics in the Imperial University of Tokyo, first visited Rutherford's laboratory at the University of Manchester, and then toured Europe, visiting a number of other physicists. On February 22, 1911, he wrote his impressions to Rutherford. In this letter Nagaoka makes the following reference to Stark: "Stark in Aix-la-Chapelle [Aachen] was propounding his 'Lichtquantentheorie'; there is some doubt whether he will succeed in explaining the interference phenomena, or not. The Germans say that he is full of phantasies, which may be partly true."[161]

E. References

[1] "The Scattering of X Rays as Particles," *Am. J. Phys.* 29 (1961):817.

[2] "Über eine neue Art von Strahlen," *Sitzber. Würz. Ges.* (1895); reprinted in "Grundlegende Abhandlungen über die X-Strahlen," *Klassische Arbeiten deutscher Physiker,* Vol. 1 (Leipzig: J. A. Borth, 1954), pp. 14–15. English translation quoted from H. A. Boorse and L. Motz (eds.), *The World of the Atom,* Vol. 1 (New York: Basic Books Inc., 1966), pp. 396–397, For a discussion of Röntgen's work, see for example Bern Dibner, *The New Rays of Professor Röntgen* (Norwalk, Conn.: Burndy Library, 1963), or W. Robert Nitske, *The Life of Wilhelm Conrad Röntgen Discoverer of the X Ray* (Tucson: University of Arizona Press, 1971).

[3] "On the Nature of the Röntgen Rays," *Proc. Cambridge Phil. Soc.* 9 (1896): 216; reprinted in *Mathematical and Physical Papers,* Vol. 5 (Cambridge: Cambridge University, Press, 1905), p. 255; E. Wiechert, "Die Theorie der Elektrodynamik und die Röntgen' sche Entdeckung," *Schriften physik.-ökon. Ges. Königsberg* 37 (1896): 1–48, especially p. 45. A. Schuster also suggested the pulse theory. See E. Rutherford, *Radioactive Substances and their Radiations* (Cambridge: Cambridge University Press, 1913), p. 83; W. H. and W. L. Bragg, *X Rays and Crystal Structure,* 4th ed. (London: G. Bell & Sons, 1924), p. 2.

[4] For a discussion of the origin and development of Thomson's ideas, see Russell McCormmach, "J. J. Thomson and the Structure of Light," *Brit. J. Hist. Sci.* 3 (1967): 362–387.

[5] Published under the title *Electricity and Matter* (New Haven: Yale University Press, 1904). See also D. L. Anderson, *The Discovery of the Electron* (Princeton: D. Van Nostrand, 1964).

[6] *Electricity and Matter,* pp. 62–63.

[7] (London, 1903).

[8] See for example, J. A. Crowther, "Research Work in the Cavendish Laboratory," *Suppl. Nature* (December 18, 1926): 58. At times, Thomson had as many as 40 research students working under him, in addition to several visiting professors who came to him

for discussions. See, for example, Lord Rayleigh, "Joseph John Thomson," *Roy. Soc. Obit. Not. (London)* 3 (1939–1941): 595. Eventually Thomson himself and seven of his students, including his son, G. P. Thomson, won Nobel Prizes.

[9] 2d. ed., pp. 321–325. Thomson used units in which $mc^2 \rightarrow m$, and consequently his expression for σ_0 did not contain the c^4 factor, which has been inserted for the sake of consistency in notation.

[10] H. S. Allen, "Charles Glover Barkla," *Roy. Soc Obit. Not. (London)* 5 (1945–1948):343. For other accounts of Barkla's life and work, see R. J. Stephenson, "The Scientific Career of Charles Glover Barkla," *Am. J. Phys.* 35 (1967):140–152, and P. Forman, "Charles G. Barkla," *Dict. Sci. Biog.*, Vol. 1 (New York: Scribner's, 1970).

[11] "Energy of Secondary Röntgen Radiation," *Phil. Mag.* 7 (1904):543–560.

[12] See John L. Heilbron, "The Scattering of α and β Particles and Rutherford's Atom," *Arch. Hist. Exact Sci.* 4 (1968):247–307.

[13] "Polarized Röntgen Radiation," *Phil. Trans. Roy. Soc. London* 204 (1905):467–479.

[14] "Polarization in Secondary Röntgen Radiation," *Proc. Roy. Soc. London* [A] 77 (1906):247–255. See also "Secondary Röntgen Radiation," *Phil. Mag.* 11 (1906):812–828.

[15] "On the Ionization of Gases by Ultra-Violet Light and on the evidence as to the Structure of Light Afforded by its Electrical Effects," *Proc. Cambridge Phil. Soc.* 14 (1907):421.

[16] Villard's work is partially translated and discussed in Boorse and Motz (eds.), *The World of the Atom,* Vol. 1, pp. 466–468. See reference 2. It was Rutherford who gave γ-rays their name.

[17] "On the Secondary Radiation caused by the β and γ Rays of Radium," *Phil. Mag.* 8 (1904):669–685. Paschen's work is discussed on p. 684. For a discussion of Eve's life and work, see J. S. Foster, "Arthur Steward Eve," *Roy. Soc. Obit. Not. (London)* 6 (1948–1949):397.

[18] "β and γ Rays," (reference 17) p. 670.

[19] *Ibid.*, p. 685.

[20] *Ibid.* Barkla made a similar observation, but attached no importance to it. See "Energy of Secondary Röntgen Radiation," (reference 11) p. 560. Also see R. T. Beatty, "Secondary Röntgen Radiation in Air," *Phil. Mag.* 14 (1907):604–613.

[21] "Ionization of Gases," (reference 15) p. 423.

[22] *Ibid.*, p. 424.

[23] See, for example, *Max Jammer, The Conceptual Development of Quantum Mechanics* (New York: McGraw-Hill, 1966), p. 7.

[24] P. Lebedev, "Untersuchung über die Druckkräfte des Lichtes," *Ann. Physik,* 311 (1910):433–458.

[25] E. F. Nichols and G. F. Hull, "The Pressure due to Radiation," *Proc. Am. Acad. Arts Sci.* 38 (1903):557–599.

[26] For a full discussion, see Roger H. Stuewer, "William H. Bragg's Corpuscular Theory of X-Rays and γ-Rays," *Brit. J. Hist. Sci.* 5 (1971):158–181.

[27] E. N. da C. Andrade, "William Henry Bragg," *Roy. Soc. Obit. Not. (London)* 4 (1942–1944):280. For additional information on Bragg, see P. Forman's article in *Dict. Sci. Biog.*, Vol. 1 (New York: Scribner's, 1970).

[28] Andrade, "Bragg," (reference 27) p. 291.

[29] "On the absorption of α-rays, and on the classification of the α-rays of radium," *Phil. Mag.* 8 (1904):719–725.

[30] "A Comparison of Some forms of Electric Radiation," *Trans. Roy. Soc. S. Aust.* 31 (1907):79–93; "The Nature of Röntgen rays," *Trans. Roy. Soc. S. Aust.* 31 (1907):94–98; reprinted as "On the Properties and Natures of Various Electric Radiations," *Phil. Mag.* 14 (1907):429–449; *Ann. Rept. Smith. Inst.* (1907):195–214. Andrade, (reference 27) p. 280, discusses the relationship between Bragg's hypothesis and Lenard's "dynamids." For Lenard's work, see "Ueber die lichtelektrische Wirkung," *Ann. Physik.* 8 (1902):149–198.

[31] Bragg, "Comparison," (reference 30) p. 90. Not until 1908 did Rutherford and Geiger show that the α-particle is doubly charged. See Rutherford, *Radioactive Substances,* (reference 3) p. 136.

[32] Bragg, "Properties and Natures," (reference 30) p. 442.

[33] H. Haga und C. H. Wind, "Die Beugung der Röntgenstrahlen," *Ann. Physik* 10 (1903):305–312; "Über die Polarisation der Röntgenstrahlen und der Sekundärstrahlen," *Ann Physik.* 23 (1907):439–444.

[34] E. Marx, "Die Geschwindigkeit der Röntgenstrahlen," *Phys. Z.* 6 (1905):768–778 and 834–835.

[35] Bragg, "Properties and Natures," (reference 30) p. 448.

[36] "The Nature of X-Rays," *Nature* 76 (1907):661–662.

[37] "The Nature of γ and X-Rays," *Nature* 77 (1908):270–271.

[38] "An Experimental Investigation of the Nature of the γ Rays," *Phil. Mag.* 15 (1908):663–675.

[39] "Secondary Radiation from a Plate exposed to Rays from Radium," *Phil. Mag.* 14 (1907):176–187.

[40] Bragg, "γ and X-Rays," (reference 37) p. 270.

[41] "The Nature of Röntgen Rays," *Nature* 77 (1908):319–320.

[42] "Note on X-Rays and Scattered X-Rays," *Phil. Mag.* 15 (1908):288–296.

[43] "The Nature of the γ and X-Rays," *Nature* 77 (1908):560.

[44] "The Nature of the γ rays," *Proc. Cambridge Phil. Soc.* 14 (1906–1908):540.

[45] "On the Different Kinds of γ Rays of Radium, and the Secondary γ Rays which they Produce," *Phil. Mag.* 15 (1908):638–663. For Kleeman's experimental setup, see p. 641.

[46] *Ibid.*, p. 652. These observations were based on the obvious application of the equation $I(x) = I(0)e^{-\mu x}$, describing the attenuation of a beam of radiation, assumed homogeneous, in an absorber of thickness x.

[47] *Ibid.*, p. 661.

[48] "Homogeneous Secondary Röntgen Radiations," *Phil. Mag.* 16 (1908):550–584. Also see "Classification of Secondary Röntgen Radiators," *Nature* 77 (1908):343–344; "Absorption of Röntgen Rays," *Nature* 80 (1909):37; "Transformations of X-Rays," *Phil. Mag.* 18 (1909):107–132; and Barkla's Nobel Lecture, "Characteristic Röntgen Radiation," reprinted in *Nobel Lectures: Physics,* Vol. 1 (Amsterdam: Elsevier, 1967), pp. 392–399.

[49] "The Nature of X-Rays," *Nature* 78 (1908):7.

[50] "The Nature of the γ and X-Rays," Nature 78 (1908):294. Barkla responded again in "The Nature of X-Rays," *Nature* 78 (1908):665.

[51] "The Nature of the gamma and X-Rays," *Nature* 77 (1908):271; "An Experimental Investigation of the Nature of γ Rays—No. 2," *Phil. Mag.* 16 (1908):918–939 (with J.P.V. Madsen).

[52] Bragg and Madsen, "Experimental Investigation—No. 2," (reference 51) pp. 934–935.

[53] *Ibid.,* pp. 934–937.

[54] *Ibid.,* pp. 936–937.

[55] "The Nature of γ and X-Rays." *Nature* 77 (1908):509–510.

[56] Bragg and Madsen, "Experimental Investigation—No. 2," (reference 51) p. 938.

[57] "On a Want of Symmetry shown by Secondary X-Rays," *Phil. Mag.* 17 (1909):863 (with J. L. Glasson).

[58] "Secondary γ Radiation," *Phil. Mag.* 17 (1909):423–448. Madsen found that the ratio mentioned in the next sentence lay between 4.5 and 6.5, and this will become important in Chapter 3.

[59] *Ibid.,* pp. 423, 431, 441.

[60] *Ibid.,* p. 445.

[61] Three letters from W. H. Bragg to A. Sommerfeld are deposited in the Archive for History of Quantum Physics (AHQP) at the American Philosophical Society Library (Philadelphia), University of California Library (Berkeley), and the Universitets Institut for Teoretisk Fysik (Copenhagen); one letter from Sommerfeld to Bragg was found by Sir Lawrence Bragg, C. H., O. B. E., M. C., F. R. S., in his father's papers, and a copy of it was kindly sent to me by Sir Lawrence.

[62] For a published account of Sommerfeld's theory, see his paper "Über die Struktur der γ-Strahlen," *Sitzber. Akad. (München)* 41 (1911):1–60. For a modern discussion of the "forward peaking," see, for example, J. D. Jackson, *Classical Electrodynamics* (New York: Wiley, 1962), p. 473.

[63] For C. T. R. Wilson's original photographs and papers, see "On a Method of Making Visible the Paths of Ionizing Particles through a Gas," *Proc. Roy. Soc. (London)* [A] 85 (1911):285–288; "On an expansion Apparatus for Making Visible the Tracks of Ionizing Particles in Gases and some Results obtained by its Use," *Proc. Roy. Soc. (London)* [A] 87 (1912):277–292.

[64] "The Consequences of the Corpuscular Hypothesis of the γ and X Rays, and the Range of β Rays," *Phil. Mag.* 20 (1910):385–416.

[65] *Ibid.,* pp. 396, 415–416.

[66] "Energy Transformations of X-Rays," *Proc. Roy. Soc. (London)* [A] 85. (1911):349–365. See also W. H. Bragg, "The Mode of Ionization by X-Rays," *Phil. Mag.* 22 (1911):222–223; "On the Direct or Indirect Nature of the Ionization by X-rays," *Phil. Mag.* 23 (1912):647–650.

[67] Bragg and Madsen, "Experimental Investigation—No. 2," (reference 51) p. 938.

[68] Bragg, "Consequences," (reference 64) p. 386.

[69] *Studies in Radioactivity* (London: Macmillam, 1912), pp. 191–192. This is a somewhat different explanation than the one he gave in "The Secondary Radiation Produced by the beta rays of Radium," *Phys. Rev.* 30 (1910):638–640.

[70] Bragg, "Consequences," (reference 64) p. 389.

[71] R. Whiddington, "The Production of Characteristic Röntgen Radiation," *Proc. Roy. Soc. (London)* [A] 85 (1911):323–332; "Characteristic Röntgen Radiation," *Nature* 88 (1911):143.

[72] *Repts. BAAS* (1911):341.

[73] *Nature* 87 (1911):501.

[74] "The Nature of γ and X-Rays," *Nature* 77 (1908):509–510.

[75] *Ibid.,* p. 509.

[76] "A Difference in the Photoelectric Effect caused by Incident and Divergent Light," *Nature* 83 (1910):311; *Phil. Mag.* 20 (1910):331–339; also see *Phil. Mag.* 22 (1911):854–864.

[77] "A Difference in the Photoelectric Effect caused by Incident and Divergent Light," *Nature* 83 (1910):339. For a full report of Kleeman's work see "On the Direction of Motion of an Electron ejected from an atom by Ultra-Violet Light," *Proc. Roy. Soc. (London)* [A] 84 (1910):92–99.

[78] "On the Asymmetry in the Distribution of Secondary Cathode Rays Produced by X-Rays; and its Dependence on the Penetrating Power of the Exciting Rays," *Phil. Mag.* 24 (1912):45.

[79] "On the Energy and Distribution of Scattered Röntgen Radiation," *Proc. Roy. Soc. (London)* [A] 85 (1911):29–43.

[80] "The Distribution of Secondary Röntgen Rays and the Electromagnetic Pulse Theory," *Phil. Mag.* 21 (1911):270–278. These results were also confirmed by H. Pealing, "The Distribution and Quality of Secondary Röntgen Radiation from Carbon," *Phil. Mag.* 24 (1912):765–783.

[81] Barkla and Ayres, "Distribution," (reference 80) pp. 276–278.

[82] "Note on the Energy of Scattered X-radiation," *Phil. Mag.* 21 (1911):648–652. Crowther felt that the number of electrons in the scatterer was numerically equal to three times its atomic weight.

[83] "On the Scattered Röntgen Radiation from different Radiators," *Proc. Cambridge Phil. Soc.* 16 (1911):365–369.

[84] "On the distribution of Scattered Röntgen Radiation," *Proc. Roy. Soc. (London)* [A] 86 (1912):478–494. See also "Theory of the Dissymmetrically-distributed Röntgen Radiation," *Proc. Cambridge Phil. Soc.* 16 (1912):534–539.

[85] "On the Scattering of Röntgen Radiation," *Proc. Cambridge Phil. Soc.* 16 (1911):161–166. For the diagram, see p. 166.

[86] Crowther, for example, attempted to account for it by borrowing concepts from both Thomson's pulse theory and Bragg's corpuscular theory. See "On the distribution" (reference 84), p. 494.

[87] "The Distribution of Scattered Röntgen Radiation," *Phil. Mag.* 27 (1914):383–385.

[88] "The Theory of the Scattering of Röntgen Radiation," *Phil. Mag.* 25 (1913):234–241.

[89] *Ibid.*, p. 235. Also see A. H. Compton, *X-Rays and Electrons* (New York: Van Nostrand, 1926, pp. 72ff.

[90] "The Theory of X-Ray Reflexion," *Phil. Mag.* 27 (1914):315–333. Darwin assumed the electrons were subject to damping and that plane waves were incident on these electrons. He found that the amount of interference to be expected depended on the ratio λ^2/ρ^3 where λ = incident wavelength and ρ = inter-electron distance. "For example, the electrons at distances 5 x 10^{-10} cm. apart would exert an effect almost proportional to the square of their number." See pp. 328–329.

[91] "Zerstreuung von Röntgenstrahlen," *Ann. Physik* 46 (1915):809–823; reprinted in *Collected Papers* (New York: Interscience, 1954), pp. 40–50. A year later Debye, together with P. Scherrer, discovered their famous powdered crystal X-ray diffraction technique. See "Interferenz an regellos orientierten Teilchen im Röntgenlicht. I.," *Phys. Z.* 17 (1916):277–283.

[92] MS. read to Royal Society and loaned to A. H. Compton. See his *X-Rays and Electrons*, pp. 72ff.

[93] Note on Scattering of X-rays and Atomic Structure," *Phil. Mag.* 31(1916):222–232, especially p. 231 (with Miss J. Dunlop).

[94] "Primary and Secondary γ Rays," *Phil. Mag.* 20 (1910):921–938. For the diagram, see p. 926.

[95] We realize today that the coefficient of absorption μ and wavelength λ are related by $\mu = A\lambda^3 + B$ (A and B are constants), so that we obtain $d\mu/\mu = 3d\lambda/\lambda$; for small relative changes in μ and λ, therefore, it is always easy to determine the change in wavelength corresponding to a given change in absorption coefficient.

[96] Reference 94, p. 935.

[97] *Ibid.*, pp. 934–937.

[98] *Ibid.*, pp. 937–938.

[99] "The Röntgen Radiation from Substances of Low Atomic Weight," *Phil. Mag.* 24 (1912):138–149. Also see C. A. Sadler and A. I. Steven, "Apparent Softening of Röntgen Rays in Transmission Through Matter," *Phil. Mag.* 21 (1911):659–668.

[100] Sadler and Mesham, "Röntgen Radiation," (reference 99) p. 140.

[101] *Ibid.*, pp. 148–149.

[102] J. A. Gray, "The Scattering and Absorption of the Rays of Radium," *Phil. Mag.* 26 (1913):611–612.

[103] *Ibid.*, p. 612.

[104] *Ibid.*, pp. 614–616.

[105] *Ibid.*, pp. 617–618.

[106] "Secondary γ Radiation," *Phil. Mag.* 27 (1914):225–244; see also "Scattering of γ-Radiation," *Phil. Mag.* 28 (1914):363–367.

[107] "Secondary γ Radiation," (reference 106) pp. 241–243.

[108] "Über einen die Erzeugung und Verwandlung des Lichtes betreffenden heuristischen Gesichtspunkt," *Ann. Physik* 17 (1905):132–148; translated in chapter 36, "Einstein's Legacy," in Boorse and Motz (eds.), *The World of The Atom,* Vol. 1 (New York: Basic Books, Inc., 1966), pp. 545–557.

[109] "Photo-electricity," *Proc. Roy. Inst.* 25 (1928):341.

[110] *Studies in Radioactivity,* (reference 69) pp. 192–193.

[111] *Ibid.*

[112] Quoted in Martin J. Klein, "Einstein's First Paper on Quanta," in *Nat. Phil.* Vol. 2 (New York: Blaisdell, 1963), p. 59. Most of my analysis and discussion of Einstein's work closely follows Klein's lead, as developed in this and the following papers: "Einstein and The Wave-Particle Duality," in *Nat. Phil.* Vol. 3 (New York: Blaisdell, 1963), pp. 3–49; "Einstein, Specific Heats and the Early Quantum Theory," *Science* 148 (1965):173–180; and "Thermodynamics in Einstein's Thought," *Science* 157 (1967):509–516. See also M. Born, "Albert Einstein und das Lichtquantum," *Naturwiss.* 11 (1955):425–431.

[113] Martin J. Klein, *Paul Ehrenfest,* Vol. 1 (New York: American Elsevier, 1970), pp. 249–250.

[114] Quoted from Boorse and Motz (eds.), *The World of The Atom,* Vol. 1 (New York: Basic Books, Inc., 1966), p. 553. One symbol (N_0) has been changed in this and later expressions to maintain a consistent notation. J. Dorling demonstrated that Einstein's argument is stronger than an argument by analogy in that it can be shown to be a deduction from his basic premises. See his paper, "Einstein's Introduction of Photons: Argument by Analogy or Deduction from the Phenomena?" *Brit. J. Phil. Sci.* 22 (1971):1–8.

[115] *The World of The Atom,* Vol. 1, (reference 114) pp. 555–556.

[116] See reference 31. Lenard also found that the energy of the photoelectrons was independent of the intensity of the incident radiation, an observation later confirmed by P. D. Innes, "On the Velocity of the Cathode Particles emitted by Various Metals under the Influence of Röntgen Rays, and its Bearing on the Theory of Atomic Disintegration," *Proc. Roy. Soc. (London)* [A] 129 (1907):442–462. In 1925, Arthur H. Compton, following W. H. Bragg, explained the boldness of Einstein's predictions regarding the energy transformations in terms of the following metaphor: "There was once a sailor on a vessel in New York harbor who dived overboard and splashed into the water. The resulting wave, after finding its intricate way out of the harbor, at last reached the other side of the ocean, and a part of it entered the harbor at Liverpool. In this harbor there happened to be a second sailor swimming beside his ship. When the wave reached him, he was surprised to find himself knocked by the wave up to the deck." See "Light Waves or Light Bullets?" *Sci. Am.* 4 (1925):246. Bragg evidently used this analogy first in a Robert Boyle Lecture. See S. Russ, "The Release of Electrons by X-rays," *Nature* 111 (1923):534.

[117] Quoted from Boorse and Motz (eds.), (reference 114) p. 555.

[118] "Plancksche Theorie der Strahlung und die Theorie der spezifischen Wärme," *Ann. Physik* 22 (1907):180–190; 800.

[119] Quoted in Klein, "Thermodynamics," (reference 112) p. 513. See also C. Seelig, *Albert Einstein, A Documentary Biography,* translated by M. Savill (London: Staples Press, 1956), p. 87. The intensity of Einstein's work at this time is also discussed in P. Frank, *Einstein, His Life and Times,* translated by G. Rosen (New York: Knopf, 1947), p. 98.

[120] "Zum gegenwärtigen Stand des Strahlungsproblems," *Phys. Z.* 10 (1909):185–193—hereafter cited as "Present State." For the full calculation of one expression in this paper, see "Statistische Untersuchung der Bewegung eines Resonators in einem Strahlungsfeld." *Ann. Physik* 33 (1910):1105–1115 (with L. Hopf).

[121] "Present State," p. 185.

[122] Einstein's papers have been reprinted in R. Fürth (ed.), *Investigations on the Theory of the Brownian Movement,* translated by A. D. Cowper (New York: Dover, 1956).

[123] See Klein, "Wave-Particle Duality," (reference 112) pp. 5ff., for a complete discussion. Einstein's work has been recently verified experimentally. See G. W. Kattke and A. van der Ziel, "Verification of Einstein's Formula For Fluctuations in Thermal Radiation," *Physica* 49 (1970):461–464.

[124] Einstein, "Present State," p. 190.

[125] See Max Jammer, *Conceptual Development of Quantum Mechanics,* (reference 23) p. 38.

[126] Einstein, "Present State," (reference 120) p. 191.

[127] See Klein, "Specific Heats," (reference 112) pp. 176–178.

[128] Einstein, "Present State," p. 191.

[129] For a full discussion, see Armin Hermann (ed.), *Die Hypothese der Lichtquanten, Dokumente der Naturwissenschaft (Physik),* Vol. 7 (Stuttgart: Ernst Battenberg 1965), p. 17, and Armin Hermann, *The Genesis of Quantum Theory* (Cambridge, Mass.: M.I.T. Press, 1971), pp. 74–75. See also Armin Hermann, "Albert Einstein und Johannes Stark: Briefwechsel und Verhältnis der beiden Nobelpreisträger," *Arch. Ges. Med.* 50 (1966):267–285.

[130] Hermann, *Hypothese,* p. 16; "Briefwechsel," p. 281. (See reference 129.)

[131] "Briefwechsel," p. 274.

[132] "Über Röntgenstrahlen und die atomistische Konstitution der Strahlung," *Phys. Z.* 10 (1909):585.

[133] Published as "Über die Entwicklung unserer Anschauungen über das Wesen und die Konstitution der Strahlung," *Phys. Z.* 10 (1909):817–826—hereafter cited as "Development."

[134] *Ibid.,* p. 817. Translation by Klein, "Wave-Particle Duality," (reference 112) p. 5.

[135] *Ibid.,* p. 820.

[136] *Ibid.,* p. 821.

[137] *Ibid.,* pp. 823–824; also see Klein, "Wave-Particle Duality," p. 15.

[138] Einstein, "Present State," (reference 120) p. 192; also see Klein, "Thermodynamics," (reference 112) p. 514.

[139] Einstein, "Development," (reference 133) p. 824.

[140] *Ibid.* [Discussion], p. 825.

[141] Hermann, *Genesis,* (reference 129) p. 68.

[142] *Ibid.,* p. 826.

[143] *Ibid.*

[144] "Zur experimentellen Entscheidung zwischen Ätherwellen und Lichtquantenhypothese. I. Röntgenstrahlung," *Phys. Z.* 10 (1909):902–913.

[145] *Ibid.,* p. 903.

[146] *Ibid.,* p. 905.

[147] Hermann, *Genesis,* (reference 129) pp. 82–84.

[148] "Zur experimentellen Entscheidung," (reference 144) pp. 912–913.

[149] For an example, see P. Forman, "The Discovery of the Diffraction of X-Rays by Crystals; A Critique of the Myths," *Arch. Hist. Exact Sci.* 6 (1969):54.

[150] By 1922 Stark apparently regarded his theoretical work of 1909–1910 as a necessary but unpleasant phase in his life. In this regard, the following passage (pp. 28–29) from his polemic *Die Gegenwärtige Krisis in der Deutschen Physik* (Leipzig: J. A. Barth, 1922) is particularly relevant:

> Already during the years I worked at the University in Aachen, where I learned to esteem my numerous colleagues in engineering for their scientific achievements, I thought my exclusive concern with cathode, canal, and x-rays, and the processes of emission and absorption of light, rather onesided. But to reach my goal, I had to acquaint myself, during long persevering years of experimental work, with all of the details of my chosen area and, by means of systematic experiments, to bring this research to a conclusion. After this was achieved, it was a great relief for me to start experimental work in another area of physical research, which is of importance for industry. The raw materials and industries of the region closely surrounding my home in Oberpfalz led me to pursue technical research with powdery substances, and, as a first project, I embarked on physical-technical research involving the most important raw materials of ceramics, namely kaolin and clay.

Also see Armin Hermann's discussion in "Briefwechsel," (reference 129) pp. 281–284. I know of no evidence to suggest that Compton was influenced by Stark at all.

[151] "Über die Verteilung der Intensität bei der Emission von Röntgenstrahlung," *Phys. Z.* 10 (1909):969–970.

[152] D. L. Webster, "Problems of X-Ray Emission," *Bull. Natl. Res. Counc.* 1 (1920):449. See also his "Quantum Emission Phenomena in Radiation," *Phys. Rev.* 16 (1920):31–40.

[153] Quoted in Forman, "Discovery of Diffraction of X-Rays," (reference 149) p. 52. For Sommerfeld's paper, see reference 62.

[154] Quoted in Hermann, *Hypothese,* (reference 129) p. 21. See also A. Hermann, "Die frühe Diskussion zwischen Stark und Sommerfeld über die Quantenhypothese (1)," *Centaurus* 12 (1967–1968):45.

[155] *Hypothese,* (reference 129) p. 20; "frühe Diskussion," p. 49.

[156] "Zur experimentellen Entscheidung zwischen der Lichtquantenhypothese und der Ätherimpulstheorie der Röntgenstrahlen," *Phys. Z.* 11 (1910):24–25.

[157] *Ibid.*, p. 30.

[158] Hermann, *Hypothese,* p. 21.

[159] "Zur experimentellen Entscheidung zwischen Lichtquantenhypothese und Ätherwellentheorie. II. Sichtbares and ultraviolettes Spektrum," *Phys. Z.* 11 (1910):179–187, especially p. 186.

[160] For a brief discussion, see Forman, "Discovery of Diffraction of X-Rays," (reference 149) p. 54.

[161] L. Badash, "Nagaoka to Rutherford, 22 February 1911." *Phys. Today* 20 (1967): 59. Nagaoka himself is best known for his pre-Bohr "Saturnian Atom." See for instance Max Jammer, *Conceptual Development of Quantum Mechanics,* (reference 23) p. 76.

CHAPTER 2

Waves and Quanta

A. Introduction

The general attitude of physicists toward Stark's work as expressed by Nagaoka was no doubt accurate: Before 1910 Einstein's revolutionary light quantum hypothesis, if known, either was misunderstood—Einstein's statistical mechanical arguments were unfamiliar to most of his contemporaries—or was regarded with deep skepticism. Thomson and Bragg, for example, while aware of Einstein's hypothesis (even if only by reading or skimming Stark's papers), were not noticeably influenced by it. Bragg saw no deep connection between it and his own corpuscular theory of X-rays, even though most of his own experiments, and many of the experiments stimulated by his controversy with Barkla, were compatible with Stark's interpretation of the particle-quantum interaction.[1] It was as if two independent streams of thought existed: one developing around the pulse-particle debates of the English physicists, the other developing around the quantum arguments of the Continental physicists. Little cross-fertilization of ideas took place.

This picture changed rather dramatically between 1910 and 1912. Physicists in different countries became much more aware of their mutual interests, and the quantum theory itself became much better known, especially after Walther Nernst was convinced of the importance of Einstein's theory of specific heats and succeeded in organizing the enormously fruitful 1911 Solvay Conference in Brussels.[2] Einstein's light quantum hypothesis was still so controversial that no paper on it was presented at the conference; nevertheless, the hypothesis cropped up during the discussions, and in the succeeding months a sense of its importance gradually developed. It began to take on the status of a hypothesis to be reckoned with, though not necessarily espoused. Was it, perhaps, only a manifestation of a deeper reality? Could the phenomena be explained in a less radical way? These became some of the basic questions that were asked; and they were asked by physicists in England, on the Continent, and in America. Interaction, rather than independence, became the keynote.

Several developments provided the focus for this interaction, but two were salient. First, the key importance of the photoelectric effect to Einstein's light quantum hypothesis was recognized; as a result a number of experiments were carried out, and the correct interpretation of the photoelectric effect became the subject of intense discussions. Secondly, Laue's discovery of the diffraction of X-rays by crystals profoundly affected contemporary debates on the nature of X-rays. After a discussion of these two developments, other pre-1917 research will be analyzed in an attempt to assess the attitude of physicists toward the nature of X-rays that prevailed when Arthur Holly Compton entered this field.

B. Non-Einsteinian Theories of the Photoelectric Effect (1910–1913)

1. *Einstein's Speculations on Quanta and Lorentz's Theory*

Early in 1909, Einstein, convinced that no one could bring more profound criticism to bear on his developing ideas about radiation than H. A. Lorentz, initiated a correspondence with Lorentz by sending him a copy of his paper "Zum gegenwärtigen Stand des Strahlungsproblems"[3] ("On the Present State of the Radiation Problem"), which we discussed in the last chapter. Lorentz evidently responded in detail, but his letter has been lost and its general contents must be inferred from subsequent correspondence and published papers. In any event, Lorentz had now been drawn directly into the discussion and he became the first prominent theoretical physicist, apart from Planck, to display some of the apparent shortcomings of Einstein's light quantum hypothesis—and to propose an alternate interpretation of the photoelectric effect.

Lorentz's most general criticism, that independently moving, localized light quanta could not give rise to ordinary optical phenomena such as interference and diffraction, was not new, but the incisiveness with which he drove it home was. He argued,[4] in the first place, that Lummer and Gehrche's interference experiments, which involved path differences up to roughly 80 cm, proved that that distance represented a lower limit on the *longitudinal* extension of quanta. Second, he argued that G. E. Hale's new reflecting telescope on Mount Wilson, which had a mirror 150 cm in diameter to provide high resolution, proved that 150 cm represented a lower limit on the *lateral* extension of quanta. How then, Lorentz asked, could a quantum this monstrously large pass through the pupil of an eye, or interact with a single electron without—by hypothesis—being subdivided? And since—again by hypothesis—the quantum experiences a change in frequency when subdivided, how could an incident quantum be split up into a reflected and refracted quantum at a boundary between two different media, with no resulting change in frequency?[5]

Einstein prefaced his reply to Lorentz[6] by expressing his deep gratitude and "great joy" at Lorentz's criticism; it indicated to Einstein how carefully Lorentz had considered the light quantum question. It was, in fact, difficulties of precisely the kind Lorentz had raised that convinced Einstein, as he told Lorentz, that quanta could *not* be small, localized, independently moving entities—a remark that fully justified Einstein's later characterization of himself as "nicht der orthodoxe Lichtquantler."[7] Einstein then outlined his thoughts—actually only speculations at the time—to Lorentz in considerable detail. They amounted to an elaboration of his earlier idea that in analogy to electrons which are surrounded by electrostatic fields, light quanta are singular points (not necessarily mathematical singular points) which are surrounded by extended vector fields, diminishing with distance and somehow capable of superposition. A single light quantum therefore was to be conceived of as a singularity, perhaps mathematically identical to that characterizing the electron, without which the vector field could not exist. The composite motion of all the light quantum singularities would completely characterize the radiation field, with the total field energy, at least at low radiation densities, connected in some way to the total number of singularities present. Absorption of radiation would occur only when a singularity disappeared and the radiation field supported by it degenerated. At an interface between two different media, the singularities present before reflection and refraction would have to disappear and new ones would then have to be formed.

Einstein considered the essence of the theory to be not the assumption of singular points, however, but rather the assumption of linear and homogeneous field equations whose solutions permitted the propagation of small, localized, and directed energy bundles at velocity c. This goal, he felt, should be attainable by slightly modifying Maxwell's theory, and, although he had not yet succeeded in finding this modification, he suggested at the end of his letter to Lorentz that a possible starting point might be the fourth-order differential equation

$$\Delta\phi - \lambda^2\Delta\Delta\phi = 0 \qquad (2.1)$$

where Δ is defined as $\partial^2/\partial x^2 + \partial^2/\partial y^2 + \partial^2/\partial z^2$. One solution of this equation, namely,

$$\phi = \epsilon \, \frac{1 - e^{-r/\lambda}}{r} \tag{2.2}$$

is the only solution that goes over to the Coulomb potential ϵ/r at large distances r and has no (mathematical) singularities at $r = 0$. A system of four such equations, each generalized to include time dependence by replacing Δ by $\Delta - (1/c^2)(\partial^2/\partial t^2)$, would possibly yield an electrodynamic system of equations encompassing both light quanta and electrons. There were a number of difficulties, however (ϵ for instance was still left arbitrary), and Einstein admitted that the path might lead nowhere. Perhaps, Einstein suggested, he, Lorentz, could see at a glance that his hopes and efforts were completely futile, and if so, would do him [Einstein] the great favor of telling him and sparing him further efforts along these lines.

Lorentz's reply to Einstein has been lost, but one thing is certain: Lorentz was not persuaded of the necessity of introducing Einstein's light quantum hypothesis into physics—at least not for the moment. Lorentz soon emphasized its difficulties in print in early 1910, and again in his Wolfskehl Lectures at Göttingen in October of the same year.[8] He told his Göttingen audience that "The speaker would not like to quarrel with the heuristic value of this hypothesis, but he would like to defend the old theory [of Maxwell] as long as possible."[9] There were of course serious difficulties with Maxwell's theory also. Energy considerations, for example, proved convincingly that very large amounts of time would be required before photoelectrons could accumulate enough energy from incident electromagnetic waves to be ejected from an atom. Einstein's hypothesis met this difficulty, and others, simply and naturally. Nevertheless, Lorentz believed that the disadvantages associated with Einstein's hypothesis outweighed its advantages.

Lorentz's unwillingness to embrace Einstein's light quantum hyothesis led Lorentz to search for an alternative interpretation of the photoelectric effect, which he discussed at length at Göttingen. He saw as one possible starting point the "somewhat risky" conceptions of A. E. Haas.[10] According to Lorentz, Haas envisioned that an electron in a Thomson atom was set in motion by a light wave incident upon it. As long as the electron remained inside the positive sphere, the incident wave would only be scattered or dispersed. If, however, the electron received more than some threshold energy, it would be ejected from the positive sphere and a discrete amount of energy would be subtracted simultaneously from the incident wave. Planck's constant h entered into Haas's theory—Haas assigned the quantum energy $h\nu$ to the electron at the boundary of the positive sphere—and Lorentz noted with satisfaction that application of Haas's ideas to a gas, Argon, yielded order-of-magnitude agreement with Planck's own value for h. Lorentz concluded: "In

Haas's hypothesis, the riddle of the energy elements is combined with the question of the nature and action of positive electricity, and it may be that these different questions will, for the first time, together find their complete solution."[11]

The crux of Lorentz's position was that it was far more reasonable to assume the photoelectric effect would eventually be understood on the basis of atomic structure considerations, than to believe it entailed abandoning Maxwell's theory. This was both reasonable and natural—reasonable because in 1910 little of a definite nature was known about the dynamics of the atom,[12] and natural because it was not without precedent: As early as 1902 Lenard had proposed, as we saw in Chapter 1, that photoelectrons are expelled from their parent atoms because of disruptions "triggered" by the incident radiation.[13] Although this specific hypothesis was later rejected (Lenard himself abandoned it by 1911[14]), its general approach (focusing on the structure of the atom rather than on the structure of the radiation) was not —except, of course, by Einstein. To some degree, therefore, Einstein's interpretation must have appeared radical to his contemporaries on two counts: In focusing on the radiation rather than the atom, he not only explicitly rejected a Maxwellian approach, he also implicitly rejected a Lenardian approach. This, however, appeared unjustified or unnecessary to Lorentz—as well as to J. J. Thomson, Arnold Sommerfeld, and O. W. Richardson, all of whom also developed alternative, non-Einsteinian interpretations of the photoelectric effect between 1910 and 1913.[15]

2. *Thomson's Theories*

Thomson's initial reaction to Lenard's "trigger" hypothesis, judging from his discussions of it in 1905 and 1906,[16] was acceptance. However, after learning that Ladenburg in 1907 had found evidence for a definite relationship (actually a *quadratic* relationship) between the energy of the ejected photoelectrons and the frequency of the incident ultraviolet light, Thomson regarded Lenard's hypothesis as "exceedingly improbable."[17] By 1910 he had formulated his own theory. Interestingly enough, Thomson (in common with Lorentz, Sommerfeld, and Richardson) took the validity of the *linear* relationship between photoelectron energy and light frequency for granted, although this relationship remained in doubt, as we shall see, until 1915.

Thomson published his theory of the photoelectric effect,[18] actually the first of two theories, a few months before Lorentz delivered his Wolfskehl Lectures in Göttingen. His basic assumption (which he had stated earlier already) was that electric doublets of moment M are dispersed throughout the atoms constituting the photosensitive metal.[19] At least some of these doublets, Thomson then argued, must have electrons ("corpuscles") circulating around their lower, positively charged ends with angular frequency $\dot{\phi}$, so that the vector **r** to a given electron traces out a cone of half-angle θ with apex at the midpoint of the doublet AB (see Fig. 1). Setting up the electron's equations of motion, simplifying them for the steady-state condition ($\dot{r} = \dot{\theta} = 0$),

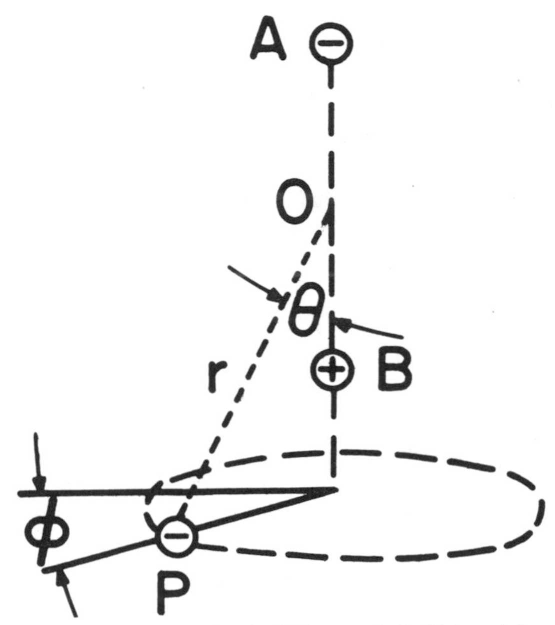

FIG. 1. Author's sketch of Thomson's doublet model.

and solving them simultaneously, Thomson easily proved that the half-angle θ of the cone is fixed at $\theta_o = \tan^{-1}\sqrt{2} = 55°$, and that the kinetic energy of the electron (charge e, mass m) is given by

$$\tfrac{1}{2} mv^2 = (\sqrt{Mme} \cos^{3/2} \theta_o)\, \dot{\phi} \sim \dot{\phi} \tag{2.3}$$

This direct proportionality between the electron's kinetic energy and its frequency of revolution (which followed directly from Thomson's model, but which is obvious from simple considerations), was all Thomson required for his theory of the photoelectric effect. It suggested to him that in the photosensitive metal there were no doubt many electrons circling about their respective doublets at many different frequencies $\dot{\phi}$, which in some cases were equal to the frequency ν of the incident radiation. Resonance might then occur, and the "electric forces in the wave might do work upon the doublet, twisting its axis so as to alter the angle it makes with the radius to the corpuscle."[20] A sufficiently large alteration might then involve the "liberation" of the corpuscles. This would occur at a time when their kinetic energy was proportional to their frequency of revolution $\dot{\phi}$, and hence proportional to the frequency of the incident light ν. Granted that this picture was correct, Thomson concluded, we "cannot regard Ladenburg's experiments as a proof of the unitary structure of light. . . . [My] theory enables us to explain the electrical effects produced by light, without assuming that light is made up of unalterable units, each containing a definite and, on Planck's hypothesis, a comparatively large amount of energy, a view which it is exceedingly difficult to reconcile with well-known optical phenomena."[21]

J. H. Jeans criticized[22] Thomson's doublet-electrons on the grounds that they were dynamically unstable, but Thomson answered[23] that their long-term stability was unnecessary for the success of his theory. Perhaps, however, Thomson did not entirely forget Jeans' criticism, since twice during the next three years he returned to the problem, modifying his theory. Initially (May 1912), he only qualified it somewhat by proving that it was unnecessary for the doublet to be "twisted" to release its electron at the proper energy—any method of releasing the corpuscles would suffice.[24] Fifteen months later, however (October 1913), he completely rejected his earlier theory and formulated a new one[25] based on nothing less than an entirely new atomic model—a piece of theoretical surgery which is understandable only when one takes Thomson's general attitude toward atomic models into account. As his biographer Lord Rayleigh remarked: "J. J. was not inclined to be dogmatic about his atomic theories, and indeed he was quite prepared to change them, sometimes without making it altogether clear that he had wiped the slate clean. . . ."[26] If a model did not obviously conflict with known facts, it was logically conceivable—and should be exploited. Thomson concentrated on what a theory *would* explain and not on what it would *not*.

In his new atomic model, Thomson postulated the coexistence of two forces: "(1) A radial repulsive force, varying inversely as the cube of the distance from the centre, diffused throughout the whole of the atom," and "(2) A radial attractive force, varying inversely as the square of the distance from the centre, [and] *confined to a limited number of radial tubes in the atom.*"[27] An electron finding itself inside such a tube at a distance r from the center of the atom would then be subjected to *both* forces, and hence would move according to the equation

$$m\dot{r} = \frac{Ce}{r^3} - \frac{Ae}{r^2} \qquad (2.4)$$

where C and A are constants. Thomson proved that it was possible for the electron to be in equilibrium at some distance $r = a$ by proving that a small displacement x away from the equilibrium position is opposed by a restoring force $-kx$. Thus, he substituted $r = a + x = a(1 + x/a)$ into equation (2.4), expanded the denominators to first order, and used the fact that at equilibrium $\ddot{r} = 0$, or $Ce/a^3 = Ae/a^2$, to obtain the new differential equation

$$m\ddot{x} = \frac{Ce}{a^4}x\,. \qquad (2.5)$$

This equation shows that the electron will oscillate about its equilibrium position at $r = a$ at a frequency $f = (1/2\pi)\sqrt{k/m} = (1/2\pi)\sqrt{Ce/ma^4}$.

Once again this was all Thomson required, since if an electromagnetic wave of frequency $\nu = f$ were now incident on the electron, resonance would occur; and if some "casual magnetic force" would then move the electron laterally out of the tube, it would come under the "uncontrolled action" of the repulsive force and be expelled from the atom. The kinetic energy T acquired during its expulsion would be given by the equation

$$T = \int_a^\infty \frac{Ce}{r^3}dr = \frac{Ce}{2a^2} = \pi\sqrt{Cem}\,f\,. \qquad (2.6)$$

According to Thomson, this result fixed the constant C at $C = h^2/me\pi^2$, since then, by substitution, $T = hf = h\nu$. Completely ignoring this blatantly *ad hoc* feature in his theory, Thomson concluded: "Thus we see that the kinetic energy with which the corpuscle is expelled is proportional to the frequency of the light and is equal to the frequency multiplied by Planck's constant. This is the well-known law of Photo-Electricity."[28]

Thomson's ideas on the structure of light (see Chapter 1) and his theories of the photoelectric effect attracted some attention and stimulated some important experimental work, particularly at the Cavendish Laboratory. Especially noteworthy were G. I. Taylor's interference experiments (1909) using an extremely weak source of light, and N. R. Campbell's interferometric correlation experiments (1909–1910), which proved incapable of deciding between wavelike and particlelike theories of radiation.[29] But the *ad hoc* character of many of Thomson's basic assumptions in his theories of the photoelectric effect must have been apparent to many of his contemporaries. W. H. Bragg, for example, regarded certain features of Thomson's first theory as "fantastic."[30] In general, Bragg considered attempts like Thomson's to squeeze quantum manifestations out of classical theories to be retrogressive. Thomson's second theory, in fact, took on an even more contrived cast a few weeks after he had proposed it when he found it necessary to replace the radial attractive cones with radial attractive cylinders.[31] However, although Thomson's theories rested on insecure grounds, a basic point should not be lost. Like Lorentz, Thomson proved that it was possible to envision atomic models upon which a quantitative theory of the photoelectric effect could be based. This very same point was made in 1911 in a more sophisticated way by Arnold Sommerfeld.

3. *Sommerfeld's Theory*

We have already seen from Sommerfeld's earlier, negative reactions to Stark's and Bragg's theories[32] that in 1911 he shared Lorentz's and Thomson's skepticism toward Einstein's light quantum hypothesis. He had, however, already become convinced that a quantum theory of *matter* was unavoidable. He contended that an "electromagnetic or mechanical 'explanation' of [Planck's constant] h seems to me to be just as unnecessary and unpromising as a mechanical 'explanation' of Maxwell's equations. It would be much more useful to pursue the h-hypothesis in its various consequences and trace other phenomena back to it."[33] One of these "other phenomena" was the photoelectric effect.

Sommerfeld developed his theory of the photoelectric effect in collaboration with his assistant and former student, Peter Debye,[34] and first presented it at a meeting of the Gesellschaft Deutscher Naturforscher und Ärtze in Karlsruhe on September 25, 1911. It would stand, he explained, "approximately midway between, on the one hand, Lenard's idea, according to which the ripping off of the electron is a resonance phenomenon in the atom, with the photoelectron's energy arising out of the atom; and, on the other hand, Einstein's light quantum hypothesis, according to which the energy of the incident radiation is depleted, the structure of the radiation, however, being changed in a way that contradicts common experience."[35] Guided by the dimensional equivalence of Planck's constant h and the integral $\int(T - V)dt$, which expresses Hamilton's principle in classical mechanics, Sommerfeld postulated that in "every purely molecular process a definite and universal amount of action" is taken up or given up. He introduced Planck's constant by fixing the exact amount by the condition

$$\int_0^\tau (T - V)dt = \frac{h}{2\pi} \qquad (2.7)$$

where τ is the time during which the process takes place. For the photoelectric effect, Sommerfeld later[36] estimated τ to be on the order of 10^{-5} second, a small, and in 1911 unobservable delay, but of course one which was nonetheless inconsistent with Einstein's interpretation.

Sommerfeld brought his general postulate to bear on the photoelectric effect by assuming that the incident radiation—electromagnetic radiation of amplitude E—sets a bound electron into oscillation according to the equation

$$m\ddot{x} + kx = eE \tag{2.8}$$

and thereby (in time τ) supplies it with enough energy to escape from the atom. By substituting the electron's kinetic energy $T = (1/2)m\dot{x}^2$ and its potential energy $V = (1/2)kx^2$ into equation (2.7), by transforming the result by partial integration, and finally by introducing the electron's equation of motion (2.8), Sommerfeld obtained the equation

$$\frac{1}{2}mx\dot{x} - \frac{1}{2}e\int_0^\tau xEdt = \frac{h}{2\pi}. \tag{2.9}$$

The first term, as may be seen by calculating

$$\frac{d}{dt}\left(\frac{h}{2\pi}\right) = 0 = \frac{d}{dt}\int_0^\tau (T - V)\,dt$$

is equal to $T/2\pi\nu_0$, where $\nu_0 = (1/2\pi)\sqrt{k/m}$ is the natural frequency of the electron. The second term represents the time virial of the external force and goes to zero in the special case of complete resonance ($\nu = \nu_0$) between the frequency of the incident radiation and the natural frequency of the electron. In that event the above equation reduces to $T/2\pi\nu_0 = h/2\pi$, or $T = h\nu_0 = h\nu$. And that, concluded Sommerfeld, is "Einstein's law."[37]

One difference between Sommerfeld's result and Einstein's may be seen immediately: the absence of an additive constant, the "work function," in Sommerfeld's expression. This difference, however, did not trouble Sommerfeld. As he later remarked, "This is obviously foreign to the pure molecular process and would appear only if we were to follow the photoelectron [after its ejection from the atom] along on its subsequent path through the surface of the metal."[38] This is a statement with which Thomson would also have agreed. The key difference between Sommerfeld and Einstein was a conceptual one: Sommerfeld, like Thomson, envisioned a resonance phenomenon.

This difference had observable consequences. Whereas Einstein's theory predicted that the electron's energy should be entirely independent of any atomic frequencies, Sommerfeld's theory (and Thomson's also, of course) predicted that a plot of energy versus frequency "should possess, for each natural frequency of the atom, a maximum." Which theory fit the existing data better, Sommerfeld's or Einstein's? Sommerfeld had the answer. J. R. Wright in R. A. Millikan's laboratory at the University of Chicago had recently measured the potential difference required to just stop the emission of photoelectrons from aluminum when using ultraviolet light of wavelength 2166 Å.[39] Wright found that this potential difference was 14 volts; this was in contrast to $h\nu/e = 5.8$ volts, which Sommerfeld took to be the Einsteinian value. Moreover, Sommerfeld wrote that Wright's measurements seemed to show "with certainty," first, that the "maximum photoelectron energy does not vary approximately linearly with the frequency," and secondly, as R. Pohl and P. Pringsheim had recently found,[40] that the photoelectric effect depends upon the plane of polarization of the incident radiation. "With respect to both points," Sommerfeld concluded, "our theory is in better accord with Wright's measurements than Einstein's light quantum theory."[41]

At the end of October 1911, one month after Sommerfeld had reached that conclusion, he had a chance to discuss his theory personally with Einstein at the first Solvay Conference in Brussels. In his Brussels paper,[42] Sommerfeld treated the more general case of near-resonance ($\nu \approx \nu_0$) and found that the electron's kinetic energy was then given by the equation

$$T = \frac{h\nu_0}{1 + \dfrac{\epsilon - \sin\epsilon}{1 - \cos\epsilon}} \tag{2.10}$$

where ϵ is a convenient abbreviation for $2\pi(\nu - \nu_0)\tau$. At resonance ($\nu = \nu_0$, or $\epsilon = 0$) this expression reduces to $T = h\nu_0$, so that "Einstein's law" was recovered.

Nevertheless, there was an apparent difficulty: at $\epsilon = -\pi/2$, or at $\nu - \nu_0 = -1/4\tau$, which is a *negative* frequency difference, the above expression predicts $T = 2h\nu_0/(4 - \pi) = 2.34\ h\nu_0$, a *positive* kinetic energy. Therefore the radiation should be able to eject electrons even when its frequency ν is less than the electron's natural frequency ν_0, a clear violation of Stokes' law of fluorescence. Sommerfeld met this difficulty by pointing out that at $\epsilon = -\pi/2$ the relative difference in frequencies, $(\nu-\nu_0)/\nu$, is equal to $-1/4\nu\tau$, an "extraordinarily small" and even unmeasurable quantity for typical values of ν and τ. It was therefore an open question as to whether or not Stokes' law was violated.

In fact, there was good reason to assume that it *was* violated. Wright's experiments had indicated that 2166 Å radiation ejected 14 eV electrons from aluminum. Now, since for this radiation $h\nu = 5.8$ eV $\approx h\nu_0$, one obtains $T/h\nu_0 = 2.44$, which Sommerfeld compared with the result predicted by equation (2.10); that is, $2/(4 - \pi) = 2.34$. This close agreement, he told his audience in Brussels, "certainly cannot be accidental."[43]

Although we now know that the agreement was accidental, Sommerfeld found it most encouraging. He freely admitted, however, that there were certain apparently arbitrary aspects and limitations to his theory. First, the factor $1/2\pi$ multiplying h in his basic postulate had been chosen only because it yielded $T = h\nu_0$ ("Einstein's law") at resonance; other considerations actually indicated that 1/4 was preferable. Secondly, the difference $(T - V)$ had been chosen over the sum $(T + V)$ simply because the difference, and not the sum, allowed the integral to be transformed into a tractable expression by partial integration. Both choices, therefore, found their ultimate justification in the results they produced, one of which seemed particularly striking to Sommerfeld. This was his conclusion that the maximum amplitude of the electron's motion turned out to be roughly as large as the diameter of a Thomson atom. It was, in fact, proportional to $\sqrt{h}$, so that h determinded the scale of atomic dimensions, and this was a prediction that Sommerfeld considered to be "remarkable" and "worthy of attention."

In the discussion following Sommerfeld's Brussels paper, Einstein was the first to respond. His remarks indicated that he was at best lukewarm toward Sommerfeld's theory. First, he raised an objection against Sommerfeld's basic postulate. Second, he noted that according to Sommerfeld's theory "the number of electrons liberated per second would not be proportional to the intensity of the [incident] light"[44]—a question which Einstein regarded as not yet conclusively settled experimentally.[45] In general, Einstein's remarks at the conference were in keeping with those in his recent letters to Lorentz. On the one hand, he "insisted" on the "provisional character" of the light quantum hypothesis, "which does not appear to be reconcilable with the experimentally verified consequences of the wave theory." On the other hand, he was convinced that "it is necessary . . . for us to introduce . . . a hypothesis such as that of quanta along side the indispensable equations of Maxwell."[46]

4. *Einstein's Photodecomposition Papers*

The impact of the Solvay Conference on its participants was profound. Brillouin, in his concluding remarks, characterized their common conviction as follows: "Henceforth, it appears completely certain that it will be necessary to introduce into our physical and chemical conceptions a discontinuity, an element varying by jumps, of which we had no idea some years ago. How will it be introduced? This I do not see so clearly. Will it be in the first form proposed by M. Planck, in spite of the difficulties which it raises, or in the second form? Will it be in M. Sommerfeld's form, or in another yet to be found?"[47] Individual physicists from different countries, many of whom had met each other for the first time, left the conference inspired. The most poignant example was that of Henri Poincaré, who shortly before his death in 1912 pointed out some of the profound physical and philosophical implications of the quantum theory, and, in an intense period of creative activity, proved mathematically that the quantum hypothesis alone is compatible with a finite energy for cavity radiation. Poincaré's proof became a key factor in convincing physicists from other countries, notably J. H. Jeans, of the need to accept the quantum theory.[48]

The Solvay discussions were also stimulating to Albert Einstein, who after returning to Prague from Brussels found yet another argument—one characteristically rooted in thermodynamics—for his light quantum hypothesis.[49] Einstein's argument rested on his analysis of the decomposition and recombination of gas molecules in thermodynamic equilibrium with monochromatic radiation in a cavity of definite volume. He assumed that radiation of energy NE can decompose a single molecule of a given gas into two other molecules according to the reaction $m_1 \rightarrow m_2 + m_3$, and that in the inverse reaction, $m_2 + m_3 \rightarrow m_1$, the energy NE is returned to the radiation. To carry out his analysis, he required three further assumptions: (1) The number of decompositions per unit time of the first gas is unaffected by the presence of the other two gases. (2) The probability of decomposition of the first gas is proportional to the radiation density ρ. (3) The recombination, or inverse process, proceeds according to the law of mass action. He then proved, by a direct but somewhat lengthy calculation, that these three assumptions were compatible with thermodynamic equilibrium between the gas and radiation only if the radiation density ρ were given by Wien's law. He felt, however, that the "most important consequence" of his analysis was his conclusion that under these conditions the energy E must necessarily be equal to $h\nu_o$, that is, that "a gas molecule, which decomposes under the absorption of radiation of frequency ν_o, absorbs (on the average) energy $h\nu_o$ from the radiation."[50]

An important question, however, still remained unresolved. Was the frequency ν_o to be associated with the radiation, or with some characteristic frequency of the gas molecule? After all, both Thomson and Sommerfeld had in effect presented arguments that this frequency was determined by the structure of the atom and was only indirectly (through resonance) dependent on the radiation. Einstein saw the crucial importance of this question, and shortly after publishing the above results he published a supplementary paper[51] in which he addressed himself to it, directly and unambiguously.

Rather than assuming that the radiation was monochromatic, he now assumed that it consisted of various frequencies $\nu^{(1)}$, $\nu^{(2)}$, $\nu^{(3)}$, . . . , and that these corresponded to various radiation densities $\rho^{(1)}$, $\rho^{(2)}$, $\rho^{(3)}$, . . . , which together form a continuous distribution. To carry his analysis further, he saw that to his three original assumptions he now had to add the assumption that the total number of gas molecules decomposed was simply equal to the sum of those decomposed by the individual frequencies $\nu^{(i)}$. This additional assumption then enabled him to prove that the frequency $\nu^{(1)}$ must be associated with the energy $E^{(1)}$ through the relationship $E^{(1)} = h\nu^{(1)}$, with analogous relationships holding for all other frequencies $\nu^{(i)}$. Einstein therefore concluded that "the energy absorbed per molecular decomposition does not depend upon the characteristic frequency of the absorbing molecules, but rather upon the frequency of the radiation that brought about the decomposition."[52]

While Einstein's papers drew a sharp response from Johannes Stark,[53] who attempted to initiate a priority dispute with Einstein, they apparently had little impact on Sommerfeld. Sommerfeld did not react to Einstein's important conclusion, and with Debye's help he continued to pursue the consequences of his general postulate—Max Born later called it Sommerfeld's "wild adventure"[54]—well into 1913. At that time he sent off a 58-page paper to the editor of *Annalen der Physik*.[55] All the ideas on the photoelectric effect which Sommerfeld expressed in this paper were either identical with those he had expressed earlier, or were direct outgrowths of them. He considered, for example, the influence of a damping term in the electron's equation of motion, and he discussed the relationship between the time τ and the coherence properties of the incident radiation. Given the nature of his model, it is not surprising that he drew two new conclusions—(1) that the kinetic energy of the ejected electrons should possess no upper limit, and (2) that the intensity of the radiation should be unrelated to the number of electrons ejected from the metal—that once again directly contradicted the corresponding predictions of Einstein's theory. Actually, what seemed to impress Sommerfeld most at this time was the possible relationship between his general theory of atomic processes and Planck's "second quantum theory," according to which absorption of radiation was continuous and emission discontinuous.[56] In the final analysis, Sommerfeld tended to regard his basic postulate as less important for the photoelectric effect than for the general problem of atomic structure. Of course, at precisely this same time, this was also the primary concern of Niels Bohr.[57]

5. *Richardson's Theory and the Richardson-Compton Experiments*

We must return at this point to late 1911 when another, very different theory of the photoelectric effect was being developed by O. W. Richardson.[58] Although he had been born in England and educated at the University of Cambridge (D.Sc., 1904), since 1906 Richardson had been Professor of Physics at Princeton University where he fostered a thriving school of research. His students included O. Stuhlmann, Jr., C. J. Davisson, and the brothers K. T. and A. H. Compton. In Arthur Compton's judgment, it was Richardson who was "largely instrumental in making Princeton a leading center of physics research. . . ."[59] As we shall see, it was also Richardson—and not Einstein, Lorentz, or Sommerfeld—who exerted the most influence on Arthur Compton personally, although he and Richardson overlapped for only one term at Princeton.

Richardson's interest in the photoelectric effect arose naturally out of his interest in the subject he himself christened "thermionics."[60] Both phenomena dealt with the emission of electrons, the former as a consequence of irradiating a substance and the latter as a consequence of heating it. This connection, however, might be viewed differently: Heating a substance generally causes it to emit both electrons and radiation, a fact which implied to Richardson that electrons and radiation were intimately related to each other. "A little consideration," he wrote, "will show that in its relations to matter and to temperature, electronic emission is in precisely the same general position as electromagnetic 'aethereal' radiation."[61]

This symmetry lay at the heart of Richardson's theory of the photoelectric effect. In some respects, it is the most interesting one of the period, since in his attempt to achieve complete generality, he adopted not a microscopic, but a macroscopic, approach. As he wrote: "For the present I wish to avoid discussion of the vexed question of the nature of the interaction between the material parts of the system and the aethereal radiation and to confine my remarks to the conclusions which may be drawn from the existence of a statistically steady condition of the aethereal and electronic radiations."[62] Thus, Richardson would adopt the methods of thermodynamics, which explicitly avoids hypotheses on the micro-world.

In the photoelectric effect, high-frequency radiation incident on a metal plate causes the emission of electrons from it, and ordinarily one does not worry about the possibility of any electrons returning to the plate and being reabsorbed by it. Richardson, however, did worry about these returning electrons: His goal was to determine the equation describing equilibrium between the rates of electron emission and electron absorption. To analyze this problem, Richardson carried out a thought experiment using the most common of all thermodynamic thought apparatus—a piston in a cylinder. He assumed that the only photosensitive surface in the system was the bottom of the cylinder, and he further assumed that the piston was both completely impervious to electrons and completely transparent to radiation. Suppose, now, that the cylinder is filled with radiation and that the piston, initially in contact with the bottom of the cylinder, is slowly lifted. The radiation will then pass through the piston from above and eject electrons from the bottom of the cylinder. The space below the piston will fill with electrons, which Richardson assumed would behave like an ideal gas. By slowly moving the piston up and down, the electrons may be pumped into and out of the bottom of the cylinder in a reversible manner.

It was this process that Richardson analyzed in detail thermodynamically. First, he calculated the number of electrons per unit volume below the piston by considering the infinitesimal changes in energy and entropy associated with the piston's movement. He then used this result and kinetic theory to determine the number of electrons (and hence the total energy) *absorbed* by the photosensitive surface per unit time. Finally, he equated this result to the rate of electron *emission,* as expressed in terms of parameters characterizing the enclosed radiation. His final result was the following equilibrium equation:

$$\frac{c}{4}\int_0^\infty \epsilon F(\nu)\rho(\nu)d\nu = AT^{1/2}\,e^{\int \frac{w}{RT^2}dT} \qquad (2.11)$$

where c is the velocity of light *in vacuo;* the product $\epsilon F(\nu)$ is the number of electrons ejected per unit time from the photosensitive surface (emissivity ϵ); ν is the frequency and $\rho(\nu)$ the energy density of the radiation; A is a constant; T is the absolute temperature; R is the ideal gas constant; and w is "the latent heat of evaporation reckoned for a single electron."[63] The structure of Richardson's equilibrium equation reflects the macroscopic nature of his analysis.

To achieve concrete predictions for the rate of emission of the photoelectrons and for their energy, Richardson had to specify the energy density $\rho(\nu)$. He did this in a second paper of 1912,[64] choosing Wien's law, which for the moment he regarded as a sufficiently close approximation to Planck's law. In addition, as a "convenient approximation," he set $w = w_o + (3/2)RT$, where, in Richardson's interpretation, w_o was the latent heat of evaporation per electron at absolute zero. He realized that w_o played the same role in his theory as the work function P played in Einstein's. Substituting these results into equation (2.11), Richardson obtained a new equation:

$$\int_0^\infty \epsilon F(\nu)h\nu^3 e^{-\frac{h\nu}{RT}}d\nu = A_1T^2e^{-\frac{w_o}{RT}} \qquad (2.12)$$

where A_1 is a new constant.

Two highly significant results followed from this equilibrium equation. First, as one may easily verify by direct substitution, it predicts that the rate of electron emission $\epsilon F(\nu)$ is given by

$$\epsilon F(\nu) = \begin{cases} 0 & 0 < h\nu < w_o \\ \dfrac{A_1 h}{R^2\nu^2}\left(1 - \dfrac{w_o}{h\nu}\right) & w_o \leqq h\nu < \infty \end{cases} \tag{2.13}$$

Thus, no electrons whatsoever will be emitted if the frequency of the radiation is less than some critical frequency $\nu_o = w_o/h$, while a steady emission will occur for frequencies $\nu \geqq \nu_0$. Second, since at equilibrium the kinetic energy T_ν of the emitted electrons must be proportional to the energy $2RT$ returned to the surface through electron impacts,[65] one may multiply the integrand on the left-hand side of equation (2.12) by T_ν, and the right-hand side by $2RT$, to obtain a new equilibrium equation (which need not be written out). This equation, as again may be verified by direct substitution, has the solution:[66]

$$T_\nu = \begin{cases} \text{physically meaningless} & 0 < h\nu < w_o \\ h\nu - w_o & w_o \leqq h\nu < \infty \end{cases} \tag{2.14}$$

It is obvious that these results are formally identical to Einstein's, and Richardson was fully aware of this significance of his theory. The above equations, he wrote, "have been derived without making use of the hypothesis that free radiant energy exists in the form of 'Licht-quanten,' unless this hypothesis implicitly underlies the assumptions:—(A) that Planck's radiation formula is true; (B) that, *ceteris paribus,* the number of electrons emitted is proportional to the intensity of monochromatic radiation. Planck . . . has recently shown that the unitary view of the structure of light is not necessary to account for (A) and it has not yet been shown to be necessary to account for (B). It appears therefore that the confirmation of [the above equations] . . . by experiment would not necessarily involve the acceptance of the unitary theory of light."[67]

This conclusion was extremely important to Richardson, because he and his former student K. T. Compton were concurrently carrying out experiments on the photoelectric effect.[68] In Richardson and Compton's judgment, all earlier experiments were

. . . very contradictory. For instance, von Baeyer and Gehrts, and Klages found, in the case of several metals, that the maximum initial velocity of the electrons was independent of the metal used. Ladenburg and other physicists have decided that the electronegative metals give off electrons with the greatest velocities; whilst Millikan and Winchester concluded that the initial velocity bears no relation to the Volta series. Most physicists who have investigated the subject believe that the maximum initial kinetic energy is a linear function of the frequency of the light; but some have obtained results supporting the view that the maximum velocity varies as the first power of the frequency.[69]

In an effort to correct this lamentable state of affairs, Richardson and Compton carefully designed and constructed an apparatus employing a source of continuously variable wavelengths, and one which enabled them to directly correct for contact potential differences between the photosensitive metal and the collector (a spherical bulb). Their experiments were unquestionably the most accurate and refined to date. In all, they used eight different photosensitive metals and subjected each to between three and five different ultraviolet wavelengths, generally ranging between 2000 and 2700 Å. At each wavelength both the maximum and mean kinetic energies of the electrons were determined. Plotting these values against frequency (for each metal), Richardson and Compton found "a linear relation between the corresponding variables." However—and this was one aspect of their work which immediately opened it up to criticism—their measurements involved certain experimental difficulties which inclined them to "place more reliance on the measurements of the mean than on those of the maximum values."[70] The maximum energies, of course, were the ones that were significant theoretically.

Nevertheless, Richardson and Compton took their experiments to confirm the linear relationship. In spite of his theoretical work, this came as a surprise to Richardson: "When these experiments were started," he later wrote, "I thought it improbable that [Einstein's] . . . equation would turn out to be correct, on account of the very grave objections to the form of quantum theory on which it had up to that time been based by Einstein."[71] But whose *theory* of the photoelectric effect did Richardson consider to be confirmed? Einstein's equation, he noted, "evidently has a wider basis than the restricted and doubtful hypothesis used by Einstein."[72] He soon made the implication explicit: His and Compton's experiments, he wrote, could be taken to "confirm the theory of photoelectric action which was recently developed by one of the writers."[73]

It is no doubt unfair to generalize at this point and conclude that whenever a scientist has a choice between a traditionally rooted or boldly revolutionary theory, he will choose the former (especially if he himself has developed it!), but that was clearly the case here. Actually, of course, Richardson's choice hinged more on a desire or lack of desire to avoid hypotheses on the micro-world. But just because Richardson had succeeded in deriving Einstein's equation without explicitly invoking Einstein's light quantum hypothesis, this did not automatically establish the validity of that equation. The experimental issue was by no means settled. As a matter of fact, it was in response to Richardson and Compton's *experimental* work that Pohl and Pringsheim of the University of Berlin's *Physikalisches Institut* published in 1913 a penetrating analysis of all post-1905 work.[74] They summarized the situation at that time as follows: "Joffé . . . pointed out that the values for the photoelectric initial velocities published by E. Ladenburg in 1907 could be just as well represented by Einstein's formula as by the formula originally adopted by Ladenburg, according to which the velocity itself, and not the square of the velocity, is proportional to the frequency. Since that time a great many papers have appeared in which this question of proportionality was thought to have been conclusively settled. In our opinion, however, no great advance in this direction has been made over the results of Joffé and of Ladenburg." They added: "Nor do the results which Richardson & Compton in their joint paper give for, say, Al[uminum] lead in our opinion to a definite conclusion."[75]

Pohl and Pringsheim argued that owing to the restricted range of frequencies over which Richardson and Compton's experiments had been carried out, a number of different functional relationships fit their data reasonably well. In fact, they pointed out in a footnote that "far better agreement" with experiment could actually be obtained by using a logarithmic relationship, $T = k\log\nu - h\nu_o$, where k is a constant.[76] This criticism, in spite of rebuttals (for instance by A. L. Hughes[77]), would remain valid and undiminished for several years. The openness of the experimental question around 1913 is evident from Pohl and Pringsheim's conclusion: "We consider that at present the most important experimental support for Einstein's relation is the fact that an extrapolation to the probable frequencies of the Röntgen spectrum leads to velocities for the electrons liberated by Röntgen rays which agree in order of magnitude with those experimentally observed."[78]

If anything, Pohl and Pringsheim's critique spurred Richardson on to greater efforts. He continued to plan experiments, with K. T. Compton taking over more and more direct responsibility for their execution as time went on.[79] They attempted to measure, for example, the rate of electron emission as a function of incident frequency to determine if—as equation (2.13) predicts—the curve possesses a maximum at the frequency $\nu_o = (3/2)(w_0/h)$. Quite unexpectedly, they found evidence indicating that the curve for sodium possessed not one, but two, "humps." Richardson attempted to account for these two (actually spurious) "humps" by postulating a second solution $\epsilon F'(\nu)$ of his equilibrium equation, but he never succeeded in finding it.[80]

This unresolved (though ill-starred) question may have been the prime motivation for Richardson's attempt to generalize his theory in late 1913.[81] All of its basic ideas were closely tied to his past work. In the first place, he took account of the possibility that the electrons may possess some kinetic energy *inside* the atom before their emission. Second, he assumed that the radiation supplied an unknown amount of energy $\phi(\nu)$ to the electrons, and he later found this amount to be equal to $h\nu$ for all frequencies and temperatures. (He remarked that Einstein in his study of the decomposition of a gas had reached a similar conclusion.) Third, he used Planck's law instead of Wien's law in his equilibrium equation, but once again he found the linear relationship between energy and frequency. Finally, he recognized (which is apparent from his model) that his entire theory applied to uncharged as well as to charged particles.

6. Richardson's Explanation of the Beta-Ray Asymmetry

The above experimental and theoretical studies constituted the culmination of Richardson's researches on the photoelectric effect—and, indeed, his career at Princeton University—for in early 1914, after the first term had ended, he left Princeton to return to England and accept the Wheatstone Chair of Physics at King's College, London. In early 1913, however, before leaving Princeton, he was led to explore an important issue closely related to his studies on the photoelectric effect. The issue was one which bears comment here. The problem he attacked was identical to that which so concerned W. H. Bragg; that is, the question of how to explain the forward-backward asymmetry in the secondary β-rays expelled from a metal plate by highly penetrating radiation.[82] This problem must have been troubling Richardson for some time, since as early as 1910, as we have seen,[83] he had asked his student, Otto Stuhlmann, Jr., to carry out ultraviolet light experiments on it. Now, specifically citing Stark's work of 1909, Richardson noted: "The explanation then usually offered to account for these asymmetric effects in secondary emission is that the energy in the radiations is not continuously distributed, but exists in a very concentrated form in limited regions. In this way a single electron may be made to receive the major part of both the energy and momentum in a radiation unit." In view of his theory of the photoelectric effect, his next statement comes as no surprise: "As the case against such a view of the nature of light is at least arguable, it is worth while to examine the question from a rather wider standpoint, to see if an estimate of the maximum effects to be observed may be obtained without making any definite hypothesis about the structure of the radiation."[84] It is now worthwhile to briefly outline Richardson's "rather wider standpoint."

He assumed that the incident radiation supplied energy to the electron continuously until it reached a "certain condition," that is, until its total energy was equal to an integral multiple of $h\nu$. He further assumed that this condition would be "followed by an act of a different character, which we may term disruption," and which would be "succeeded by a new cycle of, in general, similar events." Any energy or momentum transferred to the "matter as a whole then takes place through the acts of the disrupted electrons." This idea, as applied to the asymmetry mentioned above, yielded the following picture: "Consider the case of X radiation or light, incident normally on a thin slab of absorbing material. . . . We shall fix our attention on the state of things which exists after the slab has been illuminated for some time; so that there is no further accumulation, in the slab, of energy abstracted from the incident beam." Assuming that there are N electrons "disrupted" per unit time, each leaving with energy $\frac{1}{2}mv^2$, we have that $Nh\nu = N(\frac{1}{2}mv^2)$. "But," continued Richardson, "the incident beam is depleted of momentum as well as energy. Without committing ourselves for the moment as to the nature of this momentum, let us suppose that when the incident beam loses an amount E of energy it loses an amount λE of momentum. On the electromagnetic theory λ is constant and equal to $1/c$. . . ." If $E = h\nu$ and "the principle of the conservation of momentum is held to apply to the whole system of radiation and matter," we also have that $Nh\nu/c = Nm\bar{u}$, where $\bar{u}$ is the average component of velocity of the disrupted electrons in the direction of the incident radiation.[85]

By simply combining these two equilibrium conditions, Richardson concluded that $\bar{u} = v^2/2c$. This equation, he wrote, is "not sufficient to determine the distribution of the velocity among the particles. But we should expect the asymmetry to be more marked the greater the value of $\bar{u}/v$ and hence the greater the value of v. Thus those radiations which give rise to the emission of electrons with the greatest velocities should exhibit the greatest degree of asymmetry in this emission. This is known to be the case."[86] And that, for the moment, sufficed.

7. *Conclusions*

What general observations may be made on the interpretations of the photoelectric effect advocated by Lorentz, Thomson, Sommerfeld, and Richardson? In the first place, these interpretations certainly did not do what their originators had believed they did, that is, undercut or circumvent Einstein's light quantum hypothesis. Neither Einstein's proof that Wien's law entails the existence of quanta, nor his proof that Planck's law entails the coexistence of waves and particles in black-body radiation, was in any way affected, logically speaking, by these non-Einsteinian interpretations of the photoelectric effect. These interpretations simply demonstrated that a specific phenomenon could be accounted for on different sets of assumptions, that a specific mathematical relationship could be derived from different hypotheses.

But this logical issue was never foremost in the minds of Lorentz, Thomson, Sommerfeld, or Richardson. The boldness of Einstein's hypothesis made these physicists, and most of their contemporaries, unreceptive to it; and the unfamiliarity of Einstein's statistical mechanical arguments tended to rob them of conviction. Furthermore, Einstein's hypothesis was constantly viewed against the background of the elegance, heuristic power, and secure experimental base of Maxwell's theory. Attempts such as N. R. Campbell's to decide between wave and particle had failed; and the one experiment, the photoelectric effect, which seemed to offer the possibility of direct proof of quanta, was found to be interpretable without their help. The way in which this entire situation would be assessed in the future would depend critically, as always, on the knowledge, insight, and conviction of the individual physicist making the assessment.

C. Laue's Discovery of the Crystal Diffraction of X-Rays and its Influence (1912–1918)

1. *Its Origin and Impact on Bragg and His Contemporaries*

It was after the establishment of all the research programs previously discussed—in fact, it was in the midst of this intense theoretical and experimental activity on the photoelectric effect—that on April 21, 1912, a dramatic discovery was made. W. Friedrich and P. Knipping, carrying out a proposal of Max Laue who was working in Sommerfeld's Institute for Theoretical Physics at the University of Munich, found that X-rays could be diffracted by a copper sulfate crystal.[87] Sommerfeld, after overcoming his initial reluctance to provide Laue with men and equipment, gave his enthusiastic support to the work immediately after the first diffraction spots had been observed. Soon, as a member of the Bavarian Academy of Sciences, he also ensured the priority of Friedrich, Knipping, and Laue by presenting their papers on the discovery to the Academy on June 8 and July 6, 1912.[88] Laue's discovery, it is worth noting, also vindicated Sommerfeld's own recent and hard-won conclusion that X-rays are slightly diffracted when they pass through a narrow slit.[89]

Laue, almost immediately after making his discovery, turned to other matters. "In the summer of 1912 he accepted a chair of theoretical physics at the University of Zürich—but even before he left Munich he had set up an experiment on the diffraction of light by a dust-covered glass plate which had nothing to do with X-ray diffraction."[90] His contemporaries took an entirely different view. As Poincaré is said to have remarked shortly before his death, if "the value of a discovery is to be measured by the fruitfulness of its consequences, the work of Laue and his collaborators should be considered as perhaps the most important of modern physics."[91]

W. H. Bragg, who probably heard of Laue's discovery from Lord Rayleigh at the Dundee meeting of the British Association in the fall of 1912, asserted that it changed "the whole aspect of the problem. . . ."[92] It was no doubt this great influence on Bragg that C. J. Davisson had in mind when he wrote in 1927 that "It is an interesting commentary on the spirit of science that no one accepted the implication of x-ray-crystal phenomena more readily than Professor Bragg."[93] Since Bragg had developed his neutral pair model of X-rays and γ-rays in entirely qualitative language, and as a purely corpuscular theory, the impact of Laue's discovery on Bragg's theory was conclusive.

Yet, the essence of Davisson's remark, that Bragg immediately abandoned his corpuscular theory and adopted the wave theory, is not accurate. Bragg's first response was to try to discover an empirical rule governing the formation of the Laue spots. He wrote that he partially succeeded "as a consequence of an attempt to combine Dr. Laue's theory with a fact which my son pointed out to me, viz. that all the directions of the secondary pencils . . . are 'avenues' between the crystal atoms"[94] through which, presumably, Bragg's corpuscular X-rays could travel. Stark, another corpuscularian, attempted to explain Laue's results with Einstein's light quantum hypothesis. Very surprisingly, Barkla—the most adamant of the supporters of the electromagnetic pulse theory—found difficulty at first in understanding Laue's discovery as an interference phenomenon.[95]

As the initial uncertainty over the correct interpretation of Laue's discovery vanished, however—after Bragg's son, W.L. (Sir Lawrence) Bragg deduced the "Bragg equation" from wave considerations—it became apparent to the elder Bragg that a purely corpuscular theory of X-rays could not be maintained. In late November 1912, only two months after the Dundee meeting, Bragg admitted forthrightly that "If the experiment helps to prove X-rays and light to be of the same nature, then such a theory as that of the 'neutral pair' is quite inadequate to bear the burden of explaining the facts of all radiation."[96]

The emphasis in the last sentence is important—it is on the word "all." For even after the two Braggs had brilliantly exploited Laue's discovery and opened up an entirely new field of research, X-ray spectoscopy—for which they shared the Nobel Prize of 1915—W.H. Bragg never relinquished his desire for a combined wave-particle theory that reflected the corpuscular behavior of X-rays and γ-rays. In April 1913, while describing Laue's discovery, he wrote: "These results do not really affect the use of the corpuscular theory of X-rays. The theory represents the facts of the transfer of energy from electron to X-ray and *vice versa,* and all the phenomena in which this transfer is the principal event. It can predict discoveries and interpret them. It is useful in its own field. The problem remains to discover how two hypotheses so different in appearance can be so closely linked together."[97] Eight years later he was still expressing similar views.[98]

A number of prominent contemporary physicists, in fact, now saw the wave-particle dilemma very clearly. J. H. Jeans, for example, who was convinced of the necessity of the quantum theory by Poincaré's celebrated proof, but who was also aware of its "enormous" difficulties, became deeply concerned with the problem of reconciling the wave theory and the quantum theory. As he remarked in September 1913 at the Birmingham meeting of the British Association: "It is hardly too much to say that the two theories appear to be in active antagonism whenever they come in contact. Everywhere the undulatory theories demand that radiation should be capable of spreading and dividing indefinitely; while the quantum theory demands the reverse, at least when there is interaction between matter and ether. The conflict is, perhaps, shown at its keenest in the case of X-rays."[99]

Jeans went even further in his influential 1914 *Report on Radiation and the Quantum-Theory*. After examining all relevant evidence, including Bohr's theory of the atom and its relationship to the photoelectric effect,[100] and after contrasting the British and Continental views regarding the nature of the ether ("a completely real substance" *versus* one "devoid of substantiality"), he concluded:

> Whatever rôle is assigned to the ether, the main physical difficulty about the quantum-theory at present seems to be the apparent impossibility of reconciling it with the established results of the undulatory theory of light. The experimental evidence—for instance, of the photo-electric effect and of interference—seems almost to indicate that both theories are true simultaneously: that radiant energy both remains concentrated and indivisible, and at the same time spreads and is divisible: the idea that these two phenomena may be concerned with two totally different kinds of radiation seems at present too fantastic for serious consideration. In any case it may be asserted with confidence that until some kind of reconciliation can be effected between the demands of the quantum-theory and those of the undulatory theory of light, the physical interpretation of the quantum-theory is likely to remain in a very unsatisfactory state. Probably the hope of most physicists is that some sort of a compromise may ultimately be effected, but at present any practical attempt at a compromise appears to require the abandonment of something which is essential to one or other of the two theories.[101]

N. R. Campbell drew similar conclusions, but one of his remarks is particularly worth quoting to emphasize the general impact of Laue's discovery: "The question is no longer whether the X-rays are like light, but what theory of light we are to adopt in order to explain the properties of both light and X-rays."[102]

The dilemma was unavoidable and widely felt. In the United States, R. A. Millikan discussed it in an article in *Science* entitled "Atomic Theories of Radiation."[103] His purpose was to evaluate the radiation problem from "the point of view of *experimental* physics alone." He discussed, in turn: (1) Planck's so-called "second quantum theory," . . . "at present the most fundamental and least revolutionary form of quantum theory"; (2) Planck's original quantum theory, one which was "more radical" than his second: (3) Professor Bragg's "frankly corpuscular" theory, which appeared to be flatly contradicted by Laue's discovery; (4) J. J. Thomson's theory of pulses traveling along stretched Faraday tubes of force; and, (5) Einstein's "discontinuous" quantum theory. After examining the evidence, Millikan appraised the situation as follows: "The facts *which have been here presented* are obviously most completely interpreted in terms of . . . [Einstein's] theory, however radical it may be. Why not adopt it? *Simply because no one has thus far seen any way of reconciling such a theory with the facts of diffraction and interference so completely in harmony in every particular with the old theory of ether waves* [Millikan's italics]. Lorentz will have nothing to do with any ether-string theory, or spotted wave-front theory, or electro-magnetic corpuscle theory. Planck has unqualifiedly declared against it, and Einstein gave it up, I believe, some two years ago. . . ."[104]

There were, of course, some contemporary physicists for whom the situation was basically nonproblematic. Barkla for instance apparently never seriously doubted the universal validity of the electromagnetic theory. In 1913 he regarded as inexplicable the observation that the velocity of β-rays ejected from substances by high-frequency radiation is independent of the intensity of the radiation.[105] (Similarly, W.F.G. Swann proposed that the forward-backward asymmetry in the intensity of secondary β-rays could be completely understood in the light of the wave theory by noting that as the electric vector of the incident pulse or wave moves the β-ray *transversely,* the magnetic vector, acting in accordance with the Lorentz force law, would move the β-ray *in* the direction of propagation, thereby producing the asymmetry.[106]) In 1916 Barkla wrote: "There is not only no suggestion of a quantum or entity of radiation, but the phenomena become absolutely meaningless on any such theory. . . . There is no evidence of absorption of X-radiation in whole quanta. X-ray phenomena do not support the quantum theory of radiation as generally understood."[107] Nothing brings out Barkla's attitude toward quantum ideas more clearly than his willingness to postulate, and to unrelentingly pursue the consequences of the existence of a new series of characteristic X-radiations, the J-series.[108] As its name implies, the lines in this series were presumably associated with a lower-lying level than the K-level, and hence the J-series was fundamentally incompatible with the Bohr atom.

2. *Certainty at Last*

The influence on later thought, the importance of an experimental discovery as dramatic and as theoretically clear-cut as Laue's, can scarcely be exaggerated. Every physics teacher who subsequently touched on X-rays could hardly fail to mention Laue's discovery and point out its significance. It is almost possible, for example, to hear Rutherford telling his students, as he wrote in 1918: "The complete proof that X-rays were a type of light wave was supplied by Laue by showing that X-rays give marked diffraction effects in passing through crystals."[109] Or, as G.W.C. Kaye wrote in his very popular book *X-Rays:* "It is only within the last few years that controversy has been stilled by the discovery that X rays can be reflected and diffracted by crystals. There can scarcely be any doubt now that X rays are identical with ultraviolet light of extremely short wavelengths. . . ."[110] These remarks fully support H.S. Allen's recollection that after Laue's discovery "many physicists regarded the question at issue as closed. The debated problem as to the nature of X-rays was for them now settled in favor of the undulatory hypothesis."[111]

D. Related Theoretical and Experimental Work (1912–1917)

1. *Millikan's Photoelectric Effect Experiments and His Interpretation of Them*

After 1912 all theoretical and experimental research bearing on the nature of X-rays, or on the nature of radiation in general (including some work which has already been discussed), took place in the shadow of Laue's discovery. Some theoretical and experimental work grew directly out of Laue's discovery; some had roots that antedated it. In the balance of this chapter, we shall survey some of the important results achieved in both cases.

Perhaps the most significant work having roots that antedated Laue's discovery was R. A. Millikan's classic experiments on the photoelectric effect. In the fall of 1912, Millikan visited a number of European physicists (some of whom had recently attended the Solvay Conference) and then returned to his laboratory at the University of Chicago. He later recalled: "I knew from my Berlin discussions how difficult it would be to get a convincing answer to the problem of this Einstein photoelectric equation, and I scarcely expected either from Planck's attitude or from our discussions at Berlin that the answer, when and if it came, would be positive. . . ." He added: "but the question was very vital and an answer of some sort had to be found." Millikan was a highly competitive man, but even he underestimated the task before him: "it occupied practically all of my individual research time for the next three years."[112]

Millikan and his students, as a matter of fact, had worked intermittently on the photoelectric effect since 1905.[113] His first significant—and unexpected—finding was that the kinetic energies of the photoelectrons were independent of temperature. By 1911, as we have seen, J. R. Wright in Millikan's laboratory had progressed further, but his experiments did not bring to light, in Millikan's words, "anything approaching a linear relationship."[114] Eventually Millikan found "the key to the whole problem."[115] While tracking down the source of a troublesome disturbing effect ("photoelectric fatigue"), he discovered that the highly electropositive alkali metals (e.g., sodium and potassium) could be used for studying the photoelectric effect over a much wider range of incident wavelengths than had ever been used before.

But many problems remained: Means for measuring contact potentials had to be found: Sources of stray light had to be eliminated: Older work had to be re-examined for errors. Step by step his apparatus increased in complexity, eventually becoming "a machine shop in vacuo."[116] The final version, however, worked beautifully: the photosensitive surface could be scraped clean, irradiated, and have its contact potential determined—all under high vacuum. Accurate photocurrent measurements at last became possible. Many years later, Millikan summarized his work as follows: "After ten years of testing and changing and learning and sometimes blundering, all efforts being directed from the first toward the accurate experimental measurement of the energies of emission of photoelectrons, now as a function of temperature, now of wavelength, now of material (contact e.m.f. relations), this work resulted, contrary to my own expectation, in the first direct experimental proof in 1914 of the exact validity, within narrow limits of experimental error, of the Einstein equation, and the first direct photoelectric determination of Planck's h."[117] He first reported his results at the April 1915 Washington meeting of the American Physical Society.

After Millikan's experiments, there could be no doubt whatsoever about the validity of Einstein's linear relationship—the accuracy with which his data points fell on the predicted straight line was truly remarkable (see Fig. 2[118]). In fact, all relevant predictions—that the slope of the line is precisely h/e, that the energy of the photoelectrons does not exceed a definite maximum for a given incident frequency, etc.—were quantitatively confirmed. Einstein's light quantum hypothesis was therefore finally vindicated. Or was it? Concerning the *theoretical interpretation* of his experiments, Millikan noted emphatically: "the semi-corpuscular theory by which Einstein arrived at his

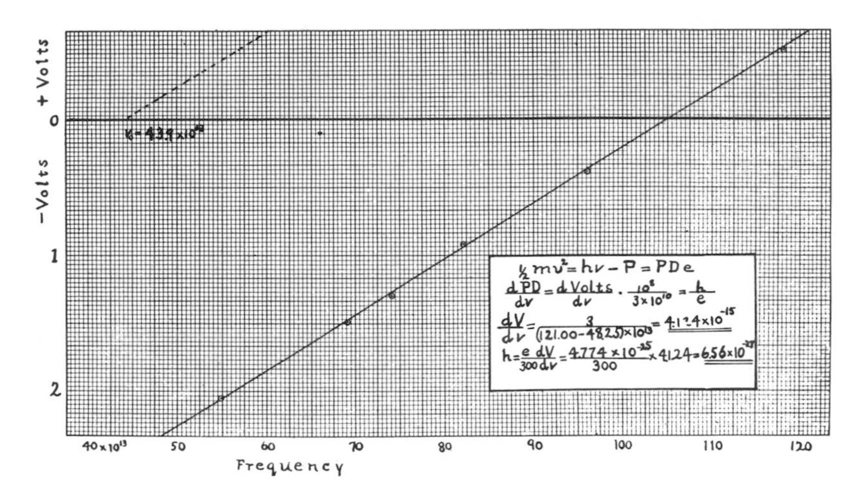

FIG. 2. Millikan's plot demonstrating the linear relationship between photoelectron energy and incident radiation frequency.

equation seems at present to be wholly untenable."[119] Einstein's "bold, not to say reckless," hypothesis "seems a violation of the very conception of an electromagnetic disturbance"—it "flies in the face of the thoroughly established facts of interference."[120]

The question therefore was: what explanation could be given as "a substitute for Einstein's theory?"[121] Millikan's attitude was identical to that expressed a few years earlier by Lorentz, Thomson, Sommerfeld, and Richardson—he, too, searched for an alternative interpretation of the photoelectric effect! He felt, however, that Thomson's "ether-string theory," for example, notwithstanding the fact that "Sir J. J. Thomson and Norman Campbell still adhere to it," would not do. In the first place, Millikan had found in the course of his recent (and equally famous) oil-drop experiments that slowly increasing the electric field produced no sudden changes in the suspended electron's velocity, which he took to prove that, contrary to Thomson's opinion, the field did not have a "fibrous structure."[122] In the second place, Thomson's theory, in common with all electromagnetic theories, predicted that very long times—several hours at least—would be required for successive wavefronts to supply to the tiny electron its observed kinetic energy. Yet, "the corpuscle is observed to shoot out the instant the light is turned on." To Millikan, there was therefore *"no alternative but to assume that the corpuscles which are ejected are already possessed of an energy almost equal to hν* [Millikan's italics]." This was possible only if the atoms in the photosensitive metal contained corpuscle-oscillators "at all times in all stages of energy loading up to the value $h\nu$." No corpuscle could radiate any of its energy until a "certain critical value" had been reached, but at this time "explosive emission" occurred.[123]

Since the corpuscle acquired its energy by resonance, Millikan's theory had a great deal in common with Sommerfeld's "accumulation" theory, although Millikan himself tried hard to separate his ideas from Sommerfeld's. He considered the latter to be only a "modification" of Planck's second quantum theory. Perhaps the most novel feature of Millikan's theory was his assumption that in the atom, in addition to corpuscle-oscillators of certain characteristic "natural" frequencies, there were "a few oscillators of every conceivable frequency." The emisssion of photoelectrons then "appears to take place to some extent at all frequencies of the incident light, though the emitted corpuscle never leaves the metal unless its energy of emission from the atom is greater than $h\nu_0$, but it takes place especially copiously when the impressed frequency coincides with a 'natural frequency'."[124]

Therefore, in common with Lorentz, Thomson, Sommerfeld, and Richardson, Millikan did not take the validity of Einstein's *equation*—which he established beyond doubt—to constitute proof of the validity of Einstein's light quantum *hypothesis*. No one expressed this point more clearly than Millikan himself, when in 1917 he wrote: "Despite . . . the apparently complete success of the Einstein equation, the physical theory of which it was designed to be the symbolic expression is found so untenable that Einstein himself, I believe, no longer holds to it, and we are in the position of having built a very perfect structure and then knocked out entirely the underpinning without causing the building to fall. It stands complete and apparently well tested, but without any visible means of support. These supports must obviously exist, and the most fascinating problem of modern physics is to find them. Experiment has outrun theory, or, better, *guided by erroneous theory*, it has discovered relationships which seem to be of the greatest interest and importance, but the reasons for them are as yet not at all understood [italics added]."[125] When Millikan at last provided an experiment that confirmed Einstein's light quantum hypothesis, he did not accept it as such.

2. *Einstein Concludes Light Quanta Possess Linear Momentum*

Millikan was equally mistaken in his appraisal of Einstein's attitude toward his light quantum hypothesis. Rather than abandoning it, Einstein in 1916, after devoting himself almost exclusively to the problem of gravitation for several years, achieved deeper insights into it. His paper, "On the Quantum Theory of Radiation,"[126] which embodied the results of his efforts, is best known for his "astonishingly simply and general" derivation of Planck's radiation law by introducing the famous "Einstein coefficients." This derivation is too well known to be repeated here. It is more important for us—and it was apparently more important for Einstein in 1916—that through his *further* analysis he was able to provide an answer to the question: "does the molecule experience an impulse when it emits or absorbs energy ϵ?" He termed his answer to this question "the main result" of his paper.[127]

Just as in 1909, Einstein arrived at his answer by analyzing the momentum fluctuations of black-body radiation in equilibrium with gas molecules. The basic equation he obtained, for the mean square fluctuation $\overline{\Delta^2}$, was identical with his earlier result, equation (1.10). He then went on to show, however, that under the assumption of the quantum theory of radiation, both $\overline{\Delta^2}$ and the "radiation friction" P could be related to the energy density $\rho(\nu)$ to obtain a new equation, provided that $\rho(\nu)$ was given by Planck's law. The "most important" aspect of this calculation was that in it he had to assume that each elementary process involving the emission and absorption of radiation was "perfectly directional";[128] otherwise, his equation would be violated. Einstein therefore regarded the following statements "as fairly certainly proved":

> If a radiation bundle has the effect that a molecule struck by it absorbs or emits a quantity of energy $h\nu$ in the form of radiation (ingoing radiation), then a momentum $h\nu/c$ is always transferred to the molecule. For an absorption of energy, this takes place in the direction of propagation of the radiation bundle, for an emission in the opposite direction. If the molecule is acted upon by several directional radiation bundles, then it is always only a single one of these which participates in an elementary process of irradiation; this bundle alone then determines the direction of the momentum transferred to the molecule.
>
> If the molecule undergoes a loss in energy of magnitude $h\nu$ without external excitation, by emitting this energy in the form of radiation (outgoing radiation), then this process, too, is directional. Outgoing radiation in the form of spherical waves does not exist. During the elementary process of radiative loss, the molecule suffers a recoil of magnitude $h\nu/c$ in a direction which is only determined by 'chance', according to the present state of the theory.
>
> These properties of the elementary process . . . make the formulation of a proper quantum theory of radiation appear almost unavoidable. The weakness of the theory lies on the one hand in the fact that it does not get us any closer to making the connection with wave theory; on the other, that it leaves the duration and direction of the elementary processes to 'chance'. Nevertheless I am fully confident that the approach chosen here is a reliable one.
>
> There is room for one further general remark. Almost all theories of thermal radiation are based on the study of the interaction between radiation and molecules. But in general one restricts oneself to a discussion of the *energy* exchange, without taking the *momentum* exchange into account. One feels easily justified in this, be-

cause the smallness of the impulses transmitted by the radiation field implies that these can almost always be neglected in practice, when compared with other effects causing the motion. For a *theoretical* discussion, however, such small effects should be considered on a completely equal footing with [the] more conspicuous effects of a radiative *energy* transfer, since energy and momentum are linked in the closest possible way. For this reason a theory can only be regarded as justified when it is able to show the impulses transmitted by the radiation field to matter lead to motions that are in accordance with the theory of heat.[129]

3. *Critical Technical Inventions*

Although Millikan was completely mistaken in believing that Einstein had abandoned his light quantum hypothesis, there is ample evidence that Millikan's attitude toward Einstein's hypothesis was typical of the time. Millikan and his contemporaries were aware of Einstein's hypothesis, saw the ease with which it accounted for, say, the photoelectric effect, but at the same time were deeply impressed with its apparent difficulties *vis-à-vis* interference phenomena. No one was prepared to take Einstein's hypothesis really seriously until these difficulties were resolved. It is not surprising, therefore, that it eventually had to be forced upon physicists. As it developed, the route by which this occurred originated not in theories or experiments which antedated Laue's discovery, but in experiments which grew directly out of it. These experiments, in turn, would not have been possible without two technical inventions perfected in 1912–1913.

The first, the Bragg spectrometer, which was invented by the Braggs shortly after Laue's discovery, has already been mentioned and will enter into our discussion again and again.[130] The second, the Coolidge X-ray tube, which was invented in 1913 by W. D. Coolidge of General Electric,[131] was less dramatic but hardly less important than the first. Rutherford termed it "a triumph of the application of the latest scientific knowledge and technique."[132] In the Coolidge tube, X-rays were produced by directing accelerated electrons onto a metal target under high vacuum. In a short time it replaced the old, erratic, gas-filled, focus-type tube as a laboratory source of X-rays. Its main advantages were that it could "be operated continuously for hours," that the intensity and penetrating power of the X-rays were "both under the complete control of the operator," and that each of these quantities could be "instantly increased or decreased independently of the other."[133] A third invention, the Wilson cloud chamber, first constructed by C. T. R. Wilson in 1911, soon became an extremely important tool for directly observing interactions between atomic and subatomic particles.[134]

It is worthwhile to emphasize that the invention of these instruments by 1913 (when Arthur Compton, incidentally, was in his first year of graduate work at Princeton University) makes that year a critical year in the history of the quantum theory of scattering. By 1913 every important and relevant experimental fact—the change in the hardness of the scattered rays with scattering angle, the lack of dependence of this change on the nature of the radiator, the forward-backward asymmetry in the secondary β-rays and X-rays (or γ-rays)—was known. It is true that they were known only qualitatively. Yet, we have just seen that by 1913 every instrument—the Bragg spectrometer, the Coolidge X-ray tube, the Wilson cloud chamber—which was required to make these facts quantitative had been developed. Moreover, the key theoretical concept, the conservation of momentum in an electron-quantum collision, had been elaborated by 1909 by Stark. These are striking facts —facts which must be kept in mind when attempting to picture the general attitude of physicists toward the nature of radiation around 1916, and consequently, when attempting to assess the difficulty of Compton's achievement.[135]

4. *Webster Undercuts the Pulse Theory Using the Duane-Hunt Law*

One important theoretical insight gained from experiments carried out subsequent to Laue's discovery concerned the status of the electromagnetic pulse theory of X-rays. While it may seem natural in retrospect to assume that the pulse theory was immediately abandoned by everyone after Laue's discovery, this was not the case. As we will see, it found adherents even after 1920. Nevertheless, in 1915 D. L. Webster, who only two years earlier had made essential use of the pulse theory in his explanation of "excess scattering," presented a strong argument against it.[136]

Webster's argument was based on a relationship recently discovered empirically by William Duane and D. L. Hunt, both of whom (like Webster himself) were then at Harvard University. Purely as a generalization from experiment, Duane and Hunt concluded that the product of the electronic charge e and the accelerating potential V of a Coolidge tube is equal to the maximum energy $h\nu_{max}$ of the X-rays emitted by the tube.[137] That is, they concluded that $h\nu_{max} = eV$. (Both Stark and Wien had reached this same conclusion earlier from theoretical considerations.[138]) That this result—which now appears to be an absolutely trivial deduction from Einstein's light quantum hypothesis—was ennobled in 1915 by calling it the "Duane-Hunt law," merely serves to underline the negative attitude of contemporary physicists toward Einstein's hypothesis. Duane himself wholeheartedly rejected Einstein's hypothesis; and even Rutherford in 1918 could say about the Duane-Hunt law: "There is at present no physical explanation possible of this remarkable connection between energy and frequency, and it is doubtful whether it involves a new and unsuspected property of the electron itself or is connected in some way with the actual structure of the atom."[139]

On the basis of the Duane-Hunt law, Webster gave a simple but shrewd argument against the pulse theory of X-rays. He noted that a pulse, when "decomposed" by Fourier analysis, is made up of *all* radiant frequencies between minus and plus infinity—all frequencies should be present in the pulse. There should be no ν_{max}, no maximum frequency of the X-rays emitted by the tube. The pulse theory of X-rays and the Duane-Hunt law were therefore mutually incompatible.

5. *X-Ray and Gamma-Ray Absorption Experiments*

A second important question investigated after Laue's discovery concerned the precise connection between γ-rays and X-rays. Certainly there was strong, perhaps almost universal feeling that γ-rays were simply X-rays of short wavelength. Yet proof was still lacking. As it developed, a great deal of information on this question was obtained from absorption experiments,[140] specifically from measurements on the variation of the linear-absorption coefficient μ (or mass-absorption coefficient μ/ρ) with wavelength λ. These experiments, superficially most prosaic ones, led eventually to the most far-reaching conclusions.

As early as 1912, E. A. Owen had allowed characteristic X-radiations from various metals to pass through absorbing gases and had measured the corresponding absorption coefficients μ. He observed that μ apparently varied as $1/A^5$, that is, as the inverse fifth power of the atomic weight *A of the metals emitting the X-rays*.[141] The following year, Moseley discovered that the frequency ν of a given characteristic X-ray varies as the square of the atomic number Z of the element emitting it.[142] C. G. Darwin drew the logical conclusion: Owen's law could be written as

$$\mu = k\lambda^{5/2}, \qquad (2.15)$$

where k is a constant.[143]

With the invention of the Bragg spectrometer, a source of continuously variable X-ray wavelengths became available, and far greater precision became possible. The first person to use this instrument (in 1914) for absorption measurements was W. H. Bragg himself, aided by S. E. Pierce.[144] Bragg and Pierce, however, did not investigate the variation of μ with λ, but rather the variation of μ with Z, the atomic number of the absorber. They concluded that μ varied as Z^4, so that Owen's law could be further transformed into

$$\mu = k'\, Z^4\lambda^{5/2} \qquad (2.16)$$

where k' is another constant.

Some doubt was cast on this result, still in 1914, when Manne Siegbahn, the well-known Swedish X-ray spectroscopist, examined the variation of μ with λ and concluded that μ apparently varied not as $\lambda^{5/2}$ but as λ^n, where n lay between 2.55 and 2.71 for various solids, and between 2.66 and 2.94 for various gases.[145] Stimulated by Siegbahn's work, W. H. Bragg performed further experiments in 1915.[146] He tentatively confirmed Siegbahn's results, concluding that for a number of substances the exponent n was apparently closer to 3 than to 5/2. One year later, A. W. Hull and Marion Rice reinforced this conclusion by finding that a good fit to their data could be obtained if the mass-absorption coefficient μ/ρ were given by

$$\frac{\mu}{\rho} = a\lambda^3 + b \tag{2.17}$$

where a and b are constants.[147] All of the above results were valid, of course, only for the regions between the so-called X-ray "absorption edges."

Mass-absorption coefficients for *γ-rays* emitted by various radioactive substances had been measured as early as 1909 by F. Soddy and A. S. Russell.[148] In general, they found that μ/ρ for γ-rays was typically two orders of magnitude smaller than it was for X-rays. That these results were not grossly in error was shown in 1917 by M. Ishino, a Japanese research student working in Rutherford's laboratory at the University of Manchester.[149] Ishino found that for RaC γ-rays passing through aluminum and lead, μ/ρ was 0.071 cm^2/g and 0.076 cm^2/g, respectively. But—in very important experiments—he went even further: He devised a somewhat involved (and not entirely unproblematic) experimental method for estimating the corresponding value of the mass-*scattering* coefficient σ/ρ, which is related to the mass-*absorption* coefficient μ/ρ, and to the mass-*fluorescent* coefficient τ/ρ by the equation

$$\frac{\mu}{\rho} = \frac{\tau}{\rho} + \frac{\sigma}{\rho} \tag{2.18}$$

or equivalently, by

$$\mu = \tau + \sigma \tag{2.19}$$

Correcting for a 15% loss due to back-scattering (compare with Madsen's 1909 estimate of 15–22%[150]), Ishino concluded that σ/ρ for RaC γ-rays scattered by aluminum, iron, and lead, was 0.045, 0.042, and 0.034 cm^2/g, respectively. These values, concluded Ishino, were "far less than . . . [those] for the ordinary X-rays."[151]

The precise values for the mass-scattering coefficient σ/ρ for X-rays had been tentatively estimated in early 1916 by Hull and Rice in connection with their previously mentioned absorption experiments. They found that for X-rays of wavelength 0.12–0.39 Å absorbed by aluminum, copper, and iron, the equation $\mu/\rho = a\lambda^3 + 0.12$ gave a good fit to their data, provided the constant a was different for each element.[152] When this result is compared with equation (2.18), it is seen that Hull and Rice's conclusion was tantamount to claiming that σ/ρ was independent of wavelength and absorber, and equal to 0.12 cm^2/g.

The work of Hull and Rice (as well as the earlier work of Soddy and Russell, and Ishino) induced C. G. Barkla and Margaret P. White to perform similar, but far more extensive X-ray absorption experiments in 1917.[153] Using the direct beam from an X-ray tube as their primary source of X-rays (which they made "remarkably homogeneous" by filtering), Barkla and White selected radiation with a wavelength range of 0.145–0.509 Å, and irradiated copper, aluminum, water, filter paper and paraffin. In all, they used 72 different primary wavelengths (although a given substance was not subjected to each wavelength). They observed that for each substance there was a systematic decrease in the mass-absorption coefficient μ/ρ with decreasing wavelength. Of all of their very extensive data, however, the direction indicated by one particular point assumed the greatest importance for the history of the Compton effect: the value of μ/ρ for 0.145 Å X-rays in aluminum. "It will . . . be noticed," wrote Barkla, "that the coefficient of absorption in aluminum sinks quite appreciably below .2 to a value .15, and that there is a tendency to lower values still. . . . [It] certainly looks as though we have in the absorption by aluminum, evidence of the beginning of a second marked deviation from the simple laws—that is, evidence that a diminution of scattering and absorption is setting in when the wave-length becomes small. This appears to be the link connecting the absorption of X-rays with the absorption of γ rays."[154]

Years later, that apparently innocuous observation, indicating as it did a highly puzzling trend, was called by Arthur Holly Compton "an observation comparable in importance with the Michelson-Morley experiment."[155]

E. References

[1] In addition to the papers by Bragg, Florance, Gray, etc., discussed in Chapter 1, see also G. W. C. Kaye, "The Emission of Röntgen Rays from Thin Metallic Sheets," *Proc. Cambridge Phil. Soc.* 15 (1909):269–272.

[2] For a discussion of Nernst's role in testing Einstein's theory, see Martin J. Klein, "Einstein, Specific Heats, and the Early Quantum Theory," *Science* 148 (1965):173–180. For other discussions of the first Solvay Conference, see Russell McCormmach, "Henri Poincaré and the Quantum Theory," *Isis* 58 (1967):37–55; L. Rosenfeld, "La première phase de l'évolution de la théorie des quanta," *Osiris,* 2 (1936):149–196, and Max Jammer, *The Conceptual Development of Quantum Mechanics* (New York: McGraw-Hill, 1966), pp. 56–61. For the reports of the conference see P. Langevin and M. de Broglie (eds.), *La Théorie du rayonnement et les quanta* (Paris: Gauthier, 1912).

[3] See Chapter 1. Einstein's initial letter to Lorentz, which is deposited in the Einstein Archives, Institute for Advanced Study, Princeton, New Jersey, is undated but was evidently written some time between March 15, 1909, the date of publication of Einstein's paper, and May 23, 1909, the date of publication of Einstein's next letter to Lorentz, which is deposited in the Algemeen Rijksarchief, The Hague, Holland. Another letter to Lorentz in the Einstein Archives is dated January 27, 1911; others in the Algemeen Rijksarchief are dated February 15, 1911, and November 23, 1911; one letter from Lorentz to Einstein in the Algemeen Rijksarchief is dated December 6, 1911. I should like to express my gratitude to both Helen Dukas for her kindness in showing me the letters in the Einstein Archives and providing me with copies of them, and A.E.M. Ribberink for providing me with copies of the letters in the Algemeen Rijksarchief.

[4] Lorentz published these arguments in "Die Hypothese der Lichtquanten," *Phys. Z.* 11 (1910):349–354, especially pp. 353–354, and in "Alte und neue Fragen der Physik," *Phys. Z.* 11 (1910):1234–1257, especially pp. 1249–1250. Other relevant arguments were also given by Lorentz in these papers.

[5] Lorentz, "Alte und neue Fragen," p. 1250, notes that Stark suggested a way out of some of these difficulties by hypothesizing that light quanta form coherent aggregates of energy, but concludes that this involves the same kind of difficulties as met in Newton's emission theory.

[6] Einstein to Lorentz, May 23, 1909. The following discussion is based on this letter. Russell McCormmach recently also discussed the Einstein-Lorentz interchanges in "Einstein, Lorentz, and the Electron Theory," in his *Historical Studies in the Physical Sciences,* Vol. 2 (Philadelphia: University of Pennsylvania Press, 1970), pp. 41–87.

[7] Einstein to Lorentz, January 27, 1911.

[8] See reference 5. The Wolfskehl Lectures were established in 1908 by Paul Wolfskehl when he bequeathed 100,000 marks to the Göttingen Academy of Sciences as an award to the first person submitting a correct proof of Fermat's famous "last theorem." Since none was forthcoming, the interest on the money was used to sponsor lectures by prominent scientists.

[9] "Alte und neue Fragen," (reference 4) p. 1250.

[10] For a description of Haas's very original atomic model and its possible influence, see Jammer, *Conceptual Development of Quantum Mechanics,* (reference, 2) pp. 40–42.

[11] "Alte und neue Fragen," (reference 5) p. 1253.

[12] For much insight into the state of knowledge, see John L. Heilbron, "The Scattering of α and β Particles and Rutherford's Atom," *Arch. Hist. Exact Sci.* 4 (1968):247–307.

[13] "Ueber die lichtelektrische Wirkung," *Ann Physik* 8 (1902):149–198, especially pp. 169–170.

[14] For relevant remarks, see J. H. Jeans, *Report on Radiation and the Quantum-Theory* (London: "The Electrician" Printing & Publishing Co., 1914), p. 62. See also Sommerfeld's remarks, reference 35.

[15] See also Roger H. Stuewer, "Non-Einsteinian Interpretations of the Photoelectric Effect" in Roger H. Stuewer (ed.), *Historical and Philosophical Perspectives of Science* (Minneapolis: University of Minnesota Press, 1970), pp. 246–263.

[16] For a discussion of Thomson's attitude, see Russell McCormmach, "J. J. Thomson and the Structure of Light," *Brit. J. Hist. Sci.* 3 (1967):362–387, especially p. 373. One can also judge Thomson's attitude from his *Conduction of Electricity Through Gases,* 2d ed. (Cambridge: Cambridge University Press, 1906), pp. 228–279.

[17] J. J. Thomson, "On the Ionization of Gases by Ultra-Violet Light and on the evidence as to the structure of light afforded by its Electrical Effects," *Proc. Cambridge Phil. Soc.* 14 (1906–1908):422. For E. Ladenburg's paper see "Über Anfangsgeschwindigkeit und Menge der photoelektrischen Elektronen in ihrem Zusammenhange mit der Wellenlänge des auslösenden Lichtes," *Ber. Deut. Physik Ges.* 9 (1907):504–515; *Phys. Z.* 8 (1907): 590–594.

[18] "On the Theory of Radiation," *Phil. Mag.* 20 (1910):238–247.

[19] Thomson made use of the doublet idea earlier in "On the Velocity of Secondary Cathode Rays from Gases," *Proc. Cambridge Phil. Soc.* 14 (1906–1908):541–545; but his present considerations seem to have grown most directly out of his analysis of how "kinks" may be produced in tubes of force emanating from electrons when the electrons interact with atomic doublets. See "On a Theory of the Structure of the Electric Field and its Application to Röntgen Radiation and to Light," *Phil. Mag.* 19 (1910):301–313.

[20] "On the Theory of Radiation," (reference 18) p. 245.

[21] *Ibid.,* p. 246.

[22] "On the Motion of a Particle about a Doublet," *Phil. Mag.* 20 (1910):380–382.

[23] Letter to the Editors, *Phil. Mag.* 20 (1910):544.

[24] "On the Unit Theory of Light," *Proc. Cambridge Phil. Soc.* 16 (1910–1912):643–652.

[25] "On the Structure of the Atom," *Phil. Mag.* 26 (1913):792–799.

[26] *The Life of Sir J. J. Thomson, O.M., Sometime Master of Trinity College, Cambridge* (Cambridge: Cambridge University Press, 1943), p. 141.

[27] "On the Structure of the Atom," p. 793.

[28] *Ibid.,* p. 795.

[29] For a discussion of this work, see McCormmach, "J. J. Thomson," (reference 16) pp. 369 and 378.

[30] Letter of Bragg to Sommerfeld, February 7, 1911. See Chapter 1, reference 61.

[31] Letter to the Editors, *Phil. Mag.* 26 (1913):1044.

[32] See Chapter 1, sections B.4 and D.4.

[33] "Das Plancksche Wirkungsquantum und seine allgemeine Bedeutung für die Molekularphysik," *Ber. Deut. Physik Ges.* 9 (1911):1092. Sommerfeld's adoption of the quantum theory may have had elements in common with his rather rapid adoption of the special theory of relativity, as described by Max Born in his obituary notice "Arnold

Johannes Wihelm Sommerfeld," *Roy. Soc. Obit. Not. (London)* 8 (1952):280. See also K. Bechert, "Arnold Sommerfeld," *Experientia* 7 (1951):477–478. Jammer has discussed this paper of Sommerfeld's in his *Conceptual Development of Quantum Mechanics,* p. 55.

[34] Although Debye helped Sommerfeld develop his theory from the start, the first paper of which he appears as co-author was the paper "Theorie des lichtelektrischen Effektes vom Standpunkt des Wirkungsquantums," *Ann. Physik* 41 (1913):873–930.

[35] "Das Plancksche Wirkungsquantum," p. 1088.

[36] "Theorie des lichtelektrischen Effektes," p. 885.

[37] "Das Plancksche Wirkungsquantum," p. 1089.

[38] "Theorie des lichtelektrischen Effektes," p. 885.

[39] "Das positive Potential des Aluminiums als eine Funktion der Wellenlänge des einfallenden Lichtes," *Phys. Z.* 12 (1911):338–343; "The Positive Potential of Aluminum as a Function of the Wave-length of the Incident Light," *Phys. Rev.* 33 (1911):43–52.

[40] "Über den selektiven Photoeffekt ausserhalb der Alkaligruppe," *Ber. Deut. Physik. Ges.* 9 (1911):474–481.

[41] "Das Plancksche Wirkungsquantum," (reference 33) p. 1091.

[42] "Application de la théorie de l'Élément d'action aux phénomènes moléculaires non périodiques," in *La théorie du rayonnement et les quanta* (reference 2), pp. 313–372; discussion, pp. 373–392.

[43] *Ibid.*, p. 356. The standard abbreviation for electron volt is eV.

[44] *Ibid.*, p. 390.

[45] *Ibid.*, p. 432.

[46] *Ibid.*, p. 443.

[47] *Ibid.*, p. 451.

[48] See R. McCormmach, "Henri Poincaré," (reference 2) especially pp. 43–50 and 53–54.

[49] "Thermodynamische Begründung des photochemischen Äquivalentgesetzes," *Ann. Physik* 37 (1912):832–838.

[50] *Ibid.*, p. 838.

[51] "Nachtrag zu meiner Arbeit: 'Thermodynamische Begründung des photochemischen Äquivalentgesetzes'," *Ann. Physik* 38 (1912):881–884.

[52] *Ibid.*, p. 884.

[53] For a brief discussion, see Armin Hermann, "Albert Einstein und Johannes Stark: Briefwechsel und Verhältnis der beiden Nobelpreisträger," *Arch. Ges. Med.* 50 (1966):279.

[54] "Sommerfeld" (see reference 33), p. 283.

[55] "Theorie des lichtelektrischen Effektes" (see reference 34).

[56] *Ibid.,* especially pp. 923–929. M. Born later maintained that the physicists "of the younger generation . . . regarded Planck's 'second quantum theory' as a weak

compromise. . . ." See his obituary of Planck in *Roy. Soc. Obit. Not. (London)* 6 (1948–1949):172. For Planck's paper see "Eine neue Strahlungs-hypothese," *Ber. Deut. Physik. Ges.* 13 (1911):138–148. See also E. Marx, "Die Theorie der Akkumulation der Energie bei intermittierender Belichtung und die Grundlage des Gesetzes der schwarzen Strahlung," *Ann. Physik* 41 (1913):161–190.

[57] For the development of Bohr's thought, see John L. Heilbron and Thomas S. Kuhn, "The Genesis of the Bohr Atom," in Russell McCormmach (ed.), *Historical Studies in the Physical Sciences,* Vol. 1 (Philadelphia: University of Pennsylvania Press, 1969), pp. 211–290.

[58] For a sketch of Richardson's life and work, see William Wilson, "Owen Williams Richardson," *Roy. Soc. Obit. Not. (London)* 5 (1959):207–215.

[59] Quoted from Marjorie Johnston (ed.), *The Cosmos of Arthur Holly Compton* (New York: Knopf, 1967), p. 199.

[60] By 1901, Richardson had already derived his famous equation for the rate of emission of electrons from a heated surface, although he later had to correct the $T^{1/2}$ factor to a T^2 factor. Richardson eventually (1928) won the Nobel Prize for his achievements.

[61] "Some Applications of the Electron Theory of Matter," *Phil. Mag.* 23 (1912):616; for abstract of this paper see, "The Application of Statistical Principles to Photoelectric Effects and some Allied Phenomena," *Phys. Rev.* 34 (1912):146–149.

[62] "Some Applications," p. 617.

[63] *Ibid.,* p. 619.

[64] "The Theory of Photoelectric Action," *Phil. Mag.* 24 (1912):570–574; essentially the same paper was published in *Science* 36 (1912):57–58.

[65] See Richardson, *The Electron Theory of Matter* (Cambridge: Cambridge University Press, 1914), p. 472. There he calculates that

$$E = \frac{n}{2}\left(\frac{m}{2\pi RT}\right)^{3/2} \int_0^\infty \int_{-\infty}^\infty \int_{-\infty}^\infty um\,(u^2+v^2+w^2)\,e^{-(m/2RT)(u^2+v^2+w^2)}\,du\,dv\,dw$$

$$= 2N_0RT.$$

[66] "Theory of Photoelectric Action," (reference 64) p. 573. The possibility that some of the electrons incident on the surface are reflected and not absorbed by it may be taken into account very simply by introducing a multiplicative factor s, where $0 \leqq s \leqq 1$, times the quantity $h\nu - w_0$.

[67] *Ibid.,* p. 574.

[68] "The Photoelectric Effect," *Phil. Mag.* 24 (1912):575–594—hereafter cited as "Photoelectric Effect."

[69] *Ibid.,* p. 575; this page also gives references to the work cited by Richardson and Compton.

[70] *Ibid.,* pp. 584–585.

[71] "The Photo-electric Action of X-Rays," *Proc. Roy. Soc. (London)* 94 (1918):269.

[72] *Ibid.*

[73] Karl T. Compton and O. W. Richardson, "The Photoelectric Effect—II," *Phil. Mag.* 26 (1913):550.

[74] "On the Long-wave Limits of the Normal Photoelectric Effect," *Phil. Mag.* 26 (1913):1017–1024.

[75] Ibid., p. 1018. W. Hallwachs, "Die Lichtelektrizität" in E. Marx (ed.), *Handbuch der Radiologie,* Vol. III (Leipzig: Akademische Verlagsgesellschaft M.B.H., 1916), pp. 537–559, lists 469 papers on the photoelectric effect between 1887 and 1912. For contemporary discussions of the photoelectric effect, see, for example, H. S. Allen, *Photo-Electricity: The Liberation of Electrons by Light* (London: Longmans, Green, and Co., 1913; 2d ed. 1925); A. L. Hughes, *Photo-electricity* (Cambridge: Cambridge University Press, 1914).

[76] *Ibid.*

[77] Letter to the Editors, *Phil. Mag.,* 27 (1914):473–475.

[78] "On the Long-wave Limits," (reference 74) p. 1019.

[79] See the paper cited in reference 73; see also "The Photoelectric Effect—III," *Phil. Mag.* 29 (1915):618–623 (with F. J. Rogers).

[80] See "The Photoelectric Effect—II," (reference 73) p. 567, for a discussion of his initial attempts.

[81] "The Theory of Photoelectric and Photochemical Action," *Phil. Mag.* 27 (1914):476–488. See also A. L. Hughes and L. A. Du Bridge, *Photoelectric Phenomena* (New York: McGraw-Hill, 1932), pp. 196–198.

[82] "The Asymmetric Emission of Secondary Rays," *Phil. Mag.* 25 (1913):144–150. See also J. Laub, "Über die durch Röntgenstrahlen erzeugten Strahlen," *Ann. Physik* 46 (1915):785–808.

[83] Chapter 1, section C.1.

[84] "Asymmetric Emission," (reference 82) p. 145.

[85] *Ibid.,* pp. 145–147.

[86] *Ibid.,* p. 148.

[87] For a detailed discussion of the origin of Laue's discovery, see Paul Forman, "The Discovery of the Diffraction of X-Rays by Crystals; A Critique of the Myths," *Arch. Hist. Exact Sci.* 6 (1969):38–71, and P. P. Ewald's adjoining comments. For references to relevant primary and secondary literature, see Forman's paper. In 1913, incidentally, Laue's father was raised to hereditary nobility; subsequent to that date, Laue always signed his name von Laue. See P. P. Ewald, "Max von Laue," *Roy. Soc. Obit. Not. (London)* 6 (1960):135 (footnote). See also Ernst Lampa, "Max von Laue," *Naturwiss.* 36 (1949):354–355, and W. Friedrich, "Erinnerungen an die Entdeckung der Interferenzerscheinungen bei Röntgenstrahlen," *Ibid.,* pp. 354–356.

[88] "Interferenzerscheinungen bei Röntgenstrahlen," *Sitzber. Akad. (München)* 1912:303–322; "Eine quantitative Prüfung der Theorie für die Interferenzerscheinungen bei Röntgenstrahlen," *Sitzber. Akad. (München)* 1912:363–373; Laue's two papers have been reprinted in, for example, his *Gesammelte Schriften und Vorträge,* Vol. 1 (Braunschweig: Friedrich Vieweg, 1961), pp. 183–218; Ostwald's *Klassiker,* Number 204, and

Naturwiss. 39 (1952):361–372; for an English translation see G. E. Bacon, *X-Ray and Neutron Diffraction* (Oxford: Pergamon, 1966), pp. 89–108.

[89] A. Sommerfeld, "Über die Beugung der Röntgenstrahlen," *Ann. Physik* 38 (1912) 473–506. See also H. Haga and C. H. Wind, "Die Beugung der Röntgenstrahlen," *Wied. Ann.* 68 (1899):884–895; *Ann. Physik* 10 (1903):305–312; *Proc. Konink. Akad. Wetensch.* 9 (1906):104–109; "Über die Polarisation der Röntgenstrahlen und der Sekundärstrahlen," *Ann. Physik* 23 (1907):439–444; and B. Walter and R. Pohl, "Zur Frage der Beugung der Röntgenstrahlen," *Ann Physik* 25 (1908):715–724; "Weitere Versuche über die Beugung der Röntgenstrahlen," *Ann Physik* 29 (1909):331–354.

[90] Ewald, "Max von Laue," (reference 87) pp. 139–140.

[91] Quoted in A. H. Compton, *X-rays and Electrons* (New York: Van Nostrand, 1926), p. 90.

[92] W. H. and W. L. Bragg, *X Rays and Crystal Structure,* 4th ed. (London: G. Bell & Sons, 1924), p. 3.

[93] "Are Electrons Waves?" *Bell Lab. Record* 4 (1927):258.

[94] "X-rays and Crystals," *Nature* 90 (1912):219.

[95] For a brief discussion, see Forman, "Discovery of Diffraction," (reference 87) p. 55.

[96] "X-rays and Crystals," *Nature* 90 (1912):360–361.

[97] "The Reflection of X-rays by Crystals," *Proc. Roy. Soc. (London),* 88 (1913):436.

[98] For more information and references to primary and secondary literature, see Roger H. Stuewer, "William H. Bragg's Corpuscular Theory of X-Rays and γ-Rays," *Brit. J. Hist. Sci.* 5 (1971):258–281.

[99] "Discussion on Radiation" in *Repts. BAAS (1913) at Birmingham*—Transactions of Section A, p. 380.

[100] *Report,* (reference 14) p. 63.

[101] *Ibid.,* p. 89.

[102] *Modern Electrical Theory* 2d ed., (Cambridge: Cambridge University Press, 1913), p. 300.

[103] *Science* 37 (1913):119–133.

[104] *Ibid.,* pp. 132–133.

[105] "The Nature of X-rays," *Eng.* 96 (1913):422.

[106] "The Pulse Theory of X Rays, γ Rays, and Photoelectric Rays, and the Asymmetric Emission of β Rays," *Phil. Mag.* 25 (1913):534–557.

[107] "On the X-Rays and the Theory of Radiation," *Proc. Roy. Soc. (London)* 92 (1916):501–504. See also "Problems of Radiation," *Nature* 94 (1915):671–672; "X-ray Fluorescence and the Quantum Theory," *Nature* 95 (1915):7.

[108] "Notes on the Absorption and Scattering of X-rays, and the Characteristic Radiations of J-series," *Phil. Mag.* 34 (1917):270–285 (with M. White).

[109] "Silvanus P. Thomson Memorial Lecture," *J. Röntgen Soc.* 14 (1918):81.

[110] 3d ed. (London, 1918), p. 236. See also Max Planck's general lecture, "Das Wesen des Lichtes," *Naturwiss* 7 (1919):903–909; translated by R. Jones and D. W. Williams and reprinted in *A Survey of Physical Theory* (New York: Dover, 1960), pp. 89–101.

[111] "Charles Glover Barkla," *Roy. Soc. Obit. Not. (London),* 5 (1945–1948):349. Barkla also exploited Laue's discovery; see "Reflection of Röntgen Radiation," *Nature* 90 (1912):435; "An X-Ray Fringe System," *Nature* 90 (1912):647 (both with G. H. Martyn).

[112] *Autobiography* (New York: Prentice-Hall, 1950), p. 100. See also L. A. Du Bridge and Paul S. Epstein, "Robert Andrews Millikan," *Natl. Acad. Sci. Biog. Mem.* 33 (1959):241–282.

[113] For a summary of Millikan's efforts, see his paper "A Direct Photoelectric Determination of Planck's '*h*'," *Phys. Rev.* 7 (1916):355–388.

[114] *Ibid.*, p. 360.

[115] *Autobiography,* (reference 112) p. 101.

[116] "Direct Photoelectric Determination," (reference 113) p. 361.

[117] "The electron and the light-quant from the experimental point of view;" reprinted in *Nobel Lectures in Physics,* Vol. II (Amsterdam: Elsevier, 1965), pp. 61–62. Copyright ©, The Nobel Foundation, 1924.

[118] Reproduced from "Direct Photoelectric Determination," (reference 113) p. 373.

[119] *Ibid.*, p. 382.

[120] *Ibid.*, pp. 355, 384.

[121] *Ibid.*, p. 385.

[122] *Ibid.*, p. 384.

[123] *Ibid.*, p. 385.

[124] *Ibid.*, pp. 387–388.

[125] *The Electron;* facsimile reprint (Chicago: Phoenix, 1963), p. 230. Copyright ©, The University of Chicago, 1917 and 1924. It is rather shocking to compare this statement with what Millikan wrote in 1950 (*Autobiography,* pp. 100–101) concerning his "complete verification of the validity of Einstein's equation." "This seemed to me," he wrote, "as it did to many others, a matter of very great importance, for it rendered what I will call Planck's 1912 explosive or trigger approach to the problem of quanta completely untenable and proved simply and irrefutably, I thought, *that the emitted electron that escapes with the energy $h\nu$ gets that energy by the direct transfer of $h\nu$ units of energy from the light to the electron* [Millikan's italics] and hence scarcely permits of any other interpretation than that which Einstein had originally suggested, namely that of the semi-corpuscular or photon theory of light itself."

[126] "Zur Quantentheorie der Strahlung," *Phys. Z.* 18 (1917):121–128; translated and reprinted in B. L. Van Der Waerden (ed.), *Sources of Quantum Mechanics* (New York: Dover, 1967), pp. 63–77, as well as in D. ter Haar, *The Old Quantum Theory* (Oxford: Pergamon, 1967), pp. 167–183. For a discussion of this paper in the context of Einstein's work, see Martin J. Klein, "Einstein and the Wave-Particle Duality," *Nat. Phil.,* Vol. 3 (New York: Blaisdell, 1963), pp. 3–49.

[127] Van Der Waerden, p. 65.

[128] *Ibid.*

[129] *Ibid.*, pp. 76–77.

[130] For some early papers by Bragg on X-ray spectroscopy, see, for example, "The Reflection of X-rays by Crystals," *Proc. Roy. Soc. (London)* [A] 89 (1913):246–248; "X-Rays and Crystalline Structure," *Eng.* 97 (1914):814–815. Also see Bragg's Nobel Lecture, "The Diffraction of X-Rays by Crystals," reprinted in *Nobel Lectures: Physics*, Vol. 1 (Amsterdam: Elsevier, 1967), pp. 370–382, as well as his book *X Rays and Crystal Structure* (London: G. Bell & Sons, 1915).

[131] "A Powerful Röntgen Ray Tube with a Pure Electron Discharge," *Phys. Rev.* 2 (1913):409–430.

[132] "Silvanus P. Thomson Memorial Lecture," (reference 109) pp. 83–84.

[133] Coolidge, "Powerful Röntgen Ray Tube," (reference 131) p. 430.

[134] For references to Wilson's papers, see Chapter 1, reference 63. See also P. M. S. Blackett, "Charles Thomson Rees Wilson," *Roy. Soc . Obit. Not. (London)* 6 (1960):269–295. It is also worthwhile to note that H. Geiger invented the "Geiger counter" in 1913. See "Über eine einfache Methode zur Zählung von α- und β-Strahlen," *Ber. Deut. Physik. Ges.* 15 (1913):534–539.

[135] I know of no evidence to suggest that Compton was in any way influenced by Stark's work. Nor, for example, did Kallmann and Mark in an extensive 1926 review article (see Chapter 7, reference 23) refer at all to Stark's ideas—they went back only to Einstein's 1917 paper for the concept of light quantum momentum.

[136] "The X-Ray Spectrum of Tungsten at a Constant Potential," *Phys. Rev.* 6 (1915):56.,

[137] "On X-Ray Wave-Length," *Phys. Rev.* 6 (1915):166–177. See also F. C. Blake and W. Duane, "X-ray Determination of *h*," *Phys. Rev.* 10 (1917):624–627; Webster later pointed out that he played a key role in the discovery of this law. See transcript of interview with him on deposit at the Center for History of Physics, American Institute of Physics, New York City.

[138] See Armin Hermann (ed.), *Die Hypothese der Lichtquanten, Dokumente der Naturwissenschaft (Physik)*, Vol. 7 (Stuttgart: Ernst Battenberg, 1965), p. 18.

[139] "Silvanus P. Thomson Memorial Lecture," (reference 109) p. 84.

[140] For a general discussion, see A. H. Compton and S. K. Allison, *X-Rays in Theory and Experiment* (New York: Van Nostrand, 1935), pp. 533ff.

[141] "The Passage of Homogeneous Röntgen Rays through Gases," *Proc. Roy. Soc. (London)* [A] 86 (1912):426–439.

[142] For a full discussion, see John L. Heilbron, "The Work of H. G. J. Moseley," *Isis* 57 (1966):336–364.

[143] Compton and Allison, p. 533.

[144] "The Absorption Coefficients of X rays," *Phil. Mag.* 28 (1914):626–630.

[145] "Über den Zusammenhang zwischen Absorption und Wellenlänge bei Röntgenstrahlen," *Phys. Z.* 15 (1914):753–756.

[146] "The Relation between certain X-ray Wave-lengths and their Absorption Coefficients," *Phil. Mag.* 29 (1915):407–412.

[147] "The Law of Absorption of X-Rays at High Frequencies," *Phys. Rev.* 8 (1916):326–328.

[148] "The γ-Rays of Uranium and Radium," *Phil. Mag.* 18 (1909):620–649; see also F. and W. M. Soddy and A. S. Russell, "The Question of the Homogeneity of γ-Rays," *Phil. Mag.* 19 (1910):725–757.

[149] "The Scattering and Absorption of Gamma Rays," *Phil. Mag.* 33 (1917):129–146.

[150] See Chapter 1, reference 58.

[151] "Scattering and Absorption," (reference 149) p. 145.

[152] "Law of Absorption," (reference 147) pp. 327–328.

[153] See reference 108 for exact reference to their paper "Notes on Absorption."

[154] *Ibid.*, p. 277.

[155] Unpublished letter dated April 28, 1958, to Professor Franklin Miller, Jr., Kenyon College, Gambier, Ohio, deposited at the Center for History of Physics, American Institute of Physics, New York City. Quoted by permission of Mrs. A. H. Compton. In private correspondence, Professior R. S. Shankland told the author that neither he nor Professor Miller, both of whom worked in Compton's laboratory in Chicago, ever got the impression that the *single* measurement of Barkla assumed the great importance which Compton later attributed to it in his 1958 letter to Professor Miller. Indeed, Chapter 3 will show, in agreement with Professor Shankland's recollections, that it was rather the *trend* indicated by Barkla's measurements that was of such great importance for Compton's thought. It is also worth noting that Compton was obviously implying that there was a genetic relationship between the Michelson-Morley experiment and Einstein's discovery of special relativity, which, as the researches of Gerald Holton have shown, is a very problematic contention.

CHAPTER 3

The Large Ring Electron

A. Introduction

To shift from a study of the work of many physicists to that of one, Arthur Holly Compton, is to shift from a portrayal of a relatively broad historical panorama to an illumination of some of its significant details. Of course, even here not all subjects can be treated; selection principles were and will be at work. They operated in the first two chapters to eliminate, for example, virtually all mention of the profound changes that took place in our knowledge of atomic structure between 1895 and 1916. They will operate in this and succeeding chapters to confine our discussion to a detailed analysis of the research program pursued by Compton between 1916 and 1923. To offset some of the obvious disadvantages entailed by this restriction, there is at least one great advantage: It will enable us both to delineate the conceptual hurdles Compton had to leap and to show the relative importance Compton attached to various theories and experiments as his thought evolved. It will enable us, in a word, to explore the anatomy of one of the most important discoveries in modern physics.

B. Arthur Holly Compton (1892–1962)

1. Early Life and Work at Wooster and Princeton

Arthur Holly Compton was born September 10, 1892, in Wooster, Ohio, where his father Elias was Professor of Philosophy and later Dean of the College of Wooster. The religious, intellectual, and intimate home atmosphere created by the parents, Elias and Otelia Compton, and its enormous influence on the lives of their four children, Karl Taylor (1887–1954), Mary Elesa (1889–1961), Wilson Martindale (1890–1967), and Arthur Holly (1892–1962), have already been described by James R. Blackwood.[1] Most, if not all, of the traits that later characterized Arthur Compton's personality and approach to scientific problems had already emerged before he left Wooster for graduate work at Princeton University in 1913. His profound sense of wonder when contemplating natural phenomena and his unbounded patience in observation were already apparent by age 12 when he spent hour after hour observing the sun and stars with a Sears, Roebuck telescope his father had allowed him to purchase from his savings. His great personal confidence, single-mindedness, inventiveness, and love of working with his hands became obvious by age 16 when he researched the early literature on airplanes, criticized certain aspects of it in print, and then convinced his father to allow him to construct and fly a glider with a 27-foot wingspan. At the same time (May, 1910), he took a photograph of Halley's Comet that he kept for the rest of his life. Three years later, he published a mature article in *Science* describing a clever laboratory method he had discovered for demonstrating the earth's rotation.[2]

Arthur Compton's choice of career was influenced most by his father and brother Karl. He was undecided, because of the strong religious atmosphere in which he had been reared, whether to become a minister or scientist after he graduated from Wooster College in 1913. His father advised him: "If I am not greatly mistaken, it is in science that you will find that you can do your best work. Your work in this field may become a more valuable Christian service than if you were to enter the ministry or become a missionary."[3]

Compton was particularly receptive to this advice, because years earlier, during the winter of 1908–1909, his brother Karl had already taken him under his wing and had taught him some techniques of experimental physics at Wooster. Karl had enlisted Arthur's help while investigating for his Master's thesis the physical reasons for the "popping" sound emitted by the so-called Wehnelt interrupter, which was part of the X-ray equipment Professor Bennett had purchased for the College shortly after Röntgen's discovery.[4] This was the first X-ray equipment Arthur Compton ever handled. Around 1911 he used it again, rather extensively, in his own laboratory research at Wooster. This experience, he recalled, "was of substantial help to me when I went to Princeton a few years later to do my graduate study."[5]

Compton had chosen Princeton University largely because his brother Karl had already gone there for graduate work. Karl, in turn, had chosen Princeton because of its outstanding reputation for research in physics. Princeton's President Woodrow Wilson had fostered a physics department with six full professors, the "lodestar" of which was O. W. Richardson.[6] Karl soon became Richardson's best student and began assisting Richardson in his studies on the photoelectric effect. We have already seen the fruits of that collaboration in Chapter 2. We have also seen that Richardson left Princeton University for the University of London in early 1914—just after Arthur Compton had completed his first term of graduate study there. In an extraordinary gesture, Richardson turned over his X-ray equipment to Arthur Compton when he left,[7] and hence it is not surprising that even this brief contact with Richardson exerted immense influence on him. As he recalled in 1918:

> After I had completed my studies in college, I felt a great interest in the work that was being done in modern physics. I had an ambition to follow this work to the limit in a graduate school. But I was unable to see that such study would enable me ever to do work of any considerable value in the world. After learning all the physics that I could, I might teach the subject to someone else, who might in turn study and teach. But I wanted to prepare myself to do some kind of work that would be of material value to my brother men. As a result I laid my plans to spend a year of graduate study in physics for a thorough foundation, and then to enter an engineering college. I am certainly thankful that by the time I had completed my first year of graduate study I had come to see in physics research a field of the greatest possible promise so far as achievements of real and permanent value are concerned.[8]

In 1915 Arthur Compton followed in Karl's footsteps and won the coveted Porter Ogden Jacobus Fellowship at Princeton. Concurrently with his own studies, he had the immense pleasure of directly observing Karl (who meanwhile had become assistant professor of physics) organize a research program for studying the passage of electrons through gases. At one point, when Karl required a highly sensitive electrometer, Arthur showed him a prototype of one he had carefully constructed by trial and error for use in his own research. Karl then made a thorough theoretical study of the instrument, and together the two brothers eventually designed and constructed a new and very sensitive model, which they later described in the only scientific paper on which they collaborated.[9] Arthur soon displayed his inventiveness once again by developing an improved version of the Bragg spectrometer.[10]

In 1916 Arthur Compton received his Ph.D. degree at Princeton. His thesis research, suggested by Richardson but completed under H. L. Cooke,[11] involved studying the intensity of X-rays as a function of their angle of reflection from crystals, primarily to see how it depended upon the distribution of the electrons in the crystal atoms. The resulting publication[12] clearly reveals Compton's deep commitment to X-ray research as well as his mastery of all relevant physical and mathematical concepts—many of which proved very useful to him in his subsequent research. Later in life he was fond of recalling how, at the last moment, he discovered that C. G. Darwin in England had worked on almost exactly the same problem as he had, which cast doubt upon whether he could use it as a thesis topic. "In working it

FIG. 1. The University of Minnesota physics faculty and staff, 1916–1917. From left to right: A. Johnsrud, C. J. Johnson, O. Rognley, A. H. Compton, F. Vik, A. Zeleny, H. A. Erikson (Chairman), P. D. Foote, J. T. Tate, L. W. McKeehan, E. O. Dietrick, C. Dowe, and P. A. Klopsteg.

over, however," he smiled, "I was able to make a few additions and changes that made it acceptable."[13]

2. *First Experiments at Minnesota (1916–1917)*

Immediately after finishing his Ph.D. degree Compton married Betty McCloskey, who had been a classmate at Wooster College. Shortly thereafter, having already accepted a position as Instructor in Physics at $1500 per year at the University of Minnesota, he and his wife moved to Minneapolis, arriving there together with Paul D. Foote and John T. Tate (see Fig. 1). At Minnesota, Compton helped out in a general physics course for engineers; he was also allowed to follow his own dominant interests of that time by offering a graduate course in the mathematical theory of electricity and magnetism. Furthermore, he later recalled that at Minnesota he had had an experience "that proved to be one of the most illuminating experiences of my scientific career."[14]

As his first postdoctoral research, Compton had decided to test a prediction of Weber's theory of the elementary magnet. According to that theory, magnetization of a substance was accompanied by a reorientation of its atoms, and Compton had concluded from his experience with X-rays that he should be able to detect this reorientation by reflecting X-rays from a magnetite crystal, at the same time observing the anticipated shift in the Laue spots as the crystal was changed from an unmagnetized to a magnetized state. Months of work followed while Compton assembled or constructed all of the required spectrometers, electrometers, and electromagnets. No small effort was invested in Compton's experiment by other members of the Physics Department. At last the experiment was ready.

". . . I turned on the magnet," Compton recalled, "and nothing at all happened. The spot of light stayed right where it was. Again and again I tried, checking the instruments one by one. Still nothing happened. I could hardly believe the fact before my eyes. My heart sank." Just then the Chairman of the Department, Henry Erikson, walked in and asked Compton how his experiment was coming along. Compton ruefully replied, " 'It isn't.' . . . Even then it seemed unbelievable that the beautiful theory I had hoped to verify could be false." Eventually, Compton drew some remarkable conclusions from this negative result, but at the time it was a great disappointment, and Erikson could not help observing Compton's long face. He reacted by giving Compton a friendly slap on the back, saying, "Well, Compton, . . . the way things are is always more interesting than the way we thought they were."

That incident, wrote Compton, ". . . was one of the best lessons in the understanding of science that I have ever had. The mistaken notion is to get some idea and then try to prove it. . . . The real thing that a scientist tries to do when he is faced with a phenomenon is to attempt to understand it. To do that he tries all the possible answers that he can think of to see which one of them works best."[15] Compton's dictum was followed by no one more closely than Compton himself; during the next six years he indeed tried "all the possible answers" he could think of to understand the scattering of X-rays.

C. Compton's Large Electron Theory of Scattering (1917)

1. *Its Origin and Fruitfulness*

Compton embarked directly on the research program that ultimately proved so fruitful sometime after mid-1917 and after he had decided to leave the University of Minnesota to accept a lucrative position as a research engineer in the Lamp Division of the Westinghouse Electric and Manufacturing Company in East Pittsburgh, Pennsylvania.[16] He was always explicit on what triggered his thoughts. It was his reading of Barkla's 1917 paper, in which Barkla had reported the observation Compton later referred to as "comparable in importance with the Michelson-Morley experiment."

That judgment, of course, was made with hindsight. At the time, Barkla himself regarded his observations as somewhat surprising and worthy of attention, but was led to no profound conclusions of permanent value from them. Compton also regarded them as puzzling—but at the same time he sensed that "Barkla had hold of something of fundamental importance."[17] Superficially, Barkla's key observation was simple enough. He found, we recall, that the mass-absorption coefficient μ/ρ steadily decreased with wavelength, and for 0.145 Å X-rays in aluminum it was equal to 0.153 cm^2/g. Why did Compton regard this result as so unsettling? The reason is easy to understand, because if equation (2.18), $\mu/\rho = \tau/\rho + \sigma/\rho$, holds, and if σ/ρ is given by Thomson's expression,

$$\frac{\sigma}{\rho} = \frac{\sigma_0}{\rho} = \frac{8\pi}{3} \frac{Ne^4}{m^2c^4\rho} \tag{3.1}$$

a serious contradiction appears. For if this expression is evaluated, assuming that the atomic number of the absorber is numerically equal to roughly one-half its atomic weight, it is a simple matter to see that σ_0/ρ is a constant, independent of wavelength, and equal to about 0.2 cm^2/g (0.188 cm^2/g is more precise[18]). Therefore, for *no incident wavelength whatsoever* should the *total* mass-absorption coefficient μ/ρ drop below about 0.2 cm^2/g.[19] Yet, γ-ray experiments such as Soddy and Russell's,[20] and Ishino's,[21] yielded values on the order of 0.03 cm^2/g, and Barkla had just recorded for short-wavelength X-rays a steady decrease to the value 0.153 cm^2/g.

That was the puzzle.[22] One possible explanation was that not all of the electrons were effective in the scattering process: Perhaps the N electrons per unit volume in Thomson's expression should be replaced by $(1/2)N$ or some other fraction of N. This would obviously produce a reduction in σ_0/ρ, and hence in μ/ρ. Ishino, however, had explicitly concluded that "the scattering of the gamma rays per atom is approximately proportional to the number of the electrons exterior to the nucleus in the atom,"[23] and Soddy and Russell had reached a similar conclusion, which to Compton meant that "all the electrons outside the nucleus are effective in producing absorption."[24] Replacing N with some fraction of N would not do.

A second possible explanation was that the incident rays were being scattered by different electrons in the atom and subsequently interfering with each other, as in "excess scattering." In fact, both Ishino and Barkla had felt that the reduced absorption could be accounted for in this manner.[25] Compton took sharp exception: "If the electron acts as a point charge there is . . . no possible grouping of the electrons which can, according to classical theory, produce a smaller absorption than that calculated according to Thomson's formula."[26] A third explanation, which Compton also found unacceptable, was that there were extremely strong electronic binding forces present in the atom.

If these explanations were untenable, what then was the correct one? It was in his answer to this question that Compton fully revealed his theoretical commitments in 1917. From his course work, research activities, and discussions at Princeton—particularly with his brother Karl and O. W. Richardson—Compton had inherited an attitude of complete confidence in the universal validity of classical electrodynamics. He revealed this attitude as early as 1914 when, deeply impressed with Laue's recent discovery, he wrote an article in which he maintained that while "there were those who still defended other hypotheses," the "true explanation of these [Laue] spots was given by Mr. W. L. Bragg."[27] He elaborated in a footnote: "The theory other than the wave theory which received most support was one put forward by Prof. W. H. Bragg in 1910. According to this theory, X-rays are corpuscles, each composed of an electron and an equal positive charge, which travel with nearly the velocity of light. This theory was made to account for the fact that the 'emergent' secondary radiation is stronger than the 'incident' radiation, when a beam of X-rays passes through matter. Even this fact, however, was shown by Prof. O. W. Richardson to be consistent with the wave theory of X-rays."[28]

One can almost hear Richardson talking when Compton, one year later, outlined what he regarded as the correct interpretation of specific heats: "The very considerable success of the quantum hypothesis in explaining the variations of the specific heat of solids with temperature has been taken as a strong confirmation of that hypothesis. Before this evidence can be considered as conclusive, however, it is necessary to see if there may not be some other satisfactory solution of the problem of specific heat, which does not involve the conception of quanta. . . . In the present paper an assumption is introduced which leads directly to an expression for the variation of the specific heat with temperature which will be shown to agree at least as well with experiment as the expression derived from the quantum hypothesis." Adopting C. Benedicks' concept of "agglomeration," Compton enunciated his non-quantum assumption as follows: "If the relative energy between two neighboring atoms in a solid falls below a certain critical value, the two atoms become agglomerated so that the degree of freedom between them vanishes; but as soon as the energy increases again above the critical value, the degree of freedom reappears."[29] Compton then derived an expression for the specific heat which depended on this critical temperature—but was independent of Planck's constant. He compared his expression with Debye's: "Debye's expression . . . has been found to be largely empirical; so to compare this agglomeration hypothesis with the quantum hypothesis my formula is rather to be compared with that of Einstein." To wit: "It is evident that Einstein's expression is much the less accurate of the two," and consequently "this agglomeration hypothesis represents more accurately the condition of the atoms in a solid than does the quantum hypothesis."[30]

The following year (1916), Compton attacked the problem of thermal conductivity, concluding that the agglomeration hypothesis "leads naturally to a satisfactory formula for the thermal conductivity while the former [the quantum hypothesis] seems to give such an expression only with the use of a highly improbable assumption. . . ."[31] A further, related paper reveals no more friendly attitude toward the quantum theory.[32] At various points Compton expressed his thanks to Professors E. P. Adams, W. Magie, and K. T. Compton for their "continued interest" and "helpful criticism."[33]

Compton's negative predisposition toward the quantum theory—as well as his positive predisposition toward classical electrodynamics—was changed neither by Rutherford's visit to Princeton around 1915,[34] nor as a result of Compton's researches at the University of Minnesota during 1916–1917. He would, in fact, give the quantum theory little serious consideration until 1921.[35] The net result was that when confronted with Barkla's puzzling 1917 observations, which involved an interaction between radiation and electrons, Compton sought its explanation *not* in the nature of the radiation, but in the nature of the electrons.

He concluded, drawing on his thorough knowledge of physical optics, that the electrons were *diffracting* the incident radiation as well as scattering it. For radiation to be diffracted, however, its wavelength must be comparable in size to the size of the obstacle, in this case, the electron. Since Barkla had just demonstrated that the reduction in the mass-absorption coefficient was still occurring for 0.145 Å X-rays, the only possible conclusion, Compton maintained, was that "the diameter of the electron is comparable in magnitude"[36] with 0.145 Å—or even perhaps an order of magnitude smaller to account for the same phenomenon with γ-rays.

This was an extremely startling conclusion, since a number of different lines of reasoning had previously suggested that the diameter of the electron was on the order of 10^{-13} cm, or 10^{-5} Å. Compton was therefore advocating an electron that was roughly 1,000 to 10,000 times as large as the old model. Furthermore, an *electron* of radius 0.1 Å was one-tenth as large as the diameter of the hydrogen *atom,* according to Bohr's theory. It seems, therefore that although Compton was familiar with Bohr's theory (even if he was not basically concerned with current research on atomic structure[37]), he was willing to countenance a very discordant assumption in order to forge ahead with his scattering theory. Many of his readers, however, must have regarded his model of the electron—his "large electron," as he repeatedly called it—to be almost as radical as Einstein's light quantum hypothesis.

Yet, qualitatively at least, radiation scattered from such an electron had very desirable properties. Consider a wavefront $S'SS''$ incident on such an electron, and imagine that two rays are scattered from two extreme points, A and B (see Fig. 2[38]). If, now, the diameter of the electron is comparable in size with the wavelength of the incident radiation, an "appreciable phase difference" will be produced between these two rays, since the path $S''BP''$ of the lower ray is much shorter than the path $S'AP'$ of the upper ray. For some scattering angle θ this path difference will result in destructive interference, and hence "the intensity of the ray scattered to P will be much reduced." This, presumably, gave rise to the reduction in intensity that Barkla had observed.

The potential fruitfulness of Compton's large electron hypothesis did not stop there, however. Compton knew, as he wrote in early 1918, "that the scattered radiation on the emergent side of the plate is much more intense than that on the incident side, both in the case of relatively soft X-rays and in the case of hard γ-rays."[39] As we have seen, in the former case this phe-

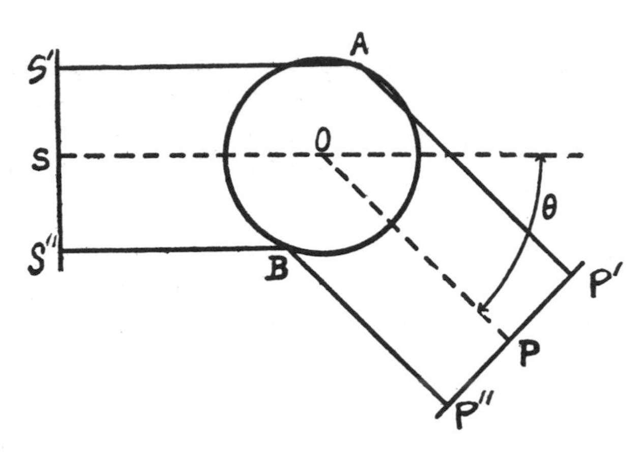

Fig. 2. Compton's diagram showing an X-ray wavefront incident on a large electron.

nomenon was called "excess scattering" and was explained by the interference of rays scattered by different electrons. In the latter case, however, Compton argued that this explanation could be ruled out, for two reasons. First, if an atom were traversed by rays of increasing hardness, then at some point Thomson's formula would hold, indicating that the electrons were scattering *independently,* and "must continue to do so for all shorter wavelengths." Second, in "excess scattering"—this was the origin of the term—the total forward scattered radiation *increased,* not decreased, as Barkla had observed. "It is thus evident," Compton wrote, "that the unsymmetrical scattering of very short electromagnetic waves is due not to groups of electrons in the atoms, but to some property of the individual electrons."[40]

Compton soon summarized very clearly what he had in mind:

> It is possible that certain assumptions regarding the conditions for scattering radiant energy, contrary to classical theory, might be made which would account for the observed low value of the scattering for very high frequencies. As long as the idea of the point charge electron is retained, however, no such assumptions can account for the observed dissymmetry between the incident and the emergent scattered radiations. Unless the theory that X-rays and γ-rays consist of waves or pulses is abandoned, the only possible explanation of this dissymmetry would seem to be that the scattering particles have dimensions comparable with the wavelength of the rays which they scatter. Since the scattering particles have been shown to be the electrons, the statement may therefore be made with confidence that *the diameter of the electron is comparable in magnitude with the wave-length of the shortest γ-rays* [Compton's italics].[41]

Thus, while momentarily toying with another idea, Compton embraced

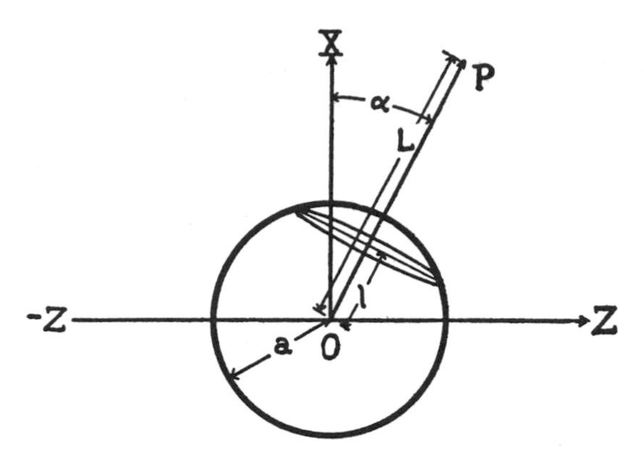

FIG. 3. Compton's diagram showing radiation being scattered from a rigid, spherical, charged shell.

his large electron hypothesis as the key to the explanation of the asymmetry. But could this explanation be made *quantitative?* Could the forward-backward asymmetry in the angular distribution $I(\theta)$, as well as the reduction in the mass-scattering coefficient σ/ρ, be *calculated?* To approach these problems a definite shape of the electron had to be assumed, since, for example, the angular distribution of radiation scattered from a cube would obviously not be the same as that scattered from a sphere. Compton illustrated the general nature of the problem—which had a number of features in common with his thesis research[42]—by considering how radiation of wavelength λ is scattered from a rigid, spherical, charged shell of finite radius $a \sim \lambda$.

Suppose that electromagnetic radiation of wavelength λ is incident from the $-Z$ direction on such an electron, and suppose that it is scattered in the direction $OP = L$ (see Fig. 3[43]). If the amplitude of the incident radiation is A, then the electric field E_0 at any point z on the Z-axis is given by $E_0 = A\cos[(2\pi/\lambda)(\Delta - z)]$, where Δ is the path difference of the waves with respect to $z = 0$. Defining η to be the charge on a shell between two planes infinitely close together and perpendicular to the Z-axis, the total force on the electron will then be given by

$$F = \int_{-a}^{a} E_o \eta dz = A\eta \int_{-a}^{a} \cos \frac{2\pi}{\lambda} (\Delta - z) dz = \frac{A\eta\lambda}{\pi} \cos \frac{2\pi\Delta}{\lambda} \sin \frac{2\pi a}{\lambda}. \qquad (3.2)$$

This force determines both the acceleration $\ddot{x} = F/m$, which the electron *as a whole* experiences perpendicular to the Z-axis, and the magnitude of the reradiated (or scattered) electric field $E(L,\alpha)$ at the point $P(L,\alpha)$. The latter quantity is given by $(e\ddot{x}/Lc^2) \sin \alpha = (eF/Lmc^2) \sin \alpha$, where α is the complement of the scattering angle θ. Since the charge e is distributed over the entire shell, however, one must integrate over the distance l, where $|l| \leqq a < L$. The result is that

$$E(L,\alpha) = \frac{\sin\alpha}{Lmc^2}\int_{-a}^{a} \frac{A\eta\lambda}{\pi}\cos\frac{2\pi\Delta}{\lambda}\sin\frac{2\pi}{\lambda}(\Delta - L + l)\eta dl$$

$$= \frac{A\lambda^2\eta^2}{\pi^2 Lmc^2}\sin\alpha\sin\frac{2\pi a}{\lambda}\cos\frac{2\pi}{\lambda}(\Delta - L) \qquad (3.3)$$

Finally, one can use the fact that the intensity $I(L,\alpha)$ of the scattered radiation is proportional to the square of its amplitude $E(L,\alpha)$, as well as the definition of the scattering coefficient σ (the ratio of the total energy scattered to the total energy incident), to find that the mass-scattering coefficient σ/ρ is given by

$$\frac{\sigma}{\rho} = \frac{N}{kA^2\rho}\int_0^{\pi} I(L,\alpha)(2\pi L\sin\alpha)(Ld\alpha) = \frac{8\pi}{3}\frac{Ne^4}{m^2c^4\rho}\frac{\sin^4(2\pi\alpha/\lambda)}{(2\pi a/\lambda)^4}. \qquad (3.4)$$

Note that σ/ρ is a function of a/λ.

Two conclusions follow immediately. First, it is apparent that

$$\lim_{a/\lambda\to 0}\frac{\sigma}{\rho} = \frac{8\pi}{3}\frac{Ne^4}{m^2c^4\rho} = \frac{\sigma_0}{\rho}$$

which is just Thomson's original formula, seen here to be precisely valid only in the point charge limit. Secondly, it is apparent that

$$\lim_{a/\lambda\to\infty}\frac{\sigma}{\rho} = 0,$$

which shows that at short wavelengths (or high frequencies) this large, rigid, spherical shell electron is "able to explain at least qualitatively the decrease in the absorption for electromagnetic waves. . . ."[44]

Both conclusions were obviously encouraging. Nevertheless, this model was not entirely satisfactory. As Compton later wrote, "the hypothesis of the electron as a rigid spherical shell, incapable of rotation, though resulting in a reduced total scattering, would give rise to symmetrical scattering on the incident and emergent sides of a plate. To account for the observed dissymmetry, the further assumption must be made that the incident electromagnetic wave is capable of moving the different parts of the electron relatively to each other."[45]

That is, the shell had to be perfectly flexible instead of rigid.[46] Compton therefore carried out a second calculation, closely related to the one above, from which he concluded that for a *flexible* charged shell,

$$\frac{\sigma}{\rho} = \frac{\sigma}{\rho}\left(\frac{a}{\lambda}\right) = \frac{Ne^4}{m^2c^4\rho}\int_0^\pi \frac{(1+\cos^2\theta)}{2}\left\{\frac{\sin[(4\pi a/\lambda)\sin(\theta/2)]}{(4\pi a/\lambda)\sin(\theta/2)}\right\}^2 2\pi\sin\theta d\theta. \quad (3.5)$$

Since the scattering angle θ is the variable of integration, we note once again that σ/ρ is left as a function of a/λ only.

To compare this result analytically with that for the rigid shell electron, the integral in equation (3.5) may be evaluated either graphically or term-by-term after series expansion of the integrand. The results of this comparison may be seen most readily from the plots of σ/ρ *vs* a/λ for an aluminum scatterer ($Z = 13$) shown in Fig. 4,[47] where curve *I* describes the rigid shell electron and curve *II* the flexible shell electron. Note that qualitatively, at least, both curves show the desired behavior; the Thomson value $\sigma_0/\rho = 0.188$ cm^2/g is approached in the point charge limit ($a/\lambda \rightarrow 0$), and σ/ρ becomes greatly reduced in the short-wavelength limit ($a/\lambda \rightarrow \infty$). If the electron were assumed to be rigid but rotatable, a third curve could be obtained which would lie between the two already shown.[48]

2. *Its Further Experimental Basis*

The existing experimental data, Compton asserted, were unfortunately "too meager to submit these formulae to accurate quantitative test." Nor did he himself have any X-ray equipment on hand at Westinghouse to improve this undesirable situation.[49] Therefore, he had to rely on three experimental points which were "established with some accuracy." These three "points" have been indicated (by the author) on curve *I* as *A*, *B*, and *C* and on curve *II* as *A*, *B'*, and *C'*. They were obtained from the papers by Barkla and Dunlop (*A*), Hull and Rice (*B*, *B'*), and Ishino (*C*, *C'*).[50] Each could be used as follows to calculate a value for *a*, the radius of the electron. First, a given experimental value for σ/ρ could be located on, say, curve *I* and the corresponding ratio a/λ read off. Then, since the incident wavelength λ was known, the value of *a* could be computed. This procedure could be repeated for all experimental values of σ/ρ, and for both curves, thereby yielding a number of different (and hopefully consistent) values for the radius of the

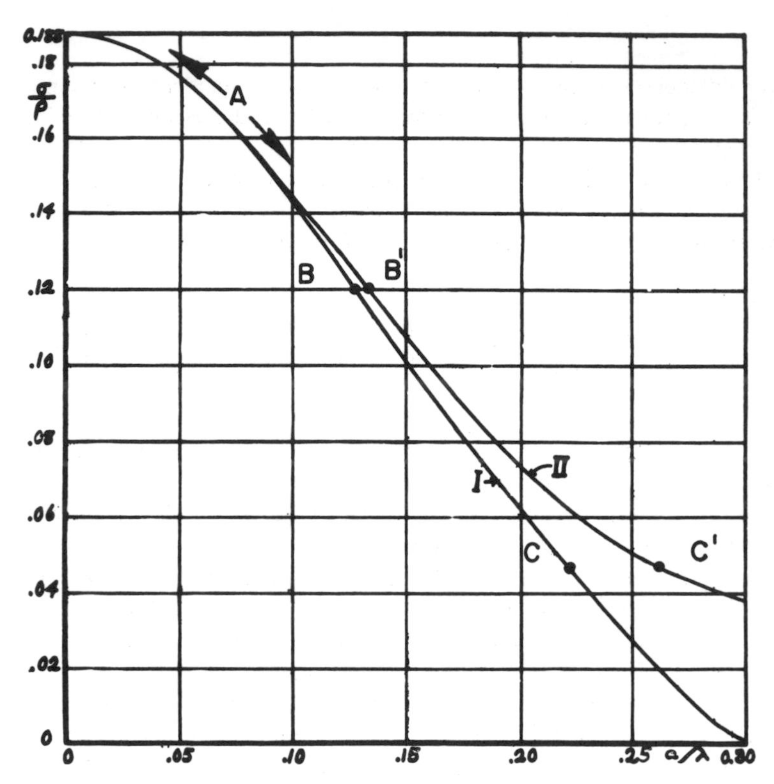

FIG. 4. Compton's plots (with points A, B, B', C, and C' added by the author) showing the behavior of the mass-scattering coefficient σ/ρ with the ratio of electron radius a to wavelength λ for an aluminum scatterer, assuming the electrons to be rigid shells (curve I) or flexible shells (curve II).

electron. The results of Compton's calculations[51] are summarized in Table 1. He concluded: "Using either formula the agreement between the two values of the radius is within the limits of probable experimental error. The unusually low absorption coefficient for γ-rays can therefore be quantitatively explained on the hypothesis that the electron is a spherical shell of electricity of radius about 2.3×10^{-10} cm."[52]

The validity of the flexible shell model could be tested in another way, by asking whether or not it predicted the desired forward-backward asymmetry in the angular distribution $I(\theta,a/\lambda)$, i.e., by asking whether or not it possessed the advantage which led to its introduction. Since, by definition,

$$\frac{\sigma}{\rho} = \frac{\sigma}{\rho}\left(\frac{a}{\lambda}\right) \sim \int_0^\pi I\left(\theta, \frac{a}{\lambda}\right) \sin\theta d\theta$$

Table 1. Summary of data and results of Compton's calculations for a, the radius of the electron

Experimentalist(s)	Scatterer	Incident λ (Å)	σ/ρ(cm²/g) experimentally determined	a(x 10^{-10} cm) predicted from curves	
				Curve	Value
Barkla and Dunlop	Light elements	Relatively long	Accurately given by Thomson formula	*A*	Point charge limit
Hull and Rice	A1	0.17	0.12	*B(I)*	2.2
Hull and Rice	A1	0.17	0.12	*B′(II)*	2.3
Ishino	A1	0.093	0.045	*C(I)*	2.1
Ishino	A1	0.093	0.045	*C′(II)*	2.5

one can see from equation (3.5) that for the flexible shell model

$$I\left(\theta, \frac{a}{\lambda}\right) \sim (1+\cos^2\theta)\left\{\frac{\sin[(4\pi a/\lambda)\sin(\theta/2)]}{(4\pi a/\lambda)\sin(\theta/2)}\right\}^2 \tag{3.6}$$

That is, the Thomson $(1 + \cos^2\theta)$ factor is modulated by the square of the (a/λ)-dependent factor in braces. Compton displayed the effects of this modulation by choosing $a/\lambda = 3\pi/4$ (which fixes λ at roughly 0.01 Å for $a = 2.3 \times 10^{-10}$ cm) and plotting the curves shown in Fig. 5.[53] Thomson's

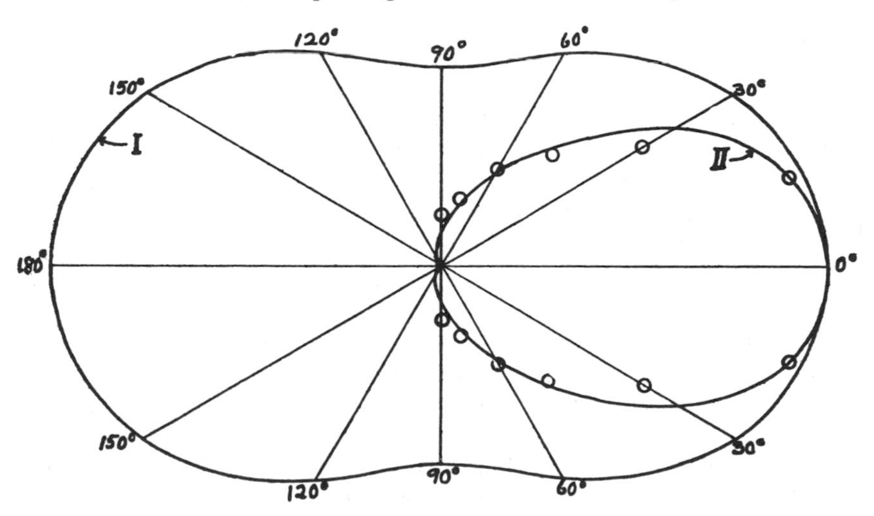

Fig. 5. Plots of $I(\theta)$ *vs* θ for Thomson's formula (curve I) and the flexible shell electron (curve II).

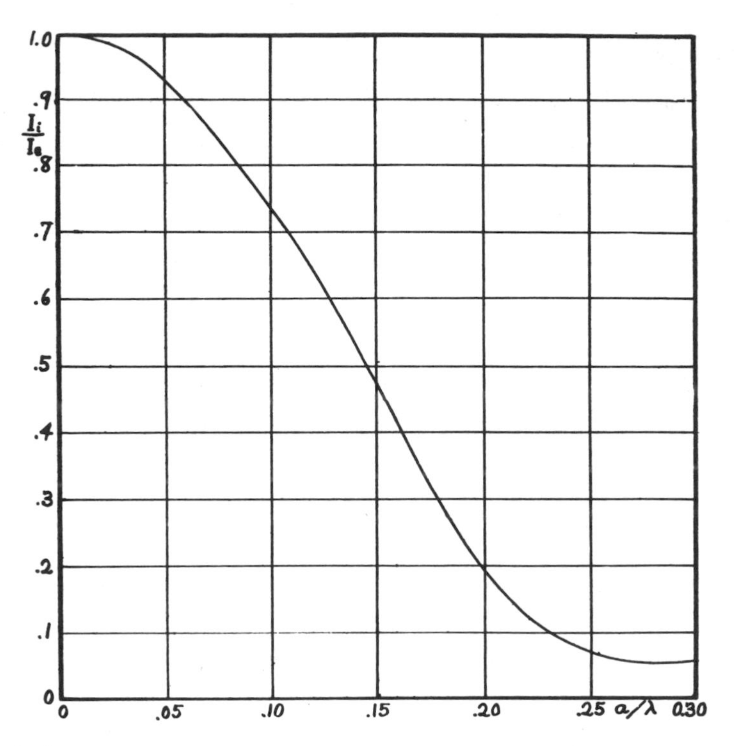

FIG. 6. Compton's plot (with one symbol changed) of σ_1/σ_e *vs* a/λ for a flexible shell electron.

expression (plotted as curve *I*) was evidently modulated in just the right way; the flexible shell model (curve *II*) predicts a strong peaking of the scattered radiation in the forward direction.

The model was also in good quantitative agreement with the experimental data represented by the small circles on the graph. To obtain these data, Compton was forced to go back in the literature to 1910—to the γ-ray experiments of D.C.H. Florance,[54] which we discussed at length in Chapter 1. Recall that Florance had used RaC as a source of γ-rays. Citing 1914 experiments of Rutherford and Andrade,[55] Compton now assumed that RaC emits a strong line at 0.099 Å and a weak line at 0.071 Å, for an "effective wavelength" of 0.095 Å. He concluded: "Inasmuch as these rays are heterogeneous, and as the softer rays are scattered relatively more strongly at larger angles, the agreement of the experimental values with curve *II* is as good as can be expected."[56]

And here we have the first hint of a surprising bonus entailed by Compton's flexible shell model. For suppose one calculates the magnitude of the forward-backward asymmetry in the scattered radiation, taking as a measure of this asymmetry the ratio of the total radiation scattered backward (toward the "incident side" of the plate) to that scattered forward (toward the "emergence side" of the plate). Denoting this ratio by σ_i/σ_e, it is apparent that

$$\frac{\sigma_i}{\sigma_e} = \frac{\int_{\pi/2}^{\pi} I(\theta, a/\lambda)\sin\theta d\theta}{\int_{0}^{\pi/2} I(\theta, a/\lambda)\sin\theta d\theta} \tag{3.7}$$

Compton could therefore substitute his flexible shell expression for $I(\theta, a/\lambda)$, equation (3.6), into the integrands, carry out the integrations, and plot the resulting expression for σ_i/σ_e *vs* a/λ. The graph he obtained is shown in Fig. 6.[57]

In the first place, note that for $a/\lambda = (2.3 \times 10^{-10}\ \text{cm})/(0.095 \times 10^{-8}\ \text{cm}) = 0.242$, one obtains $\sigma_i/\sigma_e = 8\%$, which should be compared with Ishino's value of 15%. "The agreement is hardly within the probable experimental error," Compton noted, "but the calculated value is at least of the proper order of magnitude, which is a strong verification of a flexible or a rotatable electron." A second implication of the plot was far more important, however. "This curve," Compton wrote, "explains beautifully the observation of Florance that the 'incident' scattered rays are softer than the 'emergent' and the primary rays, since it shows that the relative amount of the rays scattered backward is much greater for soft, or long wavelength, γ-rays than for hard rays."[58] Compton had therefore in effect just provided a quantitative theory for Florance's qualitative "selective scattering" interpretation of the softening.

D. Compton's Large Ring Electron Theory of Scattering (1917–1919)

1. *Its Desirable Features and Explanatory Power*

Compton's chain of reasoning seemed complete: his large, flexible, spherical shell electron accounted not only for Barkla's low mass-absorption coefficient and the forward-backward asymmetry in the angular distribution, but also for Florance's soft secondary radiation. It was an impressive achievement. Nevertheless, Compton knew that this model could not be correct: "According to electromagnetic theory," he pointed out, "it is obvious that the mass of an electron cannot be accounted for on the basis of a uniform distribution of electricity over the surface of a sphere of the size here assumed."[59] That is, assuming as Compton did that the mass of the electron was entirely electromagnetic in origin, a simple calculation shows that the self-energy of a spherical shell of total charge e and radius a is $e^2/2a$. Equating this to mc^2 and substituting $a = 2.3 \times 10^{-10}$ cm, one finds that the corresponding mass m is about 2,000 times smaller than the known rest mass of the electron. Furthermore, no advantage may be gained by assuming any other spherically symmetrical distribution of charge, such as a uniform ball of charge. All such models yield self-energies on the order of e^2/a. Hence, they possess the same fatal weakness as that of the shell electron.

Compton never suppressed difficulties; rather, he searched for ways to overcome them. In this case, he searched for a physically reasonable model of the electron. And it was then that he remembered a model which he had first encountered when attempting to explain the disappointing, negative experiment he had carried out (with Oswald Rognley[60]) at the University of Minnesota in the winter of 1916–1917. This was the ring electron model, first proposed in 1915 by the American chemist, A. L. Parson.

Parson's motivation for introducing his ring electron model was, indeed, that of a chemist: "Bohr's theory," he asserted, "gives an interesting treatment of the problem of spectrum series, but its chemical application is very meager indeed." In particular, Parson felt that Bohr's theory left the problem of atomic stability untouched. He believed, however, that this problem could be solved by postulating that the electron possesses a magnetic moment, "that the electron is itself magnetic, having in addition to its negative charge the properties of a current circuit whose radius (finally estimated to be 1.5×10^{-9} cm. . . .) is less than that of the atom but of the same order of magnitude. Hence it will usually be spoken of as the *magneton.* It may be pictured by supposing that the unit negative charge is distributed continuously around a ring which rotates on its axis (with a peripheral velocity of the order of that of light . . .); and presumably the ring is exceedingly thin."[61] Parson demonstrated that such electrons could be expected to account for both chemical bonding and the periodicity of the elements.

Its key feature, however, was that it possessed a magnetic moment. At Minnesota, Compton and Rognley had reflected X-rays from a magnetite crystal, balancing the ionization produced by the reflected beam against that produced by a standard beam (a very sensitive "null method"[62]). They had detected no changes whatsoever in the relative intensity in the higher orders when the magnetization of the crystal was changed from an unmagnetized state to various magnetically saturated states in which the magnetization was either perpendicular or parallel to one of its reflecting planes. This was the disappointing, negative result, that Compton had sought to understand.

He eventually saw that it actually confirmed an earlier conclusion of his brother Karl and E. A. Trousdale;[63] that is, that magnetization does not shift the atoms themselves in their lattice sites. Thus "the ultimate magnetic particle is not a group of atoms, such as the chemical molecule, but is the individual atom or something within the atom."[64] If it were the individual atom, other difficulties arose. Bohr's theory (in its contemporary version) ascribed the atom's magnetic moment to *co-planar* orbiting electrons.[65] However, Compton had realized ever since his thesis calculations that the reflected intensity distribution is a sensitive function of the electron distribution, and that changing the orientation of the orbital planes by 90° should produce a "very considerable" change—as high as 500 percent—in the relative intensity of the reflected beam. He concluded that, "If the atom is the ultimate magnet, its electrons are not all distributed in the same plane, as assumed by Bohr, but are arranged very nearly isotropically."[66] Even the most plausible isotropic arrangement, however, should have led to a change in intensity of at least 1 percent—a detectable amount. Therefore, Compton finally concluded that the "most obvious explanation of our negative result is that it is not the atom which is the elementary magnet, but that it is either the positive nucleus, as suggested by Merritt, or the electron, as suggested by Parson."[67] It was an impressive chain of reasoning.

Parson's ring electron model had also attracted the attention of others. In late 1917, shortly before Compton published his large electron hypothesis (January 1918), D. L. Webster[68] and C. J. Davisson[69] asked precisely the question of greatest concern to Compton: namely, whether or not Parson's ring electron was physically tenable; that is, whether or not it could possess the proper electromagnetic mass. The answer supplied by Webster was particularly revealing. He calculated that an annular charged ring of diameter $2R$ and thickness $2R'$ could indeed possess the proper electromagnetic mass, provided that

$$\frac{R'}{R} = 8 \exp\left(\frac{-\pi R m c^2}{e^2}\right). \qquad (3.8)$$

Parson thought that the ring would turn out to be "exceedingly thin." By assigning a reasonable value to R and by substituting numerical values for the other quantities, one finds that $R'/R \sim e^{-2560}$. Webster therefore did not overstate the case at all when he concluded that the ring thickness was "almost incredibly small."[70]

Nevertheless, Parson's ring electron suited Compton's purposes admirably. Compton realized from physical considerations—before he had carried out detailed calculations—that vis-à-vis his other models, "Much the same effect, so far as the scattering of γ-rays is concerned, results from the conception of the electron as a ring of electricity of diameter comparable with the wave-length of the incident beam." It seemed likely to Compton that the forward-backward asymmetry σ_i/σ_e produced by the ring electron would be "appreciably larger" than 8 percent; it would, perhaps, even approach Ishino's 15 percent. "It seems probable, therefore, that the scattering of γ-rays and X-rays may be completely explained on the hypothesis that the electron is a ring of electricity of radius about 2×10^{-10} cm., if the ring is capable of rotation about any axis."[71]

Other, independent evidence uncovered by A. H. Forman in 1916 reinforced the plausibility of this model.[72] Forman attempted to determine whether or not magnetization influences the absorption of X-rays in iron by sending a beam of X-rays through an iron absorber placed between the poles of an electromagnet. He found, he said, "an increase in the opacity of the iron when it is magnetized in a direction parallel to the Röntgen rays. . . . The effect seems to be about 5 parts in a thousand for a field of 3,500 gauss."[73]

Now, while Forman believed he was observing a reorientation of the iron *atoms*, Compton saw that Forman's results could be easily interpreted by using the ring electron model. One simply could assume, as before, that the external magnetic field interacts with the magnetic moments of the electrons, thereby orienting their orbital planes perpendicular to the direction of propagation of the X-rays. In this preferred orientation, as compared to a random orientation, the incident rays could "get a better hold" on the electrons, increase their displacement, and enhance the amount of radiation absorbed. Quantitatively speaking, Compton found that for iron, 26 percent of the electrons present—or roughly 7 electrons per atom—should be reoriented by the magnetic field. And this was almost exactly "what would be expected if it is 8 valence electrons of iron which are responsible for its ferro-magnetic properties." This interpretation of ferromagnetism, we now realize, is incorrect, but to Compton this calculation was very encouraging: "Our hypothesis of a ring electron of radius 2.3×10^{-10} cm. is therefore capable of explaining satisfactorily Forman's effect."[74]

By October, 1918, the ring electron model had attracted so much attention that H. S. Allen, Lecturer in Physics at the University of London, weighed "The Case For a Ring Electron" before members of the Physical Society of London.[75] In what became sixteen pages in print, Allen assembled thirteen different arguments in favor of the ring electron. All but three, however, dealt with atomic phenomena (chemical bonding, Zeeman effect, cohesion, etc.) or with phenomena directly related to the electron's magnetic moment (diamagnetism, paramagnetism, Planck's constant, etc.). Only as a rather incidental additional argument did Allen mention that A. H. Compton had suggested that for high-frequency radiation it was ". . . possible to explain not only the asymmetry of the scattered rays, but also the diminution of scattering with decrease of wave length."[76] In the five pages of discussion following Allen's remarks, Compton's name or work was not mentioned at all.

It is therefore evident that these physicists saw that the possible fruitfulness of the ring electron lay mostly in its application to atomic phenomena. Even here, however, the majority of Allen's audience did not share his enthusiasm. D. Owen, for instance, remarked: "Judged by its power of predicting new phenomena, the hypothesis of the ring electron is disappointing. . . . In striking contrast is the fertility of Bohr's hypothesis of the atomicity of angular momentum, leading to the prediction of a hitherto unobserved spectral series. . . ." J. H. Jeans criticized the ring electron in a similar vein. J. W. Nicholson commented: ". . . I cannot claim to be one of its supporters as a physical entity with the evidence at present available, in spite of the admiration we must all feel for some of the work to which its introduction has led." Finally, Allen himself admitted that: "Personally, he did not accept Parson's theory as it stood. . . ."[77]

2. *The Theory Becomes Quantitative*

The general skepticism of the English physicists toward the ring electron model was in sharp contrast to Compton's deep confidence in its explanatory power. While reading Allen's paper and the ensuing discussion, he must have been particularly annoyed at the following remark by Jeans: "It appears, then, that the ring electron may be welcomed mainly as giving a vivid picture of certain magnetic phenomena. There seems to be no clearly established case in which it successfully explains any phenomenon outside magnetism. . . ."[78] This was precisely the point that Compton was challenging, and not for a moment did he consider abandoning the hypothesis. On the contrary, he pursued its consequences unrelentingly, and (most likely) in late 1918 solved the scattering problem in detail. His calculation (published in March 1919[79]) is too long to be presented completely, but its main features may be readily indicated.

Suppose that unpolarized electromagnetic radiation of wavelength λ is incident from the $-Z$ direction on a charged ring of radius a with center at O (heavy ring in Fig. 7), and suppose further that the radiation is scattered through the angle θ in the direction $OP = L$. Let the line OS be the intersection of the plane of the ring with the plane POZ, and let OR bisect the angle $-ZOP$. An element of charge dq at point Q is then located in space by the angles β and γ with respect to the line OS, or by the angle ϕ with respect to the line OR, where $\cos \phi = \cos \alpha \cos \gamma - \sin \alpha \sin \gamma \cos \beta = r/a$.

These parameters define the geometry of the problem. The scattering process itself was assumed by Compton to be governed by certain physical as-

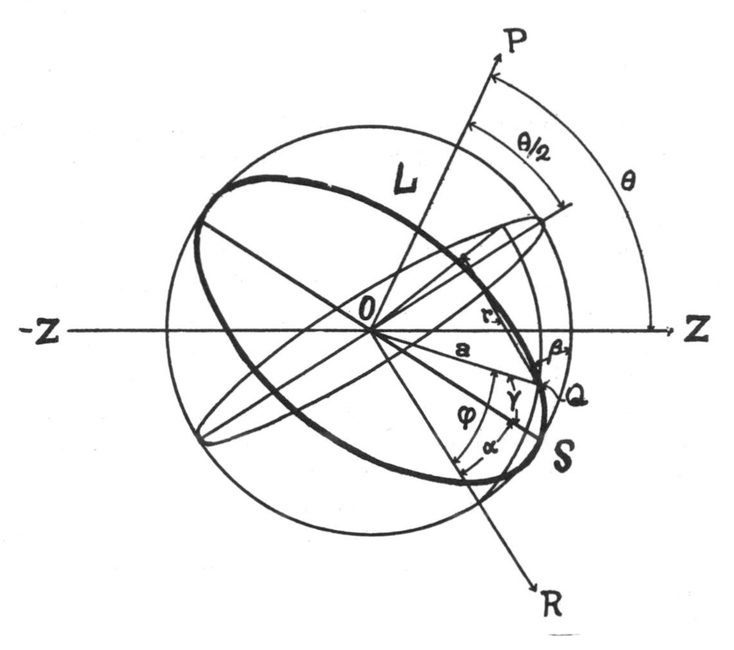

FIG. 7. Compton's diagram showing radiation being scattered from a charged ring of radius a (heavy ring).

sumptions, as follows: First, he assumed that the ring was perfectly flexible, so that different infinitesimal elements could move independently of each other. Second, to be consistent with the electromagnetic mass requirement he assumed that the ring thickness t was very small compared to its radius a. Finally, believing that a "chief factor in the complexity of the problem" was that special relativity demanded different electromagnetic masses for elements of charge accelerated in different directions, he made the problem "manageable" by assuming that "the mass of an arc element is the same in all directions, and that the velocity of the electricity in the ring is small compared with the velocity of light."[80]

Let, now, the amplitude of the incident unpolarized waves be decomposed into two components, $A_\perp$ and $A_\parallel$, perpendicular and parallel to the plane POZ. The incident intensity is then given by $I_0 = k(A_\perp^2 + A_\parallel^2)$; this, on the average, is equal to $2kA_\perp^2$, where k is a constant. One must now calculate the corresponding scattered amplitudes $E_\perp$ and $E_\parallel$. Consider first the amplitude $E_\perp$. If $\eta = e/(2\pi a)$ is the charge per unit length, and if $\mu = m/(2\pi a)$ is the mass per unit length, then an element of charge $dq = \eta a d\gamma$ accelerated at the rate $\ddot{x} = eA_\perp/m = \eta A_\perp/\mu$ will set up the field

$$dE_\perp = \frac{dq\ddot{x}}{Lc^2}\cos\frac{2\pi}{\lambda}\left(\Delta - 2r\sin\frac{\theta}{2}\right) = \frac{(\eta a d\gamma)(\eta A_\perp/\mu)}{Lc^2}\cos\frac{2\pi}{\lambda}\left(\Delta - 2r\sin\frac{\theta}{2}\right) \quad (3.9)$$

at the point $P(L, \theta/2)$. The *maximum* scattered amplitude $dE_{(m)}$, which is required for the calculation of the intensity, occurs when the reference path difference $\Delta = 0$. Consequently:

$$E_{\perp(m)} = \frac{\eta a(\eta/\mu)A_{\perp}}{Lc^2} \int_0^{2\pi} \cos \frac{4\pi}{\lambda}\left(r \sin \frac{\theta}{2}\right) d\gamma \tag{3.10}$$

It may now appear that the way to proceed further is to simply carry out the integration, square the result, and thereby obtain the intensity $I_{\perp}$. This is not as straightforward as it may seem, however, because the distance r in the integrand is actually a function of the variable γ through the relationship $r = a \cos \phi = a \cos \phi\ (\alpha, \beta, \gamma)$. Moreover, the only ultimately interesting angle is the scattering angle θ, which means that one must average over (i.e., integrate over) the angles α and β. This operation is tantamount to assuming that the ring is randomly oriented in space. Symbolically, then, the scattered intensity I corresponding to the amplitude $E_{\perp(m)}$ is given by

$$I_{\perp} = \frac{k}{\pi^2} \int_0^{\pi} d\alpha \int_0^{\pi} d\beta\, E^2_{\perp(m)} \tag{3.11}$$

To obtain the *total* intensity scattered, $I = I_{\perp} + I_{\parallel}$, one must of course also calculate the intensity $I_{\parallel} = kE^2_{\parallel(m)}$. The net result, however, is already known—one is only led to the usual ½ $(1 + \cos^2\theta)$ factor. Putting all of these ideas together, and dividing by the incident intensity $I_o = 2kA^2_{\perp}$, one obtains the following expression for the angular distribution of the scattered radiation:

$$I(\theta,L) = \frac{Ne^4\ (1+\cos^2\theta)}{8\pi^4L^2m^2c^4} \int_0^{\pi} d\alpha \int_0^{\pi} d\beta \left\{ \int_0^{2\pi} \cos\left[\frac{4\pi}{\lambda}\left(r \sin \frac{\theta}{2}\right)\right] d\gamma \right\}^2 \tag{3.12}$$

where the relationship $\eta/\mu = e/m$ has been used.

It is now a question of evaluating the integrals. Define the integral in braces (curly brackets) to be F_1. By eliminating r in favor of α, β, and γ (using the relationship $r = a \cos\phi$), and by performing a series of involved algebraic manipulations, the integral F_1 may be put into standard form and evaluated as a zero-order Bessel function:

$$F_1 = 2\pi J_0[(4\pi a/\lambda)\sin(\theta/2)\sqrt{1 - \sin^2\alpha \sin^2\beta}]$$

The square of F_1 now becomes the integrand of the second integral F_2:

$$F_2 = \int_0^{\pi} F_1{}^2 d\beta = 4\pi^2 \int_0^{\pi} J_o{}^2 d\beta$$

The only way this integration may be carried out fairly easily is to expand J_0 into a power series and evaluate F_2 term by term. This leads to the result that

$$F_2 = 4\pi^3 [M + N(\sin\alpha) + O(\sin^2\alpha) + P(\sin^3\alpha) + \ldots]$$

where $M, N, O, P, \ldots$, are known constants. Finally, F_2 becomes the integrand of the third integral F_3:

$$F_3 = \int_0^\pi F_2 d\alpha$$

Once again, by expanding by power series and evaluating the result term by term, one obtains

$$F_3 = 4\pi^4(1 - \alpha k^2 + \beta k^4 - \gamma k^6 + \delta k^8 - \ldots)$$

where $\alpha, \beta, \gamma, \delta, \ldots$ are other known constants, and $k \equiv (4\pi a/\lambda)\sin(\theta/2)$. The angular distribution of the scattered radiation is therefore finally given by

$$I(\theta,a/\lambda) = \frac{Ne^4}{L^2m^2c^4}\frac{(1+\cos^2\theta)}{2}[1 - \alpha k^2 + \beta k^4 - \gamma k^6 + \delta k^8 - \ldots] \tag{3.13}$$

The corresponding mass-scattering coefficient is, by definition,

$$\frac{\sigma}{\rho} = \frac{\sigma}{\rho}\left(\frac{a}{\lambda}\right) = \frac{1}{\rho}\int_0^\pi I(\theta, a/\lambda)\ (2\pi L^2\sin\theta)d\theta$$

$$= \frac{8\pi}{3}\frac{Ne^4}{m^2c^4\rho}[1 - \zeta\left(\frac{a}{\lambda}\right)^2 + \eta\left(\frac{a}{\lambda}\right)^4 - \ldots] \tag{3.14}$$

where $\zeta, \eta, \ldots$ are new known constants. Above all, the whole calculation pointedly illustrates the extent to which Compton in 1918 was committed to his ring electron model—and to the electromagnetic theory of light.

3. *The Ring Electron's Experimental Basis*

As Compton had guessed earlier, the predictions of his ring electron model were closely analogous to those of his other models. Comparative plots (Fig. 8[81]) of σ/σ_0 *vs* λ/a *(not* a/λ) for the rigid shell electron *(I)*, the flexible shell electron *(II)*, and the flexible ring electron *(III)*, proved that in the point charge limit ($\lambda/a \to \infty$) all scattering coefficients σ go over to the Thomson value σ_0 (i.e., $\sigma/\sigma_0 \to 1$). In the short-wavelength limit ($\lambda/a \to 0$), all approach zero.

Compton once again used these curves to calculate the radius of the electron. Since, however, he used new estimates for the incident wavelengths, he obtained not 2.3×10^{-10} cm but 2×10^{-10} cm. But, he asserted, the "important thing to notice is that if the electrons had dimensions comparable with 10^{-13} cm., as usually assumed, the scattering should be represented by the upper line . . . where $\sigma/\sigma_0 = 1.0$. The fact that experiment gives con-

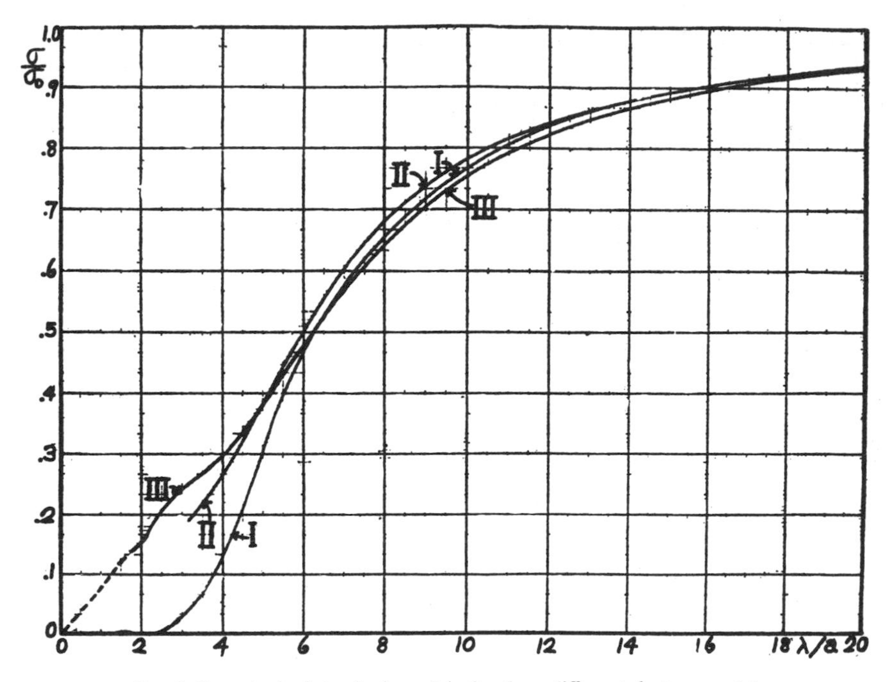

FIG. 8. Compton's plots of σ/σ_0 *vs* λ/a for three different electron models.

sistently lower values when short wave-lengths are used is sufficient proof that the electron is not sensibly a point charge of electricity."[82] He also concluded that *"the radius of the electron is the same in different atoms."*[83]

Turning to the angular distribution $I(\theta,a/\lambda)$, Compton plotted[84] the curves shown in Fig. 9. Once again he checked them against Florance's data.

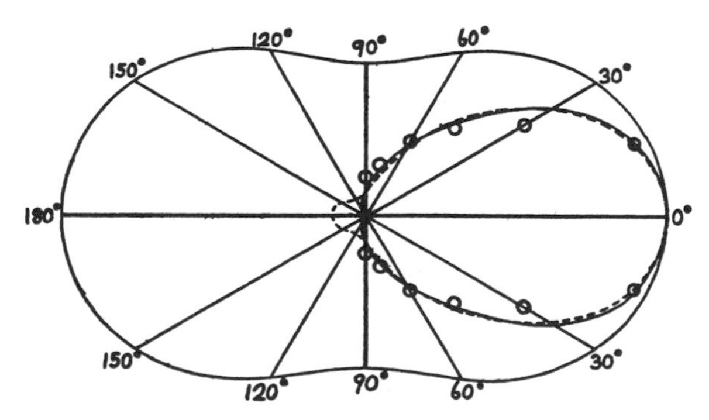

FIG. 9. Compton's plots of $I(\theta,\ a/\lambda)$ *vs* θ for the ring electron and the flexible shell electron. The Thomson variation is shown by the outside curve.

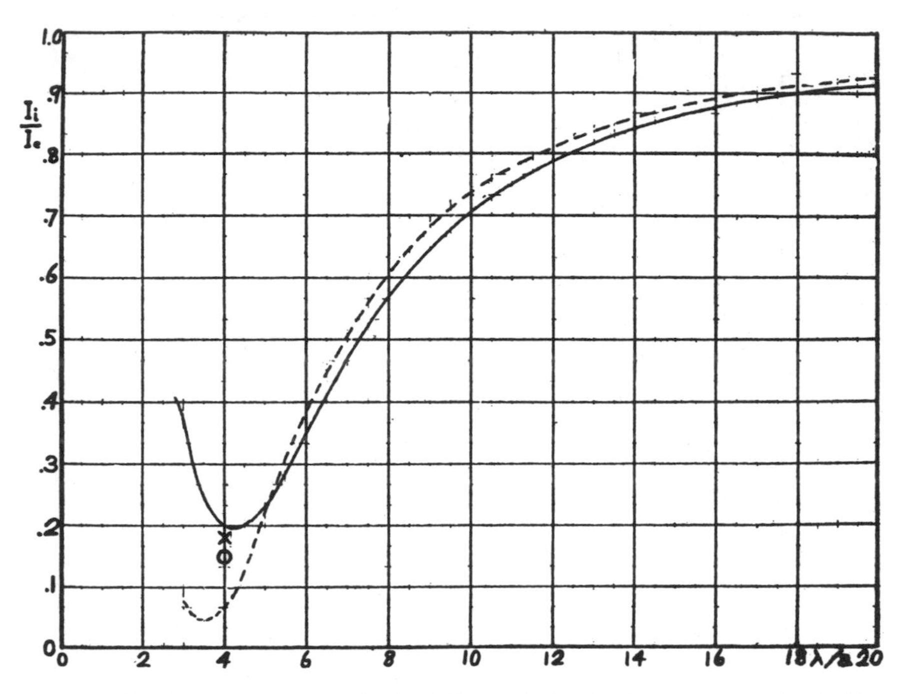

FIG. 10. Compton's plots of σ_i/σ_e (or I_i/I_e) *vs* λ/a for the ring electron and the flexible shell electron. Here the ring electron variation is given by the solid curve, whereas in Fig. 9 it was given by the dashed curve.

Note that the ring electron curve (dashed oval) is scarcely distinguishable from that of the flexible shell electron (solid oval). Obviously, therefore, these curves were not an accurate test of the relative merits of the two models: A "better quantitative test" could be obtained from comparative plots[85] of the forward-backward asymmetry σ_i/σ_e *vs* λ/a (again, *not* a/λ). The result (Fig. 10) vindicated Compton's belief that at short wavelengths the asymmetry predicted by the ring electron (solid curve) was "appreciably larger" than that predicted by the flexible shell model (dashed curve). The two lone experimental points were Ishino's 15 percent and J.P.V. Madsen's 18 percent for RaC γ-rays.[86]

One of the most significant aspects of Compton's present analysis emerged rather surreptitiously: Compton cited not only Florance's 1910 paper for the data points in Fig. 9, but also Florance's later, 1914 papcr.[87] In both papers, we recall, Florance had reported closely related experiments proving that secondary γ-rays are softer than the primary rays. By the time he had written his second paper, however, he had rather dramatically changed his interpretation of these experiments. In 1914 he no longer accepted a "selective scattering" argument, but had concluded instead that the γ-rays actually undergo a change in hardness *in the scattering process.* Since Compton now cited *both* papers, he indicated his familiarity with both arguments. Yet, he concluded that the curves of σ_i/σ_e *vs* λ/a (Fig. 10) "doubtless explain at least in part the observation of Florance that the 'incident' scattered rays are softer than the 'emergent' and the primary rays, since they show that the relative amount of rays scattered backward is much greater for soft or long wave-length γ-rays than for the harder radiation."[88] There is no way of knowing what Compton meant by the phrase "at least in part," but he very clearly believed that Florance's 1914 interpretation was incorrect, and that the ring electron theory, with its prediction of a "selective scattering effect," was correct.

4. *Initial Refinements to Determine the Precise Radius of the Ring*

Between the time Compton first became convinced of the physical reality of Parson's ring electron and the time he was able to base a full-fledged theory of scattering on it, somewhat less than two years—mid-1917 to early 1919—had elapsed. The terminal date is known with some accuracy because on February 2, 1919, Compton made the first entry in the first volume of his surviving laboratory notebooks,[89] and all subsequent entries assume a knowledge of the ring electron mass-scattering coefficient, equation (3.14). These notebooks therefore enable us to see how Compton, once he had made his basic model quantitative, probed various avenues of research suggested by it, and successively refined his scattering theory in the succeeding months.

His first efforts were directed at determining as accurately as possible the characteristic parameter of the ring electron, its radius. These efforts alone prove that Compton was fully convinced of its physical reality. His basic procedure was essentially a "curve-fitting" process: He varied certain adjustable parameters until he obtained a "best fit" to the existing experimental data, which consisted of the X-ray data of Bragg and Pierce (1914), Hull and Rice (1916), Rutherford (1917), Owen (1918), and Williams (1918), as well as the γ-ray data of Ishino (1917).[90] Surveying all of this work, Compton saw that the measurements had been taken over the widest range of wavelengths for aluminum and copper, and these were the data on which he based most of his subsequent analysis.

His point of departure amounted to a critique of equation (2.18), $\mu/\rho = \tau/\rho + \sigma/\rho$, relating the mass-absorption coefficient μ/ρ to the mass-fluorescent and mass-scattering coefficients, τ/ρ and σ/ρ. All of his earlier work had been tantamount to proving that the latter coefficient σ/ρ is not constant and equal to σ_0/ρ, as Thomson had assumed, but is actually a function of the ratio a/λ—symbolically, that $\sigma/\rho = (\sigma_0/\rho)f(a/\lambda)$, where $f(a/\lambda)$ represents the series expansion in equation (3.14). Compton now argued[91] that the former coefficient τ/ρ, instead of varying simply as λ^3 as had been generally assumed, also depended on the ratio a/λ. As a first approximation, he took this dependence to be identical with that governing the mass-scattering coefficient. Moreover, he regarded the exponent of λ to be uncertain, and hence replaced the 3 by an unknown n, not necessarily an integer. In sum, he concluded—not explicitly, but unquestionably by implication—that equation (2.18) should be replaced by the following expression:

$$\frac{\mu}{\rho} = k\frac{\sigma}{\rho}\lambda^n + \frac{\sigma}{\rho} = k\frac{\sigma_0}{\rho} f\left(\frac{a}{\lambda}\right)\lambda^n + \frac{\sigma_0}{\rho} f\left(\frac{a}{\lambda}\right) \qquad (3.15)$$

where k is a constant.

His program was now clear: for given wavelength λ, determine the set $\{k, n, a\}$ which provides the best fit to the existing absorption data. This set would automatically fix the radius a of the ring electron.[92] Almost before he had begun, however, he ran into trouble: He found that the set

$$\{k, n, a\} = \{39.7, 2.5, 0.0193 \text{ Å}\}$$

which yielded a good value of μ/ρ for aluminum at *short* wavelengths, yielded a bad value at a selected *long* wavelength (0.615 Å). He concluded: "∴ formula not good for extrapolat." Then, doubly underlined, "Revise formula. Instead of σ/ρ as coefficient of λ^n, we should use the square [of the] force with which the radiation acts on the electron. This is approximate[ly] $\sim \sqrt{\sigma}$. . . ."[93] He was probably led to this conclusion by observing that since the amplitude of the impressed force F varies as the incident wavelength λ, and since σ varies as λ^4 (as clearly seen in the case of a spherical shell electron), one has that $F \sim \lambda \sim \sigma^{1/4}$, or $F^2 \sim \sqrt{\sigma}$. In any event, right at the outset, Compton felt that disagreement with experiment forced him to replace σ/ρ with $(\sigma/\rho)^{1/2}$. He therefore modified equation (3.15) to read

$$\frac{\mu}{\rho} = k\left(\frac{\sigma}{\rho}\right)^{1/2}\lambda^n + \frac{\sigma}{\rho} \qquad (3.16)$$

where, as before, $\sigma/\rho = (\sigma_0/\rho)f(a/\lambda)$.

Compton's method for now determining the best set $\{k, n, a\}$ was as follows. Rearranging equation (3.16), and taking the logarithm of both sides, one obtains

$$\log(\mu/\rho - \sigma/\rho) = \log k + (1/2)\log(\sigma/\rho) + n\log\lambda \qquad (3.17)$$

Now, for a given value of the electron radius a, σ/ρ becomes a function of λ only. Hence, at a given experimental point $(\mu_1/\rho, \lambda_1)$ equation (3.17) takes the form

$$\alpha_1 = \log k + \beta_1 + n\gamma_1 \tag{3.18}$$

where α_1, β_1, and γ_1 are constants. A second experimental point $(\mu_2/\rho, \lambda_2)$ yields a second, similar equation:

$$\alpha_2 = \log k + \beta_2 + n\gamma_2 \tag{3.18'}$$

Solving these two equations simultaneously, one may obtain a value of n corresponding to the assumed radius a. Choosing different pairs of experimental points $(\mu/\rho, \lambda)$, any number of n-values may be obtained and an average taken. Changing to a second value for a, the entire process may be repeated. Finally, when a small spread in the n-values results, both n and a may be considered fixed, and the corresponding k-value may be found, which completes the calculation and determines the best set $\{k, n, a\}$.

"Let us first evaluate a from Barkla's data on paraffin $C_{22}H_{46}$,"[94] Compton began, and arbitrarily tried a = 0.02 Å. Midway through his calculation, however, he saw that this was "obviously not right." Trying a = 0.015 Å, he found a huge spread in the six n-values he calculated. For a = 0.017 Å, he found no more promising results, and he cryptically remarked: "nothing satisfactory." These discouraging results—obtained for a light hydrocarbon—were offset by his observation that for wavelengths below 0.34 Å, the mass-scattering coefficient σ/ρ, as before, still dropped below Thomson's value. Furthermore, he obtained much more satisfying results when he turned to a relatively heavy element, aluminum. Assuming first that a = 0.02 Å, then that a = 0.019 Å, and finally that a = 0.0187 Å, he found that the spread in the corresponding n-values successively decreased. The average of the last set (of six) was n = 2.88, for which the corresponding k-value was 26.45, yielding {26.45, 2.88, 0.0187 Å} as the best set $\{k, n, a\}$.

Compton's calculations were therefore apparently complete. Just at this point, however, he became convinced from "recent researches with homogeneous radiation of known wavelength" that "the value of the exponent is more nearly equal to 3." He therefore immediately abandoned all of his previous work and abruptly changed tack: For a given value of the radius a, he now assumed n = 3 and looked for consistency in the corresponding k-values. These calculations[95] were very fast. Knowing that the earlier set had given a good fit to the data, he first tried a = 0.0187 Å. Too much spread in the computed k-values resulted. Trying a = 0.019 Å, the spread increased, so he moved in the other direction and chose a = 0.0185 Å, with much improved results. Momentarily, he thought he could do better with a = 0.0184 Å, but midway through his calculation he abandoned the idea, crossing his work out. He was satisfied with the set $\{k, n, a\}$ = {31.9, 3, 0.0185 Å}. He readily convinced himself that the resulting expression

$$\frac{\mu}{\rho} = 31.9\left(\frac{\sigma}{\rho}\right)^{1/2}\lambda^3 + \frac{\sigma}{\rho} \qquad [a = 0.0185 \text{ Å}] \qquad (3.19)$$

gave a far superior fit to the aluminum data than either a simple λ^3 or a $(k\lambda^3 + \sigma_0/\rho)$ dependence.

Compton's success in fitting the aluminum data removed all hesitancy from his mind when he turned next to the copper data.[96] He assumed from the outset that n was equal to 3, and that the radius a was equal to 0.0185 Å. All that was necessary was to determine corresponding k-values and calculate their average, a very easy matter. Result: the spread he found in the k-values was a bit too large to satisfy him, and he backed off entirely, trying another calculation in which he assumed $n = 5/2$. Curiously, even though a smaller spread resulted, he had enough confidence in the "recent researches" to set $n = 3$ and conclude that the set $\{k, n, a\} = \{350, 3, 0.0185 \text{ Å}\}$ provided the best fit to the copper data. He again showed the superiority of the resulting equation for μ/ρ over the two older forms given above. Finally, more or less as a footnote, he tried to fit Hull and Rice's data for lead. Using $n = 3$ and $a = 0.0185$ Å, he found he could not obtain satisfactory consistency in the corresponding k-values. He remarked: "probably $\sqrt{\sigma}$ is not accurate value for coeff."[97]—which was the first hint concerning the direction his future calculations would take.

Meanwhile, however, Compton concluded:

> The formula just proposed on the basis of a large electron . . . obviously is in good agreement with the experimental values of the absorption both of X-rays and γ-rays.
>
> The form of the theoretical curve changes rapidly with small changes in the value assumed for the radius of the electron. In fact, unless there is some consistent error in the figures given by Hull and Rice, those figures suffice to determine the radius of the electron within 2 per cent.
>
> It is thus possible to account quantitatively for the absorption of very high frequency radiation if the electron is a ring of electricity of radius $a = 1.85 \pm .04 \times 10^{-10}$ cm.[98]

5. *Subsequent Studies Provide Deeper Understanding of the Fluorescent Absorption Term*

Still not entirely satisfied with his theory, Compton spent most of March, April, and May of 1919 refining it by carefully scrutinizing the fluorescent absorption term. His point of departure was a thorough examination of a pulse-theoretical calculation of J. J. Thomson's, which appeared in the second edition of Thomson's *Conduction of Electricity Through Gases* (1906), and which Compton copied verbatim into his notebooks.[99] Thomson had focused his attention on the acceleration imparted by an electromagnetic pulse to an elastically bound electron, and after first calculating the electron's displacement from equilibrium, he computed its elastic potential energy and divided it by the incident electromagnetic energy. This ratio was by definition proportional to the *atomic* fluorescent absorption coefficient $\tau/\nu = (\rho/\nu)(\tau/\rho)$, where ν is here the number of atoms per unit volume in the absorber. In Compton's notation,[100] Thomson concluded that

$$\frac{\tau}{\nu} = \frac{4\pi^3 e^2 d^3}{mc^2}\sum_k \frac{n_k}{\lambda_k^{\,2}} \tag{3.20}$$

where d is the width of the incident pulse, n_k is the number of electrons per atom of the type k, and λ_k (which is simply related to the electron's mass and elastic binding constant) is the "wave-length of the free vibration of the [k-th] corpuscle."

Immediately, on March 10, 1919, Compton convinced himself that Thomson's pulse-theoretical calculation predicted values of τ/ν (or τ/ρ) that were too high. He therefore attempted a "modification of Thomson's theory of true absorption considering incident rays to be long trains."[101] Specifically, Compton set up and attempted to solve the familiar differential equation,

$$m\frac{d^2x}{dt^2} + b\frac{dx}{dt} + cx = Ae\cos 2\pi\nu t \tag{3.21}$$

where b and c are the damping and spring constants of the electron and A and ν are the amplitude and frequency of the incident radiation. His goal was to solve this equation for $x(t)$, calculate dx/dt, and thereby obtain the energy $F\cdot v = (b\,dx/dt)(dx/dt) = b(dx/dt)^2$ dissipated through the electron's frictional losses. He relinquished all hope of a quantitative solution, however, when he did not evaluate the two constants of integration by specifying appropriate initial conditions. Two weeks later he returned to the problem, but complicated it still further by adding the assumption that the incident wave train, after hitting the (damped) electron, is exponentially attenuated. The result was even more barren than before. After writing down the general form of the solution, Compton simply stopped calculating.

While this work therefore did not lead to concrete results, it nevertheless had important consequences for Compton's later thought. First, he kept the possibility open that resonance effects may contribute to the mass-absorption coefficient. Second, he was led to a "provisional theoretical basis" for the correct $Z^4\lambda^3$ variation of τ/ρ.[102] Finally, and of most immediate concern, this work prepared him to re-examine the fluorescent absorption term in the light of his ring electron model. As he wrote:[103]

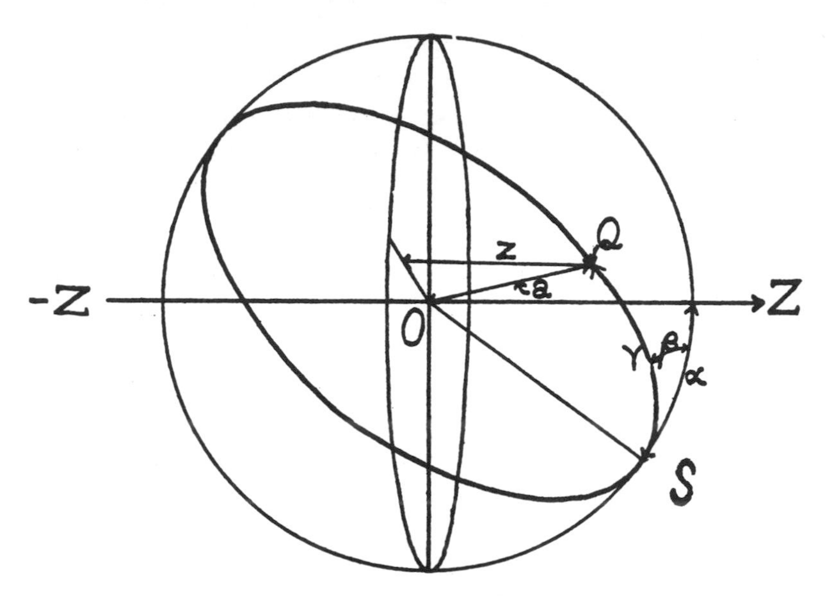

FIG. 11. Compton's simplified ring electron diagram.

. . . [Thomson] shows that the energy absorbed from the incident ray by an electron is proportional to the square of the acceleration to which the electron is subject. It would seem that this relation must hold in whatever manner the fluorescent absorption is calculated. The acceleration of a comparatively large electron will be less than that of a small electron of the same mass, however, when both are traversed by X-rays of the same intensity, since in the former case the phase of the incident ray will not be the same at all parts of the electron. The fluorescent absorption due to a large electron will therefore be less than the value $kZ^4\lambda^3$ by the factor,

$$\phi = \left\{\frac{\text{acceleration of large electron}}{\text{acceleration of small electron}}\right\}^2$$

Compton's notebooks show that he began calculating the factor ϕ on April 7, 1919; he sent his calculation to *The Physical Review* on May 24. It was based on the diagram shown in Fig. 11, which is just a simplified version of the one he drew for the ring electron scattering problem (Fig. 7). In fact, since he also had to determine the acceleration of the electron here, his whole line of argument was closely analogous to his earlier one. It is not surprising, therefore, that his final result took the form of a series expansion,

$$\phi = \phi\left(\frac{a}{\lambda}\right) = 1 - n\left(\frac{a}{\lambda}\right)^2 + o\left(\frac{a}{\lambda}\right)^4 - p\left(\frac{a}{\lambda}\right)^6 + \ldots \quad (3.22)$$

where $n, o, p, \ldots$ are known constants. Since ϕ is obviously dimensionless, Compton concluded that it should replace a similarly dimensionless quanti-

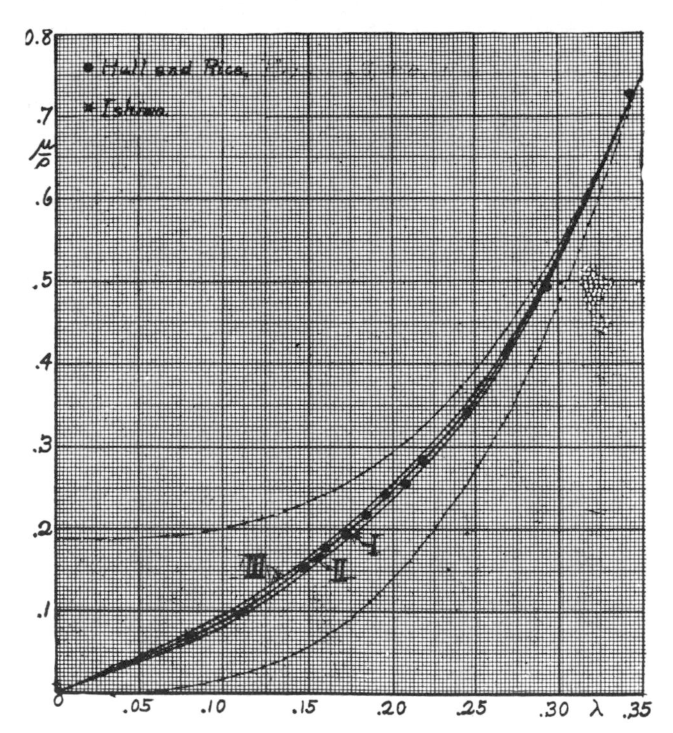

FIG. 12. Compton's plot of μ/ρ *vs* λ in the short-wavelength region for ring electrons of three different radii.

ty, σ/σ_0, in the fluorescent absorption term. On the same day on which he began his calculation, he evaluated the *first seventeen terms* of the expansion in equation (3.22).[104]

6. *Compton's Final Absorption Formula*

The culmination of Compton's refining and elaborating of his ring electron model was, therefore, the following expression for the mass-absorption coefficient μ/ρ:

$$\frac{\mu}{\rho} = \frac{\tau}{\rho} + \frac{\sigma}{\rho} = k\phi\left(\frac{a}{\lambda}\right)Z^4\lambda^3 + \frac{\sigma_0}{\rho}\, f\left(\frac{a}{\lambda}\right) \qquad (3.23)$$

To determine how accurately it agreed with experiment, he followed the same procedure as before to determine the best set $\{k, n = 3, a\}$, by first assuming three different electron radii (0.017 Å, 0.0185 Å, and 0.02 Å), and then by calculating the corresponding k-values. Since it was "at the shorter wave-lengths that the effects of the different hypotheses become evident,"[105] he plotted this region on an expanded scale (Fig. 12). He compared the ring electron variation of μ/ρ with λ for the three radii in Fig. 12 (curves *III, II,* and *I*) with that predicted by a simple $k\lambda^3$ variation (lowest curve) and that predicted by a $(k\lambda^3 + \sigma_0/\rho)$ variation (uppermost curve). After remarking on the sensitivity of the curves to small changes in the ring electron's radius, he concluded: "Unless there is some consistent error in the experimental figures, we can, on the basis of the agreement of these curves, take the radius of the electron to be $(1.85 \pm .05) \times 10^{-10}$ cm."[106] It is difficult to see how Compton could have reached a different conclusion.

7. *Uncertainty Associated With Gamma-Ray Wavelengths*

In spite of its success, Compton's ring electron model entailed one weakness which soon opened it up to criticism. Since his expressions for ϕ and σ/ρ depended not on the wavelength λ alone, but on the ratio a/λ, it soon became obvious that his estimate of the electron radius depended critically on the value he assumed for the wavelength of the incident radiation. For γ-rays this quantity was particularly uncertain in 1919.[107] The earliest quasi-reliable estimate for RaC γ-rays was made in 1914 by Rutherford and Andrade,[108] who attempted both a direct measurement and an extrapolation technique from X-ray measurements. They concluded, as we have already seen, that RaC emits a strong line at 0.099 Å and a weak line at 0.071 Å. Three years later, Rutherford, using his own absorption measurements in conjunction with the Duane-Hunt law, determined the variation of μ/ρ with λ over a broad range of X-ray wavelengths,[109] which he then extrapolated to include Ishino's RaC *γ-ray* measurements. He concluded that his earlier estimate had to be revised, that the wavelength of RaC γ-rays lay somewhere between 0.02 and 0.007 Å.[110]

This estimate was much lower than his earlier one, and the fact that Compton, right from the outset, preferred to use the earlier value in his ring electron calculations evoked immediate criticism. H. S. Allen, the most ardent English supporter of the ring electron model, asserted in 1918 that "Mr. Compton suggests that the true shape of the electron may be that of a ring, having an effective radius many times greater than that ordinarily accepted. Mr. Compton's estimate of the radius is 2.3×10^{-10} cm., but if some recent measurements by Sir Ernest Rutherford are used in the calculation, this estimate must be reduced to about one-tenth of the value stated. . . ."[111] Similar criticism had appeared in a note in *Nature* on February 28, 1918.[112]

Eventually, over a year later (in May, 1919), Compton responded: "The value of V used in Rutherford's [Duane-Hunt law] calculations was the *maximum* voltage across the tube, which obviously is considerably greater than the effective voltage. Furthermore, the filtering method of obtaining the radiation which has the shortest wave-length is uncertain in its results.

. . . Both of these sources of error led to a calculated value of the effective wave-length which is smaller than the true value."[113] Compton informed Rutherford personally of these arguments by letter. He pointed out specific reasons to believe that both errors were actually present in Rutherford's work. Rutherford replied, referring to his 1914 direct (spectroscopic) measurements: "When I recall the faintness of the radium C lines and the difficulty of fixing them . . . , I am inclined to think that a mistake could easily arise. . . . I am inclined to give a good deal more weight to the Coolidge Tube experiments than you do."[114] To which Compton responded: "As Rutherford suggests, the question is one which must finally be answered by more refined measurements of the absorption of very hard homogeneous rays of known wave-length."[115]

8. The Electron Is the "Ultimate Magnetic Particle"

Clearly, the contemporary uncertainty in γ-ray wavelengths entailed a corresponding uncertainty in Compton's value for the radius of his ring electron. However, it is also clear that this uncertainty, if repeatedly emphasized, would only have forced Compton to revise his value for the radius upward or downward. It would not, by itself, have affected the basic validity of his *model.* In fact, Compton did not regard Allen's criticism as a challenge to his model. This became evident, for instance, when Compton apparently completely ignored the following very insightful remarks of D. Owen, which accompanied Allen's paper. In attempting to explain the forward-backward asymmetry in the intensity of secondary β-rays, Owen noted that "On the view of a continuous wave-front, this imparted [electromagnetic] momentum may be insufficient to jerk an electron out of the atom. . . . On the view, however, of the existence of quanta—*i.e.*, intense condensations of activity in the wave-front—the forward momentum of the beam may be impressed like a series of hammer blows on the atoms of the target."[116]

There exists other, far more direct evidence, however, that Compton remained completely convinced of the physical reality of his ring electron: His laboratory notebooks reveal further calculations based on it. They show that by August 1919 Compton had come full circle. These calculations were directed once again at a re-examination of his negative 1916–1917 Minnesota experiment. He regarded the resulting 1920 paper[117] as the capstone of the series of three[118] which he published on the ring electron in *The Physical Review.* To a large extent, therefore, his conviction that the ring electron was the ultimate magnetic particle unified his work for several years. His present calculations,[119] which were long but straightforward, proved that his and Rognley's observations were completely consistent with this model.

Compton and Rognley had found, as Compton wrote, that "the intensity of the X-ray beam reflected from magnetite is not altered by as much as 1 per cent. in the first four orders when the crystal is magnetized to 1/3 of saturation perpendicular to the reflecting surface, nor in the third order by similarly magnetizing the crystal parallel to the reflecting face."[120] Compton now interpreted these observations quantitatively along the lines indicated earlier. He reasoned that if the crystal atoms contained ring electrons, the external magnetic field, on the one hand, should orient their planes perpendicularly to it, producing a mass-absorption coefficient

$$\left(\frac{\mu}{\rho}\right) = k\phi_{\perp} Z^4\lambda^3 + \left(\frac{\sigma}{\rho}\right) \tag{3.24}$$

On the other hand, with the crystal in an unmagnetized state (corresponding to randomly oriented ring electrons), μ/ρ should be given by the usual expression, equation (3.23).

After several trial calculations, Compton convinced himself that "the effect should not be noticeable in our experiment if the radius of the electron is less than 4×10^{-10} cm. Thus if the radius is 2×10^{-10} cm. as one of us has estimated on the basis of the scattering of X-rays and gamma rays, the hypothesis that the electron is the ultimate magnetic particle is in accord with our experimental results."[121] "Auxiliary evidence," such as the "profound effect of chemical constitution on the magnetic properties of an atom," now licensed him, he felt, to eliminate the electron's only competitor, the positive nucleus, as the ultimate magnetic particle.[122] Thus, it is well worth emphasizing that, taken as a whole, Compton's argument constitutes the first experimentally based and extended argument in the literature for assigning a magnetic moment to the electron.

Of course, Compton realized that his calculations had not provided positive proof for the correctness of his specific model, the ring electron model. They had only proved that this model was consistent with his and Rognley's Minnesota experiments. Conspicuously absent from his entire discussion, in fact, was any mention of the somewhat similar experiments of A. H. Forman, who, we recall, had concluded that magnetization *does* influence the absorption of high-frequency radiation.

E. Compton Leaves Westinghouse for the Cavendish Laboratory (1919)

In August 1919 Compton received the news he had been eagerly awaiting: he learned that he had been awarded a National Research Council Fellowship to enable him to resume pure research. He later recalled that when he had relinquished a research atmosphere at the University of Minnesota to go to Westinghouse two years earlier, "The director of the laboratory had shared my hope that in the setting of this newly created laboratory there would be an opportunity for me to do basic research in my own field of atoms and X rays, as well as to cultivate the body of knowledge from which better lamps would be built. Because of the need of the laboratory to establish itself with the company, however, it was soon evident that the former more general type of research had no essential place."[123] He had found no X-ray equipment at Westinghouse with which he could test his scattering theory, and thereby integrate his theoretical studies with an evolving experimental program. Now, most of all, he had come to understand that he wanted to devote his life entirely to research that was motivated only by "the pure love of knowledge."

He had therefore applied for a National Research Council Fellowship—and at the last moment had learned that his application had been successful. He was one of the first postdoctoral students to receive one after they had been established on R. A. Millikan's initiative after World War I.[124] Moreover, contrary to the original intent of the Council, Compton had asked to be able to study outside the United States, for there was no question in his mind where he wanted to go. Ever since the time he had heard Ernest Rutherford give a physics colloquium at Princeton, Compton had had the greatest admiration and respect for Rutherford. He had also been in correspondence with Rutherford on γ-ray wavelengths, as we have seen. To Compton, now, the "famed Cavendish Laboratory at Cambridge University was . . . simply the place where he was stationed." Actually, Compton had made up his mind to go to Cambridge with or without a fellowship. This was not an easy decision to make from a personal point of view, since he knew it would involve borrowing a considerable amount of money to enable him to take his wife and young son Arthur Alan along. "We were all set to do this, when at the last moment a fellowship was granted. As I recall it, the fellowship gave us $2,200, which covered about half our year's expenses."[125]

As soon as he received the good news, he wrote an informative letter to Rutherford, the rough draft of which has survived in his laboratory notebooks:

Within the last few weeks I have succeeded in arranging matters so that I can spend the coming year at the Cavendish Laboratory. I am expecting to arrive in Cambridge with my family about the midde of October [1919].

It is my desire to engage in some line of research in which your chief interests lie. It would of course be possible for me to undertake one of the problems which suggest themselves as a result of my recent theoretical work on the size and shape of the electron. I should prefer at this time, however, in order to secure the maximum benefit from my work, to take up an investigation which you may suggest as falling in with the research that you have in progress.

> In order that you may know of work I am prepared to undertake, I may give a brief account of my training in research, as follows:
>
> Ph.D., Princeton University, 1916; Thesis "Intensity of X-ray Reflection, and Distribut[ion] of the Electrons in Atoms."
>
> Instructor in Physics, University of Minnesota, 1916–17.
>
> Research Physicist, Westinghouse Lamp Co., 1917–
>
> My experimental work has been chiefly in connection with x-ray diffraction and reflection, high vacuum work, and discharge through gases at low pressure.
>
> I hope that conditions will be such this year as to make possible effective study and research at Cambridge.[126]

Compton's work with Rutherford, which afforded Compton the first opportunity to really integrate his experimental and theoretical studies after leaving the University of Minnesota, was so significant for his future that, as S.K. Allison has remarked, with it Compton's "career as an investigator in basic physics essentially began. . . ."[127] Allison neglected to point out that the arrangements were touch-and-go until the end, since Compton actually received an answer to his above letter, which was "to the effect that I [Compton] would be very welcome, but he [Rutherford] was afraid I would not find any place in Cambridge to live because of the great influx of students [after the war]."[128] Compton received this discouraging news, however, when Rutherford's letter was forwarded to him in Cambridge.

F. References

[1] James R. Blackwood, *The House on College Avenue: The Comptons at Wooster 1891–1913* (Cambridge, Mass.: M.I.T. Press, 1968).

[2] For more detail on these events see Blackwood, especially Chapters X and XVI, and Compton's "Personal Reminiscences," in *The Cosmos of Arthur Holly Compton,* Marjorie Johnston (ed.) (New York: Knopf, 1967), pp. 3–52. The latter work contains a bibliography (pp. 459–468) of Compton's scientific papers, which may be consulted for complete references to his early writings. His articles on airplanes, which clearly reveal the exhilaration he felt in analyzing the principles of flight and testing them with his own models, make delightful reading today. The same is true of his much more mature 1913 and 1914 articles on his laboratory method of measuring the earth's rotation. For additional accounts of Compton's life and work, see S. K. Allison, "Arthur Holly Compton," *Natl. Acad. Sci. Biog. Mem.* 37 (1965):81–110; S. K. Allison, "Arthur Holly Compton, Research Physicist," *Science* 138 (1962):796; E. U. Condon, "Prof. A. H. Compton," *Nature* 194 (1962):628–629; R. S. Shankland, "Arthur Holly Compton," *Dict. Sci. Biog.*, Vol. 3 (New York: Scribner's, 1970), pp. 366–372; "Arthur Holly Compton," *Wash. U. Mag.* 31 (1962):2–7; R. J. Stephenson, "Arthur Holly Compton, 1892–1962," *Am. J. Phys.* 30 (1962):843–844; A. A. Bartlett, "Compton Effect: Historical Background," *Am. J. Phys.* 32 (1964):120–127; and the obituaries in the *New York Times* (March 16, 1962) and the *New York World-Telegram and Sun* (March 15, 1962).

[3] Quoted by Compton in Johnston (ed.), *Cosmos,* p. 16.

[4] Blackwood, *House on College Avenue,* pp. 195–198.

[5] Johnston (ed.), *Cosmos,* p. 17.

[6] Blackwood, *House on College Avenue,* p. 211.

[7] See R. S. Shankland's introduction to the *Scientific Papers of Arthur Holly Compton: X-Ray and Other Studies* (Chicago: University of Chicago Press, 1973). In general, Professor Shankland's introduction provides a very valuable complement to my own discussion of Compton's work.

[8] "Our Nation's Need for Scientific Research," in Johnston (ed.), *Cosmos,* pp. 239–240.

[9] "An Addition to the Theory of the Quadrant Electrometer," *Phys. Rev.* 13 (1919):288; "A Sensitive Modification of the Quadrant Electrometer: Its Theory and Use," *Phys. Rev.* 14 (1919):85–98.

[10] See especially "A Recording X-Ray Spectrometer, and the High-Frequency Spectrum of Tungsten," *Phys. Rev.* 7 (1916):646–659.

[11] Johnston (ed.), *Cosmos,* pp. 184 and 20.

[12] "The Intensity of X-Ray Reflection, and the Distribution of the Electrons in Atoms," *Phys. Rev.* 9 (1917):29–57. See also "The X-Ray Spectrum of Tungsten," *Phys. Rev.* 7 (1916):498–499; "The Reflection Coefficient of Monochromatic X Rays from Rock Salt and Calcite," *Phys. Rev.* 10 (1917):95–96; and "Note on the Grating Space of Calcite, and the X-Ray Spectrum of Gallium," *Phys. Rev.* 11 (1918):430–432.

[13] Johnston (ed.), *Cosmos,* pp. 27–28. This incident led to a pleasant interchange with W. H. Bragg, and to the publication of a joint paper, "The Distribution of the Electrons in Atoms," *Nature* 95 (1915):343–344.

[14] Johnston (ed.), *Cosmos,* p. 22. From Henry A. Erikson's unpublished *History of the Physics Department of the University of Minnesota* (one copy is on deposit at the Center for History of Physics in New York), we can see that Compton's, Foote's, and Tate's positions were created by the fact that A. F. Gorton went into the ceramic industry, E. H. Kennard returned to Cornell, A. F. Kowarik was called to Yale, and O. Zobel accepted a position in the Bell Telephone Laboratories. Compton's starting salary and rank were identical to Tate's.

[15] Johnston (ed.), *Cosmos,* (reference 2) p. 23. Erikson, who had only become chairman in August 1915, recalled in his *History* that the Minnesota Physics Department purchased a 100-kilovolt Snook X-ray machine for Compton's experiment, and that the shop constructed a recording ionization spectrometer for Compton under his direction.

[16] Compton's Westinghouse offer was for $3000 per year—precisely twice as large as his Minnesota salary. Erikson recalled: "As the University did not wish to bid against an industry, this precluded any effort to retain him, an effort which would have been strenuous had the offer been of an academic character."

[17] Johnston (ed.), *Cosmos,* p. 33. Barkla's paper was "Notes on the Absorption and Scattering of X-rays and the Characteristic Radiation of J-series," *Phil. Mag.* 34 (1917):270–285 (with M. P. White).

[18] For aluminum, Z/A is 0.48. Compton later quotes $\sigma_0/\rho = 0.188$ cm^2/g, which he evidently obtained by using $N/\rho = (Z/A)N_0$, and substituting $Z/A = 1/2$, $e = 4.7 \times 10^{-10}$

esu, $m = 9.1 \times 10^{-28}$ g, $c = 3 \times 10^{-10}$ cm/sec, and $N_0 = 6 \times 10^{23}$ atoms/g•mole into equation (3.1).

[19] This is true of course even if there are other contributions to equation (2.18), as Barkla and White, in their "Notes on Absorption and Scattering," p. 275, had suggested.

[20] F. Soddy and A. S. Russell, "The γ Rays of Uranium and Radium," *Phil. Mag.* 18 (1909):620–649; "The Question of the Homogeneity of γ-Rays," *Phil. Mag.* 19 (1910):725–757.

[21] M. Ishino, "The Scattering and the Absorption of the Gamma Rays," *Phil. Mag.* 33 (1917):129–146.

[22] In 1928 Compton explicitly stated: "Historically, it was the fact that the classical electromagnetic theory is unable to account for the low intensity of the scattered X-rays which called attention to the importance of the problem of scattering." See "Some Experimental Difficulties with the Electromagnetic Theory of Radiation," *J. Franklin Inst.* 205 (1928):172.

[23] *Ibid.*, p. 142.

[24] "The Size and Shape of the Electron," *J. Wash. Acad. Sci.* 8 (1918):2—hereafter cited as "Size and Shape." For an abstract of this paper see *Phys. Rev.* 11 (1918):330.

[25] For Barkla's explanation, see "Note on the Scattering of X-rays and Atomic Structure," *Phil. Mag.* 31 (1916):231 (with J. G. Dunlop).

[26] Compton, "Size and Shape," p. 2.

[27] "New Light on the Structure of Matter," *Sci. Am. Suppl.*, 78 (1914):4.

[28] *Ibid.*, footnote 1, p. 4. As we have seen in Chapter 1, Bragg advanced his theory in 1907, not 1910.

[29] "The Variation of the Specific Heat of Solids with Temperature," *Phys. Rev.* 6 (1915):377. Benedicks' paper is also referenced here.

[30] *Ibid.*, p. 389.

[31] "A Physical Study of the Thermal Conductivity of Solids," *Phys. Rev.* 7 (1916):348.

[32] "On the Location of the Thermal Energy of Solids," *Phys. Rev.* 7 (1916):349–354.

[33] Compton, "Variation," (reference 29) p. 389; "Physical Study," p. 348.

[34] Johnston (ed.), *Cosmos*, p. 19.

[35] The only intervening mention of Einstein's name seems to be a reference Compton made to Einstein and de Haas' 1915 work on magnetism in his paper "Is the Atom the Ultimate Magnetic Particle?" *Phys. Rev.* 16 (1920):465.

[36] Compton, "Size and Shape," (reference 24) p. 2.

[37] In his 1914 paper, "New Light on the Structure of Matter" (see reference 25), Compton emphasized only those questions on atomic structure that intersected most directly with his interests in X-rays. Thus he discussed Moseley's work in considerable detail, but relegated to a footnote (p. 6) the comment that "Moseley showed that his photographs confirmed this theory [Bohr's theory as set forth in his 1913 trilogy] in a very striking manner." In a 1915 paper, "What is Matter made of?" *Sci. Am.* 112 (1915):451–452, he made a tell-tale error right at the outset when he confused William Thomson with J. J. Thomson in picturing "Lord Kelvin's theory," which he described as

based on the idea "that the positive electricity is uniformly distributed throughout a sphere whose radius is the radius of the atom. The electrons move about within this sphere under the attraction of the positive electricity and the repulsion of the other electrons. . . ." (p. 451). He then went on to discuss the work of Rutherford, C. T. R. Wilson, and Moseley, but did not mention Bohr explicitly. In his 1916 Ph.D. thesis (see reference 11), Compton concluded that his work favored Bohr's theory over models such as Crehore's, but he gave no indication that he was familiar with, for example, Sommerfeld's subsequent work. In sum, Compton consistently treated questions on atomic structure only to the extent that they impinged on his X-ray researches.

[38] This diagram, which has been slightly altered for clarity, was reproduced from Compton's paper, "The Size and Shape of the Electron. I. Scattering of High Frequency Radiation," *Phys. Rev.* 14 (1919):23—hereafter cited as "Electron I." In the interests of allowing Compton to explain his own ideas, several of the following quotations are taken from this paper, published one year after "Size and Shape"—his explanations in "Electron I" are slightly easier to follow. All of the ideas discussed, however, are present in "Size and Shape," so that this procedure is not anachronistic.

[39] Compton, "Size and Shape," p. 5.

[40] Compton, "Electron I," p. 28.

[41] *Ibid.*, p. 31.

[42] See reference 11; the similarities between the two problems become evident from an examination of the scattering equations.

[43] This diagram was reproduced from Compton, "Electron I," p. 32, and the derivation that follows is given on p. 32 and succeeding pages.

[44] Compton, "Size and Shape," (reference 24) p. 3.

[45] Compton, "Electron I," p. 28.

[46] Compton, "Size and Shape," p. 3. He regarded this condition as approximating the condition that the shell be "subject to rotational as well as translational displacements."

[47] *Ibid.*, p. 4. The "points" A, B, B', C, and C' have been added by the author for convenience in explanation.

[48] *Ibid.*

[49] Johnston (ed.), *Cosmos*, pp. 32–33.

[50] See references 21 (Ishino) and 25 (Barkla and Dunlop), as well as A. W. Hull and M. Rice, "The Law of Absorption of X-Rays at High Frequencies," *Phys. Rev.* 8 (1916):326–328.

[51] Compton, "Size and Shape," pp. 4–5. Instead of 0.093 (as in Table 1), Ishino actually gives 0.099—the difference is unimportant here.

[52] *Ibid.*, p. 5.

[53] *Ibid.*, p. 6.

[54] "Primary and Secondary γ Rays," *Phil. Mag.* 20 (1910):921–938.

[55] "The Spectrum of the Penetrating γ Rays from Radium B and Radium C," *Phil. Mag.* 28 (1914):263–273.

[56] Compton, "Size and Shape," p. 7.

[57] *Ibid.*, p. 8.

[58] *Ibid.*, p. 7.

[59] *Ibid.*, pp. 7–8.

[60] Oswald Rognley (sometimes spelled Rognlie) died in April 1922 at the age of 31 of a streptococcus infection. He was awarded the Ph.D. degree *Post obitum.* Erikson's *History* (reference 14) describes clearly what a shock his death was to his friends and colleagues.

[61] "A Magnetron Theory of the Structure of the Atom," *Smith. Misc. Coll.* 65 [No. 11] (Nov. 29, 1915):3. The paper occupies pp. 1–80. Parson evidently did not accept Bohr's rather extensive treatment of atomic and molecular stability as published in his 1913 trilogy.

[62] "The Nature of the Ultimate Magnetic Particle," *Science* 46 (1917):415–416; *Phys. Rev.* 11 (1918):132–134.

[63] "The Nature of the Ultimate Magnetic Particle," *Phys. Rev.* 5 (1915):315–318.

[64] Compton and Rognley, "Nature," (reference 62) p. 416.

[65] See, for example, Max Jammer, *The Conceptual Development of Quantum Mechanics* (New York: McGraw-Hill, 1966), pp. 89–118.

[66] Compton and Rognley, "Nature," p. 416. For estimate of 500% and comments on the cubical arrangement, see *Phys. Rev.* 11 (1918):133.

[67] Compton and Rognley, "Nature," p. 416.

[68] "The Theory of Electronmagnetic Mass of the Parson Magneton and other Non-Spherical Systems," *Phys. Rev.* 9 (1917):484–499.

[69] "The Electro-Magnetic Mass of the Parson Magneton," *Phys. Rev.* 9 (1917):570–571.

[70] Webster, "Theory," (reference 68) p. 498.

[71] Compton, "Size and Shape," (reference 24) pp. 8–9.

[72] "The Effect of Magnetization on the Opacity of Iron to Röntgen Rays," *Phys. Rev.* 7 (1916):119–124.

[73] *Ibid.*, p. 121.

[74] Compton, "Size and Shape," p. 10. Note that Compton's remark also points out the inadequacy of the contemporary theory of ferromagnetism.

[75] Allen's paper was published in *Proc. Phys. Soc. (London)* 31 (1919):49–64.

[76] *Ibid.*, p. 52.

[77] *Ibid.*, pp. 64–65, 67–68.

[78] *Ibid.*, pp. 67–68.

[79] This is the paper cited as "Electron I" (reference 38). In the diagram, which was reproduced from p. 38, the letters *L* and *S* have been added for ease in explanation. The derivation that follows is given on pp. 38–43.

[80] *Ibid.*, p. 25.

[81] *Ibid.*

[82] *Ibid.*, p. 27.

[83] *Ibid.*, p. 31.

[84] *Ibid.*, p. 29.

[85] *Ibid.*, p. 30.

[86] For Ishino's paper, see reference 21; for Madsen's, see "Secondary γ Radiation," *Phil. Mag.* 17 (1909):423–448.

[87] "Secondary γ Radiation," *Phil. Mag.* 27 (1914): 225–244; see also "Scattering of γ-Radiation," *Phil. Mag.* 28 (1914):363–367.

[88] Compton, "Electron I," p. 30.

[89] These notebooks are on deposit in the Archives of Washington University, St. Louis; a copy is also on deposit in the Center for History of Physics, American Institute of Physics, New York City. A detailed description was compiled by Roger H. Stuewer and is available at both of the above locations. Quotations are by permission of Mrs. Arthur H. Compton, whose help is hereby gratefully acknowledged. Future references to the notebooks will take the form: Compton Notebook A.1. (Wash. U., A.I.P.), p. 1.

[90] For Hull and Rice's paper, see reference 50, and for Ishino's, see reference 21. The other papers cited are W. H. Bragg and S. E. Pierce, "The Absorption Coefficients of X rays," *Phil. Mag.* 28 (1914):626–630; E. Rutherford, "Penetrating Power of the X Radiation from a Coolidge Tube," *Phil. Mag.* 34 (1917):153–162; E. A. Owen, "The Absorption of X-Rays," *Proc. Roy. Soc. (London)* [A] 94 (1918):510–524; and C. M. Williams, "On the Absorption of X-Rays in Copper and Aluminum," *Proc. Roy. Soc. (London)* [A] 94 (1918):567–575.

[91] Compton Notebook A.1. (Wash. U., A.I.P.), pp. 22–24.

[92] Note that k obviously has dimensions of λ^{-n}; the dimensions of μ/ρ and σ/ρ when not stated are understood to be cm^2/g.

[93] Compton Notebook A.1. (Wash. U., A.I.P.), p. 3.

[94] *Ibid.*, p. 4; the calculation extends to p. 10. The reference is to Barkla and White's paper cited in reference 17.

[95] *Ibid.*, pp. 11–13; also see p. 19, where the phrase "recent researches. . ." appears. This phrase was subsequently lined out in a rough draft of the abstract published as "The Law of Absorption of High Frequency Radiation," *Phys. Rev.* 13 (1919):269.

[96] *Ibid.*, pp. 14–17. Compton actually found that both $k = 385$ and $k = 350$ gave a good fit to the copper data.

[97] *Ibid.*, p. 32.

[98] *Ibid.*, p. 28.

[99] Thomson, *Conduction of Electricity Through Gases,* 2d ed. (Cambridge: Cambridge University Press, 1906), pp. 327–330; Compton Notebook A.1. (Wash. U., A.I.P.), pp. 33–37.

[100] "The Size and Shape of the Electron: II. The Absorption of High Frequency Radiation," *Phys. Rev.* 14 (1919):249—hereafter cited as "Electron II."

[101] Compton Notebook A.1. (Wash. U., A.I.P.), p. 39.

[102] Moseley's law predicted $\lambda_k^2 \sim Z^4$; if one considers the pulse thickness d to be on the order of the incident wavelength λ, one has $\tau/\rho \sim d^3/\lambda_k^2 \sim Z^4\lambda_k^2$. Compton himself called this "derivation" incomplete mainly because, as we saw in Chapter 2 Webster had demonstrated the incompatibility of the pulse theory with the Duane-Hunt law. For still other reasons, see Compton, "Electron II," p. 250.

[103] Compton, "Electron II," p. 251. Compton's capital K has been changed to a lower-

case k, and his N has been changed to a Z, for consistency in notation. Compton's diagram was also reproduced from p. 251.

[104] Compton Notebook A.1. (Wash. U., A.I.P.), pp. 43–54.

[105] Compton, "Electron II," p. 256; the plot also appears on this page.

[106] *Ibid.*, pp. 256–257. The minus sign in the exponent was omitted in the published paper, and this misprint has been corrected here.

[107] For modern values of X-ray and γ-ray wavelengths, see, for example, G. L. Clark (ed.), *The Encyclopedia of X-Rays and Gamma Rays* (New York: Reinhold, 1963); C. E. Crouthamel (ed.), *Applied Gamma-Ray Spectrometry* (Oxford: Pergamon, 1960); W. E. Burcham, *Nuclear Physics: An Introduction* (New York: McGraw-Hill, 1963); Manne Siegbahn, *Spektroskopie der Röntgenstrahlen*, 2. Aufl. (Berlin: Springer, 1931).

[108] See reference 55.

[109] Rutherford, "Penetrating Power"—see reference 90.

[110] *Ibid.*, p. 160.

[111] Allen, untitled note in *Proc. Phys. Soc. (London)* 30 (1918):143.

[112] Anonymous note in *Nature* 100 (1918):510–511.

[113] "Electron II," (reference 100) p. 257.

[114] Quoted in "Electron II," p. 258. Rutherford's letter, dated June 18, 1919, has survived and is in the Washington University Archives. Rutherford also expressed the opinion that the scattering of very short waves results from absorption or partial absorption.

[115] *Ibid.*

[116] "The Asymmetrical Distribution of Corpuscular Radiation Produced by X-rays," *Proc. Phys. Soc. (London)* 30 (1918):143. For earlier, similar conclusions see H. Moore, "On the Corpuscular Radiation Liberated in Vapors by Homogeneous X-radiation," *Proc. Roy. Soc. (London)* 91 (1915):337–345.

[117] "Is the Atom the Ultimate Magnetic Particle?" *Phys. Rev.* 16 (1920):464–476.

[118] Compton, "Electron I," (reference 38) p. 21.

[119] Compton Notebook A.1. (Wash. U., A.I.P.), pp. 55–72.

[120] Compton, "Is the Atom," p. 467. Compton's italics have been removed from the quotation.

[121] *Ibid.*, pp. 475–476.

[122] *Ibid.*, pp. 474–476.

[123] Johnston (ed.), *Cosmos*, (reference 2) p. 24. For some of Compton's Westinghouse researches, see "Cathode Fall in Neon," *Phys. Rev.* 15 (1920):492–497 (with C. C. Van Voorhis).

[124] See Millikan's *Autobiography* (New York: Prentice-Hall, 1950), pp. 124–135. Also see Johnston (ed.), *Cosmos*, p. 426.

[125] Johnston (ed.), *Cosmos*, p. 25.

[126] Compton Notebook A.1. (Wash. U., A.I.P.), pp. 73–74.

[127] "Arthur Holly Compton," *Natl. Acad. Sci. Biog. Mem.* 37 (1965):83.

[128] Johnston (ed.), *Cosmos*, p. 26.

CHAPTER 4

A New Type of Fluorescent Radiation

A. Introduction: Compton's First Experiments at the Cavendish (1919)

The Compton family arrived in Cambridge in October 1919 and found it, as Rutherford had promised, to be a city in turmoil. After the war students returned to the University in unprecedented numbers. Some, however, did not return, and of those who did many were maimed; the "class room was rare that did not have crutches leaning against a chair."[1] Suitable living space was at a premium, but eventually, with the gracious help of Lady Darwin (C. G. Darwin's mother) and Lady Thomson (J. J. Thomson's wife), Compton and his family located accommodations.

The atmosphere at the Cavendish Laboratory was as Compton had anticipated—extremely stimulating. Almost immediately he began attending lectures on a variety of subjects. J. J. Thomson, who had recently been appointed Master of Trinity College, lectured on "Positive Rays";[2] Ernest Rutherford, who had succeeded Thomson as head of the Cavendish, lectured on kinetic theory; C. G. Darwin lectured on black-body radiation; C. T. R. Wilson lectured on atmospheric electricity; and E. Cunningham lectured on electron theory.[3] "The spirit of such men as Thomson and Rutherford can only be described as contagious," Compton recalled. At Westinghouse research had been primarily directed toward achieving practical results; in contrast, at the Cavendish there was "a whole community to all of whose members the growth of understanding was the greatest possible achievement,"[4] although practical application might follow in the future.

In that atmosphere of "pure research," Compton's studies "took on a new measure of meaning. . . ." It was high adventure for Compton to come into direct contact with J. J. Thomson, whom Compton greatly admired, and whose scattering theory had become so familar to him. Rutherford, like Thomson, was internationally famous; nevertheless, he "personally went out of his way to provide for me [Compton], as a young research fellow, all the research equipment that I needed during my stay at the Cavendish." The Cavendish equipment was good, in general, although it was not as complete as that at "several of the better universities in the United States."[5] For studying radioactivity, however, Rutherford had assembled the best and most complete equipment to be found anywhere in the world, and Compton soon made good use of it. His laboratory notebooks show that within a month after arrival (on November 7, 1919), he was "Detecting Uranium Oxide by measurement of γ-rays."[6] Rutherford had therefore honored Compton's desire to engage in some line of research in which Rutherford's chief interests lay. A joint note in *Nature,* "Radioactivity and Gravitation,"[7] resulted from their collaboration.

Both Rutherford and Compton soon realized, however, that Compton was much more interested in continuing his studies on absorption and scattering than in beginning new and basically unrelated researches in radioactivity. At Westinghouse, Compton's duties as a research engineer, and the absence of X-ray equipment, had kept his researches on scattering confined to theoretical studies. At the Cavendish he was once again given the opportunity of turning to experimental work. His theoretical activity, however, remained undiminished. He repeatedly discussed his ring electron theory of scattering, and on one occasion was invited to address the Cambridge Physical Society on the subject. Rutherford, with his characteristic humor and forthrightness, introduced him as follows: "This is Dr. Compton, who is with us from the United States to discuss his work on 'The Size of the Electron.' I hope you will listen to him attentively. But you don't have to believe him!"[8] Such good-natured criticism, coupled with new knowledge gained from his experiments, had its effect: Compton's ideas on scattering and absorption changed materially during his ten months in Cambridge. We shall now examine why, and to what extent, they changed.

B. Compton's Gamma-Ray Absorption and Scattering Experiments (1919–1920)

1. *His Theoretical Preconceptions*

The most significant, or "productive"[9] experiments Compton carried out at the Cavendish[10] dealt with the absorption and scattering of γ-rays by thin metal plates. The most suggestive earlier experiments in this area had been those of D. C. H. Florance (1910, 1914) and of J. A. Gray (1913), which were discussed in Chapter 1.[11] These experiments—and Florance's and

Gray's remarkable conclusions—were fresh in Compton's mind when he reported the results of his own experiments on the subject. Just exactly *how* fresh—and how clearly Compton understood Florance's and Gray's final conclusions—may be gathered from his introductory remarks, in which he discussed in general terms the two known types of secondary radiation:

> From theoretical considerations, both scattered and flourescent radiation should undoubtedly be present in secondary gamma rays. . . .
>
> The usual method of distinguishing between scattered and fluorescent radiation is by comparing the absorption coefficients of the primary and secondary radiations. It is assumed that the scattered rays are of the same hardness as the primary rays, whereas all known high-frequency fluorescent radiations are of a less penetrating type. Gray and Florance, however, have rejected this criterion, for although they find that the secondary radiation excited by hard gamma rays is of a distinctly softer type than the primary radiation, *they conclude that the primary rays are truly scattered, but in the process of scattering are so modified as to become less penetrating* [italics added].[12]

This was the point that motivated Compton's experiments. He intended "to investigate the nature and the general characteristics of secondary gamma rays, and to study the mechanism whereby comparatively soft secondary radiation is excited by relatively hard primary radiation."[13] But before one could understand, could interpret, the experimental data, it was absolutely essential to understand "under what circumstances, if any, the hardness of the scattered rays may differ from that of the primary rays." Compton analyzed these circumstances precisely in three entirely fascinating paragraphs:

> If the scattering is due to electrons of negligible dimensions which are separated far enough to act independently of each other, there is no question but that the scattered ray will be exactly similar to the primary ray in every respect except intensity; for since the accelerations to which each electron is subject are strictly proportional to the electric intensity of the primary wave which traverses it, and since the electric intensity of the scattered ray (at a great distance) due to each electron is proportional to its acceleration, the electric vector of the scattered wave is strictly proportional to the electric vector of the primary wave. Thus the frequency, the wave-form, the damping, etc., will be the same in both beams. Radiations scattered by such electrons should, therefore, be identical in character with the primary waves.
>
> Whatever type of scattering unit is assumed, it is also clear that, if the primary wave is perfectly homogeneous . . . the scattered waves must also be homogeneous and of the same frequency. If, however, the scattering unit—whether a group of electrons or the individual electron—is of dimensions comparable with the

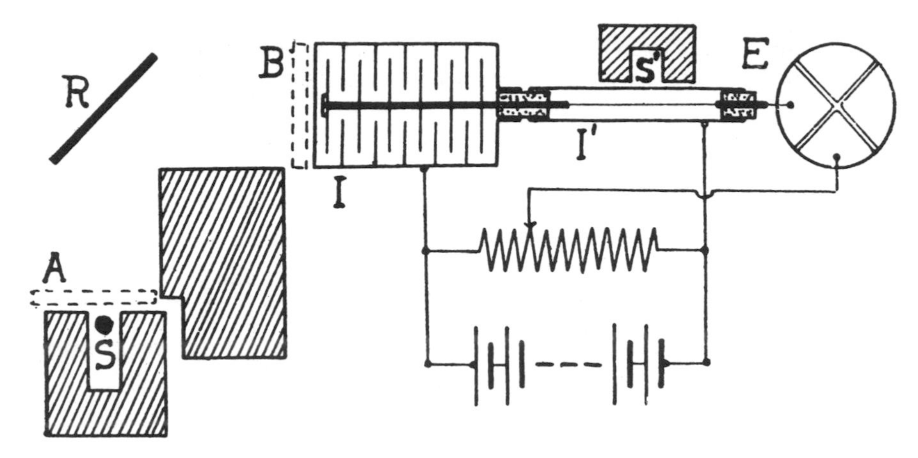

FIG. 1. Compton's Cavendish γ-ray scattering apparatus.

wave-length of the incident radiation, theory demands that the scattering, especially at large angles, shall be less for short than for longer waves. This prediction is confirmed by measurements of the scattering of X-rays and gamma rays over a wide range of frequencies. . . .

While it is possible to account in this manner for the difference in penetrating power of the primary and secondary radiation if a sufficiently heterogeneous primary beam is postulated, *it is clear that, as a result of scattering, there can be no transformation of one frequency into radiation of another frequency*. That is, the scattered rays can be no softer than the softest components of the primary rays [italics added]. . . .[14]

Every sentence in the foregoing paragraphs illustrates how completely Compton rejected both Gray's and Florance's interpretations of the softening, and how thoroughly Compton was committed to Thomson's theory of scattering, and to his own large electron modification of it. Not one word reflects considerations other than those following from classical electrodynamics. It is absolutely essential to fully appreciate Compton's total commitment to classical electrodynamics—his theoretical preconceptions—in order to understand the interpretation he placed on his Cavendish γ-ray scattering experiments.

The briefest glance at Compton's apparatus (Fig. 1[15]) reveals that it was far superior in design to any other previously used in such experiments. A sensitive quadrant electrometer E replaced the usual gold-leaf electroscope as the instrument for monitoring the current produced in the ionization chamber I. Corrections for background radiation could be made automatically by means of electrical connections to a second ionization chamber I', irradiated by a second γ-ray source S'. Finally, with a given radiator or scatterer R in place, Compton could determine the difference in hardness between the primary and secondary radiations by measuring first the ionization current I_A obtained with a given absorber in the primary beam at A, and then the ionization current I_B obtained with the same absorber in the secondary beam at B.

To see this in some detail, suppose that the primary beam is heterogeneous, consisting of various wavelengths $\lambda_1, \lambda_2, \ldots$, of intensities $I_1, I_2, \ldots$, respectively; then suppose that after being scattered, the fractions $c_1, c_2, \ldots$, enter the ionization chamber. Suppose further that with the absorber at A the fractions $k_1, k_2 \ldots$, are transmitted, and with the absorber at B the fractions $k_1', k_2', \ldots$, are transmitted. It is then evident that

$$I_A = c_1k_1I_1 + c_2k_2I_2 + \ldots = \sum_i c_ik_iI_i,$$

and similarly that

$$I_B = c_1k_1'I_1 + c_2k_2'I_2 + \ldots = \sum_i c_ik_i'I_i$$

Now, on the one hand, if each primary component is "truly scattered," it does not matter if the absorber is at position A or B. Ideally, each fraction k_i is equal to each fraction k_i', so that

$$I_A = \sum_i c_ik_iI_i = \sum_i c_ik_i'I_i = I_B, \quad \text{or} \quad I_B/I_A = 1.$$

That is, as Compton explained, *"for truly scattered radiation, the observed intensity of the secondary radiation should be approximately the same whether the absorbing plate is in the position A or B* [italics added]." On the other hand, if "the primary radiation excites in the radiator R a fluorescent radiation which is more readily absorbed than the primary rays, the observed intensity of the secondary radiation will be less when the absorption screen is in the position B; . . . the effect of any fluorescent radiation will be to make the fraction $[I_B/I_A]$ . . . less than unity."[16]

It is therefore indisputable—and startling at first to realize—that although Compton's apparatus was much more refined, the basic principle of his experiments was *precisely the same as that of Gray's 1913 experiments.* Yet, it is evident that there was a critical difference between Compton's point of view and Gray's. Gray had argued that any difference in intensity between the primary and secondary beams could only stem from a change in hardness of the radiation in the scattering process. Compton chose precisely this change in hardness as a *criterion* for distinguishing the "truly scattered" radiation from the "fluorescent" radiation. To Compton, in other words, any secondary radiation observed to be changed in hardness *must* have had its origin in the non-scattering process; it *must* be "fluorescent" radiation. The essential difference, therefore, was that Gray (and Florance) interpreted their experimental results "straight away," with no regard for possible conflicts with Thomson's classical theory of scattering, while Compton was rigidly guided in the interpretation of his experiments—precisely identical in principle to those of Gray—by a criterion *derived from* that theory. In effect Compton *defined* any secondary radiation *unchanged* in wavelength to be "truly scattered" radiation, and any secondary radiation *changed* in wavelength to be "fluorescent" radiation.

2. *His Experiments and Fluorescent Radiation Interpretation*

It was with the above dichotomy firmly in mind that Compton designed and interpreted his Cavendish γ-ray experiments. The first of these were based on the fact that the ratio I_B/I_A is a measure of the relative amounts of "fluorescent" and "truly scattered" radiation in the secondary beam. To be precise, *for large absorber thicknesses*

$$\frac{I_B}{I_A} = \frac{1}{1 + c_f/c_s} \tag{4.1}$$

where c_f and c_s are the fractions of the primary beam that become "fluorescent" radiation and "truly scattered" radiation, respectively. Equation (4.1) shows that $I_B/I_A = 1$ for a secondary beam consisting entirely of truly scattered radiation ($c_f = 0$), while $I_B/I_A = 0$ for a beam consisting entirely of fluorescent radiation ($c_s = 0$). This agrees with our intuition. In the first limit, gradually increasing the absorber thickness merely causes the secondary intensity I_B to decrease at the same rate as the primary intensity I_A, and the ratio I_B/I_A remains equal to unity. In the second limit, since the secondary radiation is of a longer wavelength than the primary, the same procedure causes the secondary intensity I_B to decrease at a more rapid rate than the primary intensity I_A, and the ratio I_B/I_A eventually becomes zero.

The results of Compton's initial experiments, in which he used an iron scatterer and lead absorbers of increasing thickness, are shown in Table 1.[17] It is especially noteworthy that Compton, probably sensing the importance of Florance's work (which was carried out under Rutherford's supervision), right from the outset took measurements at *different scattering angles,* in this case 45°, 90°, and 135°. The conclusions he drew from these measurements, which clearly display the softening of the scattered radiation, were entirely consistent with his theoretical preconceptions. They proved, he asserted, that "for gamma rays which have traversed several centimeters of lead the secondary radiation at angles greater than 90° is, except for the small probable error, all of the fluorescent type." At scattering angles much smaller than 45°, the absorption coefficient of the fluorescent radiation rapidly approached that of the primary rays. Related, but independent intensity measurements

Table 1. Compton's measurements of I_B/I_A for an iron scatterer and various lead absorbing plates at three different scattering angles.

Thickness in cm of lead absorber at B or A	I_B/I_A for iron scatterer (radiator)		
	45°	90°	135°
0	1	1	1
0.15	—	—	0.45
0.5	—	0.30	−0.02
1.0	0.52	0.13	−0.02
2.0	0.39	0.02	—
3.0	0.26	—	—
4.1	0.20	—	—

convinced Compton that "at 45° probably not as much as 15 per cent. of the whole secondary radiation consists of scattered primary rays," while at 90° the figure is roughly 2 percent and "at large angles, if there is any true scattering, it is probably less than a thousandth part of the amount predicted on the basis of the usual electron theory."[18]

This rapid angular decrease in the intensity of the "truly scattered" radiation did not worry Compton, however. It was "not impossible to account for this very low value of the scattering on the basis of the classical electrodynamics"[19]—that is, on the basis of his large ring electron theory of scattering. Instead, it was the behavior of the unusual new "fluorescent" radiation that worried Compton. No known *X-ray* fluorescent radiation varied in hardness from angle to angle. And since his large ring electron theory was a theory of *scattering*, it definitely could not be invoked to explain this variation. Compton therefore decided—one of the most vital decisions made in his entire experimental program—to undertake a systematic and detailed investigation of this newly discovered, very unusual γ-ray "fluorescent" radiation. This investigation was actually possible because he realized that he could isolate the "fluorescent" radiation from the "truly scattered" radiation by simply inserting in the primary beam an absorber for which the ratio I_B/I_A had been previously found to be essentially zero. The primary radiation destined to become "truly scattered" radiation would then not be allowed to strike the scatterer in the first place.

Insuring in this way that the secondary radiation consisted "almost wholly of fluorescent radiation. . . ," Compton first measured its angular distribution $I(\theta)$ when scattered by three different elements, aluminum, iron, and lead. His results (normalized to the intensity recorded at 90°), are shown in Table 2.[20] Obviously, for each scatterer there was a huge forward-backward asymmetry in $I(\theta)$, as both D. C. H. Florance and, more recently, K. W. F. Kohlrausch had also found. This in itself was a very surprising con-

Table 2. Compton's measurements of the angular distribution $I(\theta)$ of the "fluorescent" γ-ray radiation

	Scattering angle							
Radiator	30°	45°	60°	75°	90°	120°	135°	150°
Al	—	6.2	4.0	1.7	(1.0)	0.7	0.4	0.3
Fe	10	7.6	4.6	2.0	(1.0)	0.6	0.5	0.4
Pb	11	6.8	4.3	2.2	(1.0)	0.8	—	0.7

clusion, because it was well known that the usual *X-ray* fluorescent radiations were emitted *isotropically* from substances.

Second, Compton investigated whether or not the nature of the scatterer was important. Again making "certain that practically all of the secondary radiation was of the fluorescent type,"[21] he measured the amount emitted by five different radiators, paraffin, aluminum, iron, tin, and lead, at two different scattering angles, 45° and 135°. He found that the amount emitted per electron was roughly the same from substance to substance, and from angle to angle. This agreed with "the more quantitative experiments of Ishino," and meant that the amount emitted was essentially *independent* of the nature of the radiator. Again, this was in strong contrast to the usual *X-ray* fluorescent radiations, which were *characteristic* of the element emitting them.

Finally, Compton returned to the most puzzling by far of his preliminary observations, the apparent angular variation in the hardness of the γ-ray "fluorescent" radiation. Once again eliminating the "truly scattered" radiation, he measured the mass-absorption coefficient μ/ρ of the "fluorescent" γ-radiation emitted by the five substances above at three different scattering angles, 45°, 90°, and 135°. His results are shown in Table 3.[22] Obviously, for each substance the γ-ray "fluorescent" radiation became much softer as the scattering angle increased, although the difference in hardness between the primary and secondary rays was smaller for elements of high atomic weight. One measurement was particularly noteworthy. For iron (Fe) the value (0.50) for μ/ρ at 135° was over eight times as large as the value

Table 3. Compton's measurements of μ/ρ at three scattering angles for five different scatterers

Scattering angle	μ/ρ [cm²/g] in lead using various radiators				
	Paraffin	Al	Fe	Sn	Pb
45°	0.10	0.10	0.11	0.09	0.05 (?)
90°	—	—	0.21	—	—
135°	0.78	0.50	0.50	0.32	0.15

(0.062, not shown) he obtained for the primary γ-rays. This "presumably means," Compton noted, "that this fluorescent radiation is of very appreciably longer wave-length than the primary gamma rays." He estimated the "softest part" to have a wavelength lying between 0.06 Å and 0.12 Å, which meant that these "fluorescent" γ-rays "bridged the gap" between hard X-rays, on the one hand, and very penetrating γ-rays, on the other hand.

In sum, the properties of Compton's newly discovered "fluorescent" γ-rays were extraordinary:

> Although the secondary gamma radiation under examination seems, without doubt, to be fluorescent in nature, it differs in several important respects from the characteristic fluorescent K and L [X-]radiations excited in matter when traversed by hard X-rays. In the first place, whereas these characteristic [X-]radiations differ greatly in hardness from element to element, the secondary gamma rays, especially at small angles with the incident beam, are of nearly the same hardness over a wide range of atomic numbers. And in the second place, while the characteristic [X-]radiations are found to be distributed uniformly with regard to intensity and quality at all angles with the primary beam, the fluorescent gamma rays show marked asymmetry in both quality and quanity in the forward and reverse directions.[23]

3. *The Origin of the New Fluorescent Radiation*

The number of *dis*similarities between Compton's γ-ray "fluorescent" radiation and the usual X-ray fluorescent radiations was therefore far greater than the number of similarities. In fact, there was only *one* similarity. Both were of a longer wavelength than the primary radiation. Yet, in Compton's mind, this single similarity outweighed all of the dissimilarities. As a result, Compton was confronted with a major and difficult problem in interpretation. He did not retreat from it; but, after evidently thinking a great deal about it, concluded: "An explanation of the origin of the fluorescent radiation which appears to be satisfactory is that the high-speed secondary beta particles liberated in the radiator by the primary gamma rays excite the secondary gamma rays as they traverse the matter of the radiator. On this view the fluorescent gamma rays should be identical in character with the so-called 'white' radiation excited in the target of an X-ray tube by the impact of the cathode particles."[24]

G. W. C. Kaye's decade-old cathode-ray experiments,[25] A. S. Eve's equally old γ-ray experiments,[26] and other experiments described by Rutherford in his *Radioactive Substances and their Radiations,*[27] tended to support Compton's interpretation. We see that its focus was on the secondary β-rays liberated by the primary γ-rays; moreover, we see that Compton did not think fundamentally in terms of a γ-ray–electron *interaction* (as Stark had in 1909–1910[28]), but chose an entirely different tack. Compton concentrated once again *not* on the nature of the radiation, but on the nature of the electron.

Compton postulated that the electron, the secondary β-ray, was a kind of self-contained oscillator—an idea which D. L. Webster had also recently used in an attempt to explain the continuous X-ray spectrum.[29] Compton knew, however, that these oscillators had to be "radically different in character from those which are responsible for the K, L, and M characteristic [X-] radiations." He struggled with this dilemma, and finally resolved it:

> A question of great theoretical importance is—What kind of oscillator can give rise to radiation which not only is more intense in one direction than in another, but also differs in wave-length in different directions? Since the secondary radiation differs in frequency from the primary rays, it would seem impossible to invoke any interference between the radiation from the different oscillators to account for this phenomenon. Such an explanation is rendered the more difficult by the fact that to explain the different hardness of the rays in different directions, oscillators of different frequencies would have to be present, between which there could be no fixed phase relations. An obvious means of accounting for the observed phenomenon is to suppose that the radiator which gives rise to the secondary rays is moving at high speed in the direction of the primary beam. In this case, both the intensity and the frequency of the fluorescent radiation will be greater in the forward than in the reverse direction, as is demanded by the experiments.[30]

The forward-backward asymmetry in the *intensity* would therefore arise from the well-known forward peaking in the radiation emitted by a rapidly moving and accelerated charged particle, and the forward-backward asymmetry in the *hardness* would arise from the Doppler shift. Denoting the former asymmetry (in intensity) by R, Compton deduced the following expression[31] for it from equations given in O. W. Richardson's *Electron Theory of Matter:*

$$R = \frac{I_1}{I_2} = \left[\frac{1-\beta\cos\theta_2}{1-\beta\cos\theta_1}\right]^6 \left[\frac{(1-\beta\cos\theta_1)^2 - (1/3)(1-\beta^2)\sin^2\theta_1 + (1/3)\cos\theta_1(2\beta-\cos\theta_1 - \beta^2\cos\theta_1)}{(1-\beta\cos\theta_2)^2 - (1/3)(1-\beta^2)\sin^2\theta_2 + (1/3)\cos\theta_2(2\beta-\cos\theta_2 - \beta^2\cos\theta_2)}\right] \quad (4.2)$$

where $\beta = 1/c$ times the velocity of the β-particle, and I_1 and I_2 are the intensities observed in the directions θ_1 and θ_2 with respect to the direction of motion of the β-particle.

An exact check on equation (4.2), a "rigid calculation," was impossible because (as Wilson's cloud-chamber photographs, for example, indicated) secondary β-rays travel in highly irregular paths, and hence no precise values of θ_1 and θ_2 could be specified. The best Compton could do was to arbitrari-

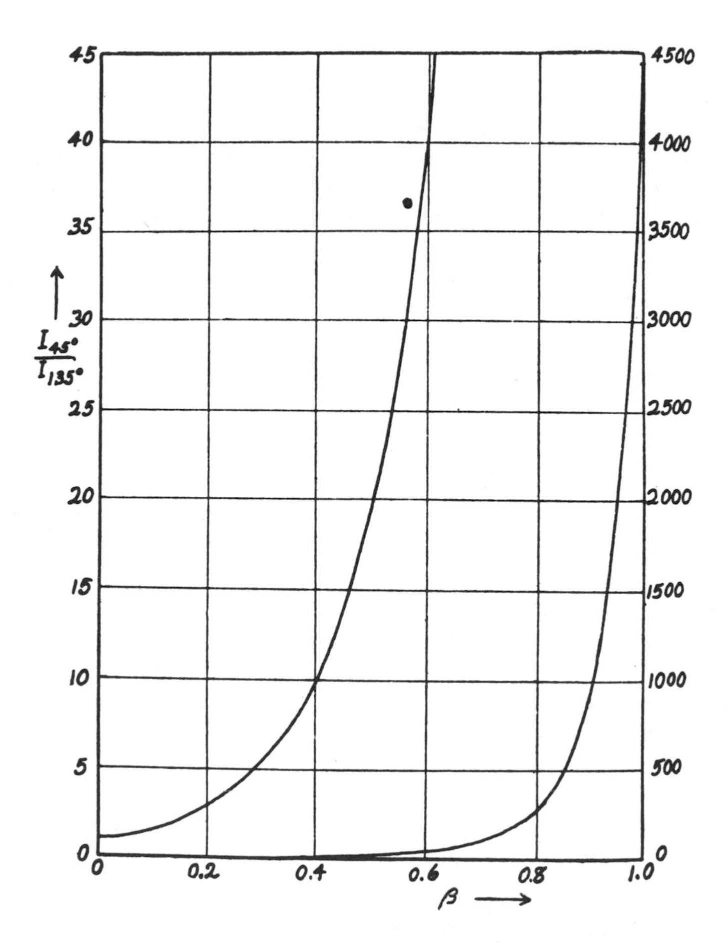

FIG. 2. Compton's plots of $I_{45°}/I_{135°}$ *vs* β, predicted by equation (4.2).

ly choose $\theta_1 = 45°$ and $\theta_2 = 135°$ and plot the calculated asymmetry as a function of β. His result is shown in Fig. 2.[32] This plot proved to Compton that "it is possible . . . to account for any reasonable degree of asymmetry of the secondary radiation. . . . The rapid increase of R with β, however, makes it reasonably certain . . . that the average speed of the oscillators which emit the secondary gamma radiation does not differ greatly from half the speed of light." Since this was several times greater than the velocity of the fastest α-particles, the oscillators (because of the mass difference) "must be individual electrons. We are thus led to the idea that it is the vibrations of the secondary beta particles themselves which give rise to the fluorescent gamma rays."[33] This conclusion was borne out by a comparison of the angular distribution $I(\theta)$ predicted for two values of β (0.5 and 0.55) with some of K. W. F. Kohlrausch's recent data.[34] The "generally satisfactory" agreement suggested to Compton "that we are working along the right line."[35]

The variation in wavelength of the fluorescent γ-radiation with scattering angle also found a ready interpretation, for "according to the [relativistic] Doppler principle, if the oscillators producing the fluorescent radiation are moving in the direction of the primary rays, the ratio of the wave-length at angle θ_1 to that at an angle θ_2"[36] is given by the following equation:

$$\frac{\lambda_1}{\lambda_2} = \frac{1 - \beta \cos \theta_1}{1 - \beta \cos \theta_2} \tag{4.3}$$

Setting $\beta = 0.5$, $\theta_1 = 45°$, $\theta_2 = 135°$, and $\lambda_2 = 0.08$ Å (which agreed with his earlier estimate of $0.06 \text{ Å} < \lambda_2 < 0.12$ Å for the "softest part" of the fluorescent radiation), Compton readily calculated λ_1 to be 0.04 Å. Since, however, this was the wavelength corresponding to $\theta_1 = 45°$, he had to produce a further argument to conclude that the wavelength of the *primary* RaC γ-rays lay between 0.02 Å and 0.03 Å, which was "not in disaccord with the calculations of Rutherford based upon the quantum hypothesis."[37] It disagreed substantially, of course, with Compton's earlier estimate of 0.09 Å, the wavelength he had used in all of his ring electron calculations.

C. Other Work by Compton at the Cavendish (1919–1920)

1. *First Indications to Doubt His Ring Electron Model*

Before Arthur Compton began his researches at the Cavendish, he had found it "natural to suppose that the electron may have the form of a ring . . . ,"[38] and he had developed a quantitative and successful theory of scattering on that hypothesis. Only after he went to the Cavendish did some very unsettling questions arise in his mind concerning the ring electron. A major reason, apparently, was his discovery that, contrary to the opinion of "Florance, Ishino, and others," if he had observed "any true scattering of gamma rays," it had represented "only a small fraction" of the "fluorescent radiation." His concern led him to carry out further γ-ray experiments at the Cavendish; the first of these were similar in principle to A. H. Forman's X-ray experiments.[39]

Forman's experiments, we recall, were designed to determine whether or not magnetization affected the absorption of X-rays in an iron plate, and his affirmative conclusion fitted in well with Compton's ring electron hypothesis. Now, Compton calculated the extent to which the mass-absorption coefficient μ/ρ might be affected by magnetization by calculating the change in both the fluorescent term $k\phi(a/\lambda)\lambda^3$ and the scattering term σ/ρ, due to a reorientation of the ring electrons. He found (in relative units) that for a 90° reorientation (axes and beam direction changed from parallel to perpendicular to each other) the former term should change from 7.2 to 0.14, while the latter term should change from 1.5 to 0.8. In other words, large changes in the mass-absorption coefficient were to be expected.

Experimentally, however, Compton now observed that "for parallel magnetization the effect on the absorption coefficient is probably less than 1 part in 5,000, and for perpendicular magnetization the effect is probably less than 1 part in 3,000."[40] That is, not only were the changes in absorption (if they existed at all) far smaller than anticipated, they were actually smaller for parallel magnetization than for perpendicular magnetization—exactly opposite to what he had anticipated from his calculations. He remarked, not without a tinge of sarcasm, that "It is . . . rather surprising that while Forman found the change in the absorption coefficient due to magnetization to increase rapidly for shorter wave-lengths, the present experiment shows no effect for the still shorter wave-length gamma rays." He did not hesitate to clearly state the implications of his experiments: ". . . it would . . . seem necessary to conclude either (1) that the ultimate magnetic particle is not a ring electron, or (2) that energy transmitted to the motion of the electron as a whole is not responsible for any considerable part of the total absorption of gamma rays." While the second possibility could not be ruled out, Compton forthrightly admitted that "the evidence seems rather opposed to the hypothesis that a ring electron is the ultimate magnetic paricle."[41]

2. *Experiments to Determine Gamma-Ray Wavelengths*

Compton's mounting doubts about the validity of his ring electron model—which Rutherford and his English colleagues undoubtedly encouraged—were perhaps largely responsible for Compton's attempt, as his last research at the Cavendish,[42] to redetermine the wavelength of RaC γ-rays. As we have seen, this quantity directly determined the radius of the electron, and Compton's initial choice of 0.09 Å over Rutherford's 1917 estimate of 0.03 Å evoked immediate criticism from H. S. Allen.[43] Moreover, Compton himself had recently found good reason to doubt his initial choice when he estimated the wavelength to lie between 0.02 Å and 0.03 Å. The whole question was therefore an open and an important one.

Compton based his experiment on an expression Peter Debye had derived in 1915 during his study of "excess scattering."[44] From Debye's expression, Compton proved that the ratio $[\sin(\theta/2)]/\lambda$ determines the angular distribution $I(\theta,\lambda)$ of the scattered radiation—or actually the ratio

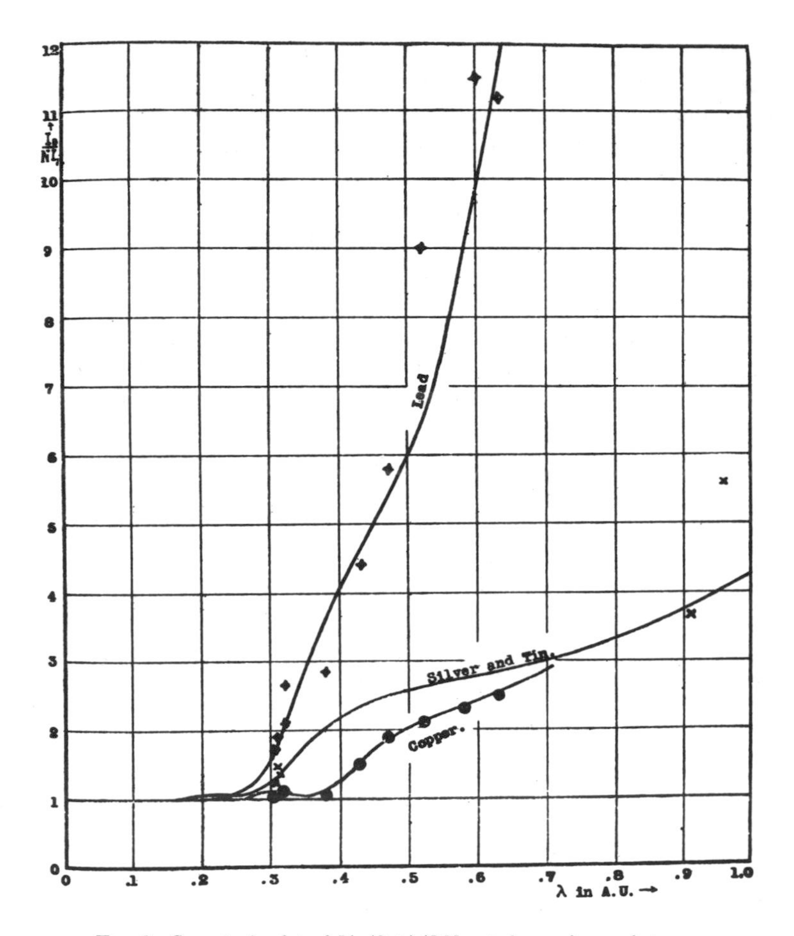

FIG. 3. Compton's plot of $I(\pi/2, \lambda)/I_1N$ *vs* λ for various substances.

$I(\theta,\lambda)/I_1N$, where I_1N represents Thomson's expression (N is the number of electrons per unit volume). If, therefore, one finds for two different elements x and y that the ratio

$$\frac{[I(\theta,\lambda)/I_1N]_x}{[I(\theta',\lambda')/I_1N]_y}$$

is some value k, it will theoretically remain at k provided only that

$$\frac{\sin(\theta/2)}{\lambda} = \frac{\sin(\theta'/2)}{\lambda'}$$

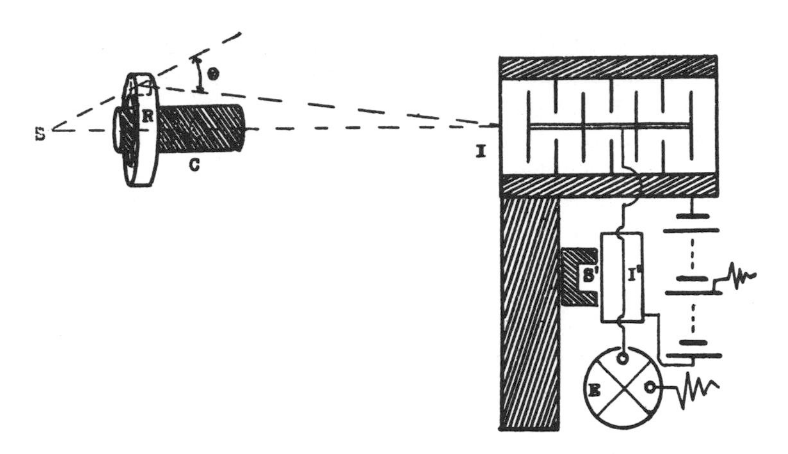

FIG. 4. Compton's apparatus for determining γ-ray wavelengths.

Now, from Barkla and Dunlop's extensive X-ray data for lead, silver and tin, and copper,[45] which were obtained for a scattering angle of $\theta = \pi/2$, Compton could plot $I(\pi/2,\lambda)/I_1N$ *vs* λ, as shown in Fig. 3.[46] For $\lambda = 0.3$ Å, for example, it is easy to see that for lead (Pb) and copper (Cu) the ratio

$$\frac{[I(\pi/2, 3\text{Å})/I_1 N]_{pb}}{[I(\pi/2, 3\text{Å})/I_1 N]_{cu}} = \frac{1.6}{1.1} = 1.45 = k$$

Therefore, Compton reasoned, if he could scatter γ-rays (wavelength λ') from both lead and copper and locate a scattering angle θ' for which the ratio

$$\frac{[I(\theta',\lambda')/I_1N]_{pb}}{[I(\theta',\lambda')/I_1N]_{cu}} = 1.45$$

he could readily calculate λ', since he would then know θ ($= \pi/2$), λ ($=$ 3Å), and θ'.

Compton's (null) apparatus is shown in Fig. 4.[47] On the left, a scattering ring R of either lead or copper is mounted coaxially with a lead shielding cylinder C, which allows the scattering angle θ' to be varied by simply changing the distance between the ring and ionization chamber I. He found that even at the smallest experimentally possible scattering angle (10°), the recorded ratio of intensities was still less than half the desired value, k = 1.45. He therefore had to be satisfied with the conclusion that "the effective wave-length of the gamma rays here employed" was less than (sin 5°/sin 45°) × (0.3) = 0.037 Å.[48]—although further arguments allowed him to reduce the figure to between 0.025 and 0.03 Å. This value of course agreed nicely with his most recent estimate, but conflicted with his initial one.

Compton recognized two possible sources of error in his work. The first, which cast a shadow over his whole procedure, was his suspicion that the radiation that he had considered as "truly scattered" in his experiments might actually have been his unusual "fluorescent" radiation. The second was more subtle. In order to argue that the wavelength of the γ-rays might be as low as 0.025 Å, Compton had to argue from *extrapolated* values obtained from the solid curves in Fig. 3. What Compton neglected to mention was that these solid curves were *theoretical* curves—and were calculated on the basis of a model of the electron other than the ring electron.

D. The Status of Compton's Thought by Mid-1920

1. *His Critique of Thomson's Basic Assumptions*

Compton had learned a great deal at the Cavendish. It is possible, in fact, to determine fairly accurately how far his thought had progressed by the time he left Cambridge by examining a paper which he wrote shortly thereafter. The object of this paper, significantly entitled "Classical Electrodynamics and the Dissipation of X-ray Energy," was "to find out whether it is possible to account for the known phenomena of X-ray absorption and scattering on the basis of the classical electrodynamics."[49] Compton—in a most revealing move—decided to examine this question systematically by adhering to his earlier, rigid dichotomy between the "truly scattered" and "fluorescent" radiation.

To Compton, the basic question regarding the "truly scattered" radiation was whether or not it could be fully explained by modifying any or all of the fundamental assumptions of J. J. Thomson's theory, *viz:* "1. That the usual electromagnetic theory is applicable to the problem. 2. That each electron in the scattering material acts independently of every other electron. 3. That there are no other forces acting on the electrons which are comparable in magnitude with the forces due to the incident beam of X-rays. And 4, that the dimensions of the electron are negligible compared with the wave-length of the incident radiation."[50] It will come as no surprise to learn that Compton rejected the second and last of Thomson's assumptions. He therefore concluded that the angular distribution $I(\theta,a/\lambda)$ could be formally written as a product

$$I(\theta,a/\lambda) = I_1\Psi\,\Phi \qquad (4.4)$$

where I_1 represents the Thomson $(1 + \cos^2\theta)$ variation for a single electron, and Ψ and Φ are the "excess scattering" and "finite electron" modulation factors arising from the breakdown of those two assumptions.

The "excess scattering" factor Ψ had been calculated by D. L. Webster, Peter Debye,[51] and other physicists—most recently by G. A. Schott,[52] who had investigated how radiation is scattered by rings of electrons (not ring electrons). Compton admitted that he had "not attempted to solve explicily the problem discussed by Schott,"[53] for two reasons. First, he saw that Schott's final result was actually incorrect, since Schott had implicitly assumed[54] that the individual rings, each containing n_s electrons, scatter independently, which implied that the maximum intensity was proportional to $\sum_s n_s^2$, and not to $(\sum_s n_s)^2$, as required by the well known N^2 variation. Second, by this time Compton himself had also solved the problem by imagining the radiation to be scattered by electrons paired off in s randomly oriented "dumbbells" of length $2\rho_s$. He had concluded[55] that

$$\Psi = \left[N + 2 \sum_1^{N/2} \left(\frac{\sin 2k_s}{2k_s} - \frac{2\sin^2 k_s}{k_s^2} \right) + 4 \left(\sum_1^{N/2} \frac{\sin k_s}{k_s} \right)^2 \right] \tag{4.5}$$

where $k_s \equiv (4\pi\rho_s/\lambda)\sin(\theta/2)$. As may easily be verified, equation (4.5) goes over to N in the short wavelength limit ($k_s \rightarrow \infty$), and to N^2 in the long wavelength limit ($k_s \rightarrow 0$), both of which are correct, the latter being the result which Schott did not obtain.

The "finite electron" factor Φ had always been Compton's major concern, and it became evident immediately from his introduction that his present discussion would contain some surprises: "We shall consider the two cases of an electron in the form of a ring, and of a solid spherical electron, one of which will be found to account in a satisfactory manner for the experiments on the scattering of radiation of very high frequency."[56]

It developed that Schott had pointed out[57] an error in Compton's ring electron calculation, which evened their scores, and which probably prompted Compton to re-examine his calculation. He eventually saw that he could solve the scattering problem simply and directly:

(1) by locating an element of charge by the angles α and γ (Fig. 5[58])

(2) by pairing off this element of charge with one diametrically opposite from it

(3) by introducing the assumption of random orientation by allowing γ to vary between 0 and π while being modulated by a probability factor $\frac{1}{2}\sin\gamma d\gamma$.

He concluded that the angular distribution $I(a/\lambda,\theta)$, which is just the "finite electron" factor Φ, was given by

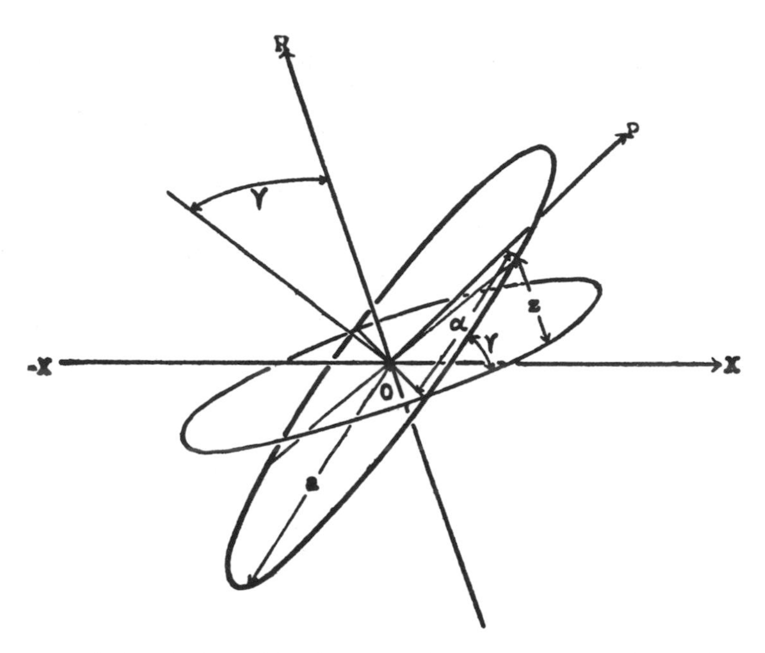

FIG. 5. Compton's simplified diagram for the ring electron scattering problem.

$$\Phi = (1/2)\int_0^{\pi} J_0^2(x\sin\gamma)\sin\gamma d\gamma = (1/x)\sum_0^{\infty} J_{2n+1}(2x) \quad (4.6)$$

where $x \equiv (4\pi a/\lambda)\sin(\theta/2)$. Similarly, Compton found that the factor ϕ in the ordinary fluorescent term was given by

$$\phi = (1/r)\sum_0^{\infty} J_{2n+1}(2r) \quad (4.7)$$

where $r \equiv 2\pi a/\lambda$. When he compared these results with equations (3.13) and (3.22), Compton saw that his mistake was not a very serious one. It originated in his randomization process and affected only the numerical values of the expansion coefficients, leaving the form of the solutions unchanged.

2. *Compton Abandons His Large Ring Electron Model—For a Large Spherical Model*

A more substantial surprise emerged during the course of Compton's treatment of an entirely new model of the electron, the solid sphere electron. His decision to treat this model was motivated no doubt by his increasing skepti-

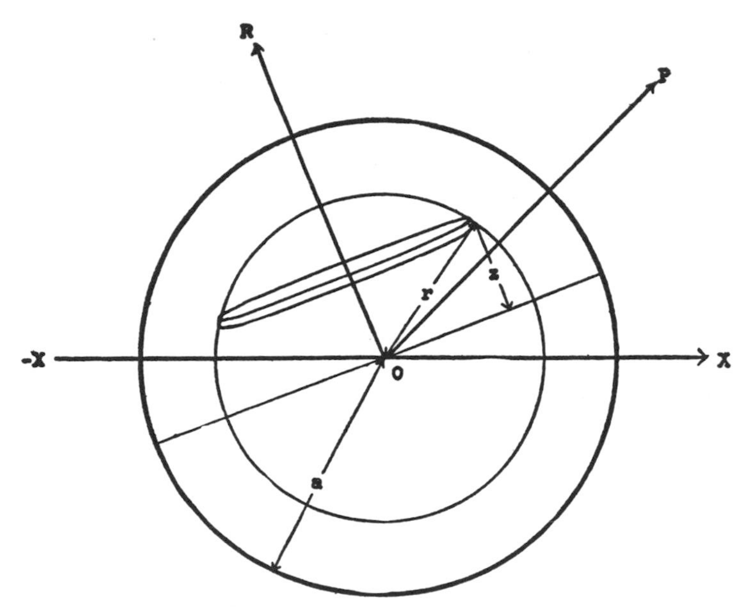

FIG. 6. Compton's diagram for calculating the "finite electron" factor Φ for a spherical electron of radius a.

cism toward the physical reality of the ring electron, which resulted from his discussions and work at the Cavendish. By late 1920 he was of the opinion, in fact, that one could accept "Parson's view of a magnetic electron of comparatively large size," but could also suppose "with Nicholson that instead of being a ring of electricity, the electron has a more nearly isotropic form with a strong concentration of electric charge near the centre and a diminution of electric density as the radius increases." He added: "While the explanation of the inertia of such a charge of electricity is perhaps not obvious, it is at least consistent with our usual conceptions. . . ."[59]

Yet, we recall that it was *precisely* the question of the inertia of the electron that originally led Compton to *reject* the spherical shell and other spherically symmetrical models in favor of the ring electron. Therefore, his decision to even consider a spherical electron is surprising, and the question immediately arises as to why he now felt that he could seriously entertain such a model. The reason, apparently, is contained in a remark he made in *Nature* (February, 1921): "With the introduction of the principle of relativity," Compton wrote, it became "clear that the variation of mass with velocity was no characteristic attribute of electrical inertia, and that therefore we have no proof that the negative electron's inertia is wholly electromagnetic in origin. In fact, the investigations of Abraham, Webster, and others have shown that there must be some mass present other than that due to the electron's electric field."[60] It seems, therefore, that in scrutinizing the whole situation once again, Compton eventually concluded that his earlier, grand objection to all spherically symmetrical models had, in fact, been unjustified.

Compton calculated the "finite electron" factor Φ for the spherical electron by referring to the diagram shown in Fig. 6.[61] Let the sphere be divided into shells of radius r and thickness dr, and let each shell in turn be divided into rings of thickness dz and of volume $2\pi r dr dz$. The amplitude of the radiation scattered by each shell through the angle $XOP = \theta$ is then given by

$$\frac{A_0 2\pi r dr \rho}{e} \int_{-r}^{r} \cos\left(\frac{2\pi}{\lambda} 2z \sin\frac{\theta}{2}\right) dz = \frac{A_0 \rho \lambda}{e \sin(\theta/2)} \sin\left(\frac{4\pi r}{\lambda} \sin\frac{\theta}{2}\right) r dr$$

where ρ is the charge density, and e the total electronic charge. If one assumes that shells of equal thickness contain equal amounts of charge, then

$$\rho = dq/dV = e(dr/a)/4\pi r^2 dr = e/4\pi a r^2$$

and one has the following expression for the amplitude scattered by the entire electron:

$$A(a/\lambda,\theta) = \left(\frac{A_0\lambda}{e\sin\frac{\theta}{2}}\right)\left(\frac{e}{4\pi a}\right) \int_0^a \frac{\sin\left(\frac{4\pi r}{\lambda}\sin\frac{\theta}{2}\right)}{r^2} r dr$$

$$= A_0 \left[1 - \frac{x^2}{3\cdot 3!} + \frac{x^4}{5\cdot 5!} - \frac{x^6}{7\cdot 7!} + \dots\right]$$

$$= A_0 \sum_0^{\infty} \frac{(-1)^n x^{2n}}{(2n+1)(2n+1)!} \qquad (4.8)$$

where $x \equiv (4\pi a/\lambda)\sin(\theta/2)$. Dividing by A_0 and squaring, one obtains the angular distribution $I(a/\lambda, \theta)$, which is just the "finite electron" factor Φ:

$$\Phi = \left[\sum_0^{\infty} \frac{(-1)^n x^{2n}}{(2n+1)(2n+1)!}\right]^2 \qquad (4.9)$$

Compton found, in the first place, that equation (4.9) agreed with Barkla and Ayres' soft X-ray data for carbon, provided that the radius of the

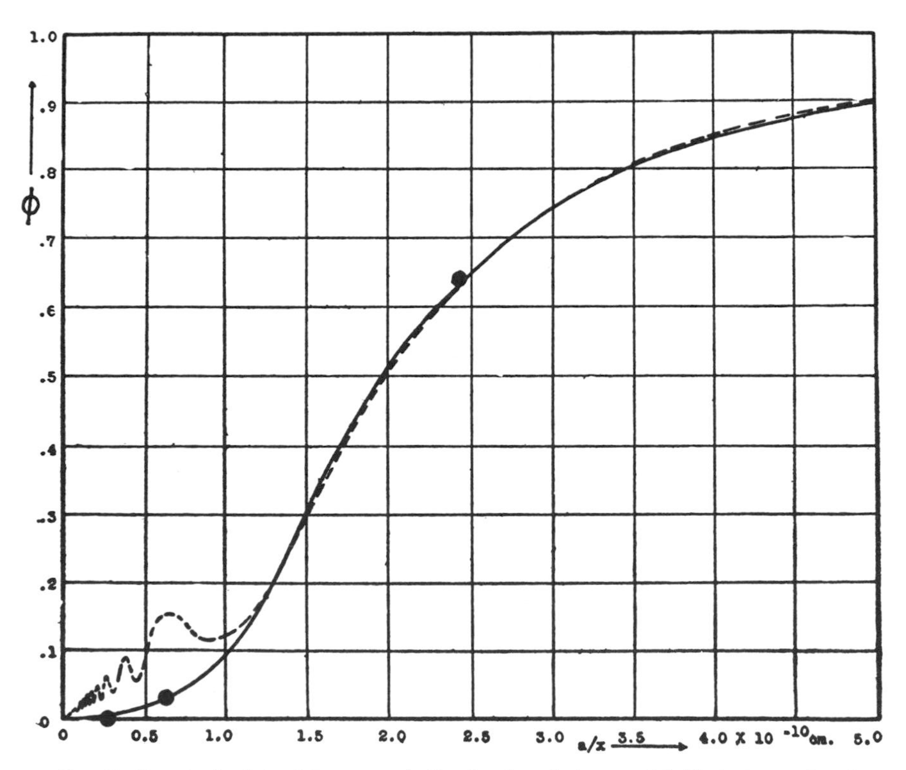

FIG. 7. Compton's plots of Φ versus a/x for the ring electron model (dashed curve) and the solid sphere model (solid curve).

sphere was assumed to be 0.05 Å.[62] Second, assuming this value for the radius, and assuming a scattering angle of 90°, he calculated curves for the ratio $\Psi\Phi/N$ *vs* wavelength—which turned out to be precisely the same curves (Fig. 3) which he had used in his Cavendish estimate of the wavelength of RaC γ-rays. This fact, it developed, was directly relevant to his key question, that is, how this spherical model compared with the ring electron model.

To answer this question, Compton restricted his discussion to the short-wavelength region (where "excess scattering" effects are negligible) and plotted Φ *vs* a/x for both models (Fig. 7[63]). The experimental data presented an immediate difficulty: Florance's and Ishino's data, Compton argued, were not entirely trustworthy because neither experimentalist had taken pains to eliminate the longer-wavelength "fluorescent" γ-radiation from the secondary beam. "The only direct measurements that have been made of the scattering of radiation of higher frequency seem to be those of the writer on the scattering of the hard gamma rays from radium C."[64] These measurements provided the two data points at low a/x ($\lambda = 0.03$ Å, with $\theta = 120°$ and $45°$, respectively). The only other data point Compton used was the one at high a/x ($\lambda = 0.15$ Å, with $\theta = 60°$) which he gleaned from Hull and Rice's work.[65] It is seen that all three fell squarely on the curve for the solid sphere electron (solid curve), but missed entirely the curve for the ring electron (dashed curve). The conclusion was inevitable: "It is clearly impossible to account for the very low value of the scattering of γ-rays on the basis of a flexible ring electron. On the other hand, the spherical electron hypothesis gives a satisfactory explanation of the meager data that are available."[66]

Compton published this conclusion in January, 1921. His first paper on the ring electron had been published in January, 1918. After a lifetime of three years—to the month—the ring electron theory was dead.[67]

But it is crucial to be absolutely clear on one point. The spherical electron theory *did* fit the data. Therefore, wrote Compton, "the present hypotheses do afford an explanation of the scattering of gamma rays and X-rays which is qualitatively correct and quantitatively as reliable as our experimental knowledge. The comparison of the present theory with experiment must therefore be taken to confirm the applicability of the classical electrodynamics to the calculation of the scattering of high frequency radiation."[68]

3. *Compton Introduces a Third Term Into the Absorption Formula*

Any challenge to classical electrodynamics, Compton sensed more and more, would come from the non-scattering process, from an attempt to understand the unusual "fluorescent" γ-radiation. He remained convinced of his Doppler shift interpretation, but in thinking further about the problem, he realized that there might be an additional source of fluorescent radiation: "If the absorption of high frequency radiation is to be explained according to the classical electrodynamics, there is another term which should appear in the expression for the total absorption. This arises from the fact that the incident radiation consists of wave-trains of finite length which presumably begin rather abruptly." This would represent a "transformation of the energy of the primary beam into another form which will probably reappear as fluorescent radiation characteristic of the electron traversed."[69]

Compton therefore resurrected the idea that had occurred to him in March, 1919, at Westinghouse.[70] Specifically, he now argued in print that the motion of the electrons should be governed by the second-order differential equation

$$\frac{d^2x}{dt^2} + 2l\frac{dx}{dt} + q^2x = \frac{e}{m}Ae^{-kt}e^{i(p_1t + \delta)} \qquad (4.10)$$

where the symbols refer both to the electron (mass m, charge e, damping constant $2lm$, force constant mq^2: $q_1^2 \equiv q^2 - l^2$) and to the incident radiation (amplitude A, phase δ, attenuation constant k, angular frequency p_1: $p_1^2 \equiv p^2 - k^2$). This equation is of the same form as equation (3.21), which Compton had entered in his notebooks almost two years earlier. But he now had its solution in hand, and hence could complete the entire calculation.[71] Symbolically, therefore, he concluded that the mass-absorption coefficient μ/ρ should be given by a sum of three terms,

$$\frac{\mu}{\rho} = \frac{\tau}{\rho} + \frac{\sigma}{\rho} + \frac{a}{\rho} \tag{4.11}$$

where a/ρ represents the additional mass-absorption coefficient.

This new coefficient a/ρ, for which Compton derived an exact expression, turned out to be on the same order of magnitude of σ/ρ, and also approximately independent of wavelength.[72] These characteristics made it difficult to check on its validity experimentally,[73] and indeed Compton's check was not very satisfactory.[74] In any event, other, much more general considerations cast doubt on the significance of his entire calculation:

> Physically . . . it does not appear that this explanation of gamma ray absorption is adequate. Experiment indicates that a large part of the X-ray energy goes into the kinetic energy of high speed beta rays; and it is difficult to understand how any large part of the 'transformed' energy, which by hypothesis is distributed nearly equally among all the electrons traversed, can show itself in the form of large amounts of energy on a comparatively small number of beta particles. For this reason I am inclined to doubt whether the new type of absorption here introduced plays any considerable part in the absorption of gamma rays.[75]

4. *Conclusions*

Roughly four and one-half years had passed since Compton had completed his Ph.D. thesis at Princeton University; three years had elapsed since he had first introduced his large electron theory of scattering; approximately one year had passed since he had discovered at the Cavendish that most of the secondary γ-radiation, especially at large scattering angles, was not "truly scattered" but a new form of "fluorescent" radiation. The point to which Compton's thought had now developed may be gathered from his answer to his basic question. Was it possible to account for the known phenomena of X-ray absorption and scattering on the basis of classical electrodynamics? Compton's answer: "There appears . . . no evidence of the failure of the classical electrodynamics in connection with measurements of the absorption of high frequency radiation, while experiment confirms the view that the classical electrodynamics is applicable to calculation of the scattering of X-rays and gamma rays."[76]

E. Compton's Initial Researches at Washington University (1920–1921)

1. His Reasons for Going There

S. K. Allison once remarked that ". . . Compton, as a physicist, was less a product of continental European universities than were many physicists of his generation in this country. He did not join the pilgrimages of the 1920's to Göttingen or Copenhagen, merely stopping off incidentally on a European tour."[77] Compton did not seek out—even deliberately avoided—the centers of greatest activity in theoretical physics, perhaps partly because he felt that the problems of most concern to him were of secondary concern to the continental theorists, but more likely largely because he sensed he could learn most by designing and carrying out his own experiments. After all, his major reason for leaving Westinghouse had been his desire to again engage in pure research in experimental physics, and his major reason for going to the Cavendish had been his conviction that Rutherford, following in J. J. Thomson's footsteps, was the foremost experimentalist of the day. And once settled in Cambridge, there was little reason for him to leave. The atmosphere created by Thomson, Rutherford, and their colleagues (Fig. 8) was extremely stimulating, and Rutherford generously provided him every facility for his research.

In mid-1920, however, as Compton's tenure at the Cavendish was drawing to a close, he had to face the decision of where to go in the fall. Many questions remained unanswered in his mind, and as he later wrote, "I came to the conclusion that my studies could be carried through to better advantage in my own laboratory, where I would be able to set up equipment especially designed for my experiments and where I could pursue my research according to the way in which one experiment leads to another."[78] Earlier he had remarked: "From the time that I first became a serious student, finding a place to do thinking has been a central problem personally for me." These twin desires—to be able to independently pursue his own research program, and to find a place "to do thinking"—were the major reasons he accepted an invitation to become Wayman Crow Professor of Physics and Head of the Department of Physics at Washington University in St. Louis:

> In 1920 Washington University, at least so far as its college of liberal arts was concerned, was a small kind of place. I went there, frankly, because I did not

Reproduced by permission of the Cavendish Laboratory

FIG. 8. Cavendish Laboratory, Physics research students, June 1920. Left to right, *back row,* A. L. McAulay, C. J. Power, G. Shearer, Miss Slater, Miss Craies, P. J. Nolan, F. P. Slater, G. H. Henderson, C. D. Ellis; *middle row,* J. Chadwick, G. P. Thomson, G. Stead, Sir J. J. Thomson, Sir E. Rutherford, J. A. Crowther, A. H. Compton, E. V. Appleton; *front row,* A. Muller, Y. Ishida, A. R. McLeod, P. Burbridge, T. Shimizu, B. F. I. Schonland.

want to become confused in being in a center where there was so much going on along the line in which I was involved that I would be led away by the thinking of the time. At Washington University there were small but reasonably adequate facilities to do the kind of work that I wanted to do, and I was far enough removed from such places as Chicago, or Princeton, or Harvard, where I felt I would have been thrown into the pattern of thought of the period and would not have been able to develop what I had conceived of as my own contribution. That was my choice as compared with what would then have been called a great university, or as compared with an industrial research laboratory, which I had left in order to get into this kind of work. Somewhat the same kind of opportunity had been presented at Cambridge, where as a research fellow I had no responsibilities other than research.[79]

To help him in his work, Compton soon brought in G.E.M. Jauncey. At the same time he "took the unprecedented step of setting up a budget of some $4000 per year for research equipment and expense."[80] We shall now see how Compton, during his first year at Washington University, used some of those research funds.

2. *"Problems to be tackled at Saint Louis."*

Compton began formulating his research plans in late summer 1920, while homeward bound on the Cunard Ship R.M.S. "Aquitania." Seven small (4½″ x 7″) pieces of ship's stationery, pointedly entitled Problems to be tackled at Saint Louis (see Fig. 9) have survived a half-century to bear witness to what was foremost in his mind as he left England.[81] Originally, he had listed at least six Problems, but unfortunately five pages, on which numbers 3 and 4 were written, have been lost. His first problem (with all of its subdivisions) was as follows:

1. Scattering of Radiation of Various hardness, up to hardest obtainable with Snook outfit. Use Snook machine & W[tungsten]-target X-ray tube, & suitable filters (Fe & Sn) to cut out soft rays. Set up something like this: [see Figure 9]. Chamber I should be on a crude spectrometer arm.

I. Study scattering at diff[erent] ∠s [angles] for diff[erent] λ's [wavelengths]. Thereby test relation:

(1) $I = f\left(\frac{\sin\ \theta/2}{\lambda}\right)$ demanded by usual e.m. theory.

(2) Also determine form of function *f*. If relation (1) is verified for v[ery] hard X-rays at small ∠s [angles], where excess scatt[ering] occurs, it will show that the diff[erent] electrons in t[he] atom cooperate, & ∴ [therefore] that all elect[ron]s are effective, & that the quantum relat[ion] does not enter. This will show ∴ [therefore] whether reduced scattering for v[ery] hard rays at *large* ∠s [angles] can be accounted for by 'constraints' on the electrons, or by interference.

1

Problems to be Tackled
At Saint Louis.

R. M. S. "AQUITANIA".

1. Scattering of Radiation of Various hardness, up to hardest obtainable with Snook outfit.

Use Snook machine & W-target X-ray tube & suitable filters (Fe & Sn) to cut out soft rays. Set up something like this:

FIG. 9. The first page of Compton's proposed St. Louis research problems written aboard ship en route to the United States from England.

The form of funct[ion] *f* will determine the size & shape [of the] e[lectron] if relation (1) holds.

II. Study relative intensity [of] the *scattered & fluorescent* radiat[ion].[82]

First and foremost, therefore, we see that Compton decided to turn from γ-ray to X-ray experiments. This was unquestionably the most far-reaching change in his experimental program since its inception. Specifically, he now decided, in a natural extension of his γ-ray work, to study the intensity of the scattered radiation as a function of scattering angle and wavelength. He reasoned that if the "excess scattering" relationship (1) held for very hard X-rays, it meant that the incident radiation was interacting with a number of different electrons, and therefore that the quantum relation "does not enter."

This is the first direct evidence that Compton was not only concerned with the "quantum relation" at this time, but that he also saw it as necessarily involving an interaction with a *single* electron, not with a group of electrons. As we will see, *Compton felt that this was a serious obstacle to its acceptance,* which future events brought to the forefront of his mind. For the moment, however, it appears that Compton expected to find that the quantum relation would *not* enter, since he dwelt at some length on this possibility, pointing out that the greatly reduced scattering at large angles for very hard rays was then to be explained either by (large electron) interference effects, or by discarding another of Thomson's assumptions, that there are no "constraints" on the electrons. Finally (Part II of problem 1), Compton, again taking a cue from his γ-ray experiments, decided to study the relative intensity of the "truly scattered" and "fluorescent" X-radiation—provided of course that such "fluorescent" X-radiation actually existed.

The second problem Compton resolved to tackle at Saint Louis, either subsequently to the one above or simultaneously with it, was a study of the "scattering of monochromatic X-rays by [the] photographic [detection] method."[83] To obtain monochromatic X-rays, Compton could use the Bragg spectrometer which he had brought along with him from England.[84] Here of course lay a primary advantage of turning from γ-ray to X-ray experiments: a precision instrument, the Bragg spectrometer, became immediately available for use. We have already seen that Compton had developed an improved version of this instrument at Princeton in 1916,[85] so that he was intimately familiar with its operation and applications.

Problem number five, following the two that are missing, dealt with the study of secondary cathode rays liberated in a Wilson cloud chamber by X-rays. In part A Compton proposed to study their trajectories in a magnetic field, while in parts B and C he proposed to study the following:

B. Length of paths for diff[erent] λ's [wavelengths].

a. See if *all* paths have same length for cath[ode] rays excited by monochrom[atic] radiat[ion] (test of nature of $h\nu$ relation)

b. See how range varies with initial velocity (assuming $h\nu$ = eV relation to hold)

C. Orientation of initial velocity of beta particle with respect to electric vector of polarized X-ray. & variation with [wavelength] λ.[86]

Once again we see that Compton had the "quantum relation" and its observable consequences on his mind, in this case in connection with the implications of the "Duane-Hunt law" $h\nu$ = eV. And once again we sense a concern with the possibility of an interaction with a single β-particle. This concern was made explicit in part 2 of his sixth and last problem, in which he proposed to study the "Manner of emission of photoelectrons, whether singly or in groups." It was his intention to "Try also for caesium (cooled in CO_2) with ultraviolet light to see if when $h\nu > 2h\nu_0$ (where ν_0 is the minimum emission frequency) more than one photoelectron is emitted at once."[87]

3. *Monochromatic X-Rays Also Excite Compton's New Fluorescent Radiation*

Since creative research is a dynamic process, it is not surprising that Compton, after becoming settled at Washington University, did not follow out his proposed experimental program in detail. Right at the outset, in fact, he modified his experimental technique in a very important respect. He apparently never used the direct beam from an X-ray tube as his primary source of X-rays, as shown in Fig. 9, but evidently concluded that it would be far better to use monochromatic, crystal-reflected X-rays.

This point bears emphasis; initially, Compton did not use his Bragg spectrometer as a *spectrometer,* that is, he analyzed no X-ray spectra with it. Rather, he first used it as a "wavelength selector"—as a means of producing a beam of monochromatic X-rays. The technique for doing this may be easily understood from the well-known Bragg equation, $n\lambda = 2d\sin\phi$, where n is the order, d is the distance between reflecting planes, and ϕ is the angle the incident beam makes with the reflecting planes. For given n and d, fixing the angle ϕ at some angle ϕ_0 selects a particular wavelength $\lambda = \lambda_0$ from the (inhomogeneous) beam incident on the crystal. Therefore, we see that the basic advantage that Compton initially derived by changing from γ-rays to X-rays was the assurance of a monochromatic primary beam.

Compton's first X-ray experiments at Washington University were carried out in early 1921 and reported in April of that year. In principle, they were a direct extension of his Cavendish γ-ray experiments: he was still deeply perplexed by Florance's and Ishino's γ-ray experiments, which "showed a considerable difference in hardness between the primary and the secondary radiation, an effect difficult to explain if the secondary radiation was truly scattered."[88] But Compton felt that Florance and Ishino, since they had been unaware of the existence of Compton's longer-wavelength "fluorescent" radiation, had mistakenly concluded that the secondary beam, especially for scattering angles greater than 90°, contained large amounts of soft "truly scattered" γ-radiation. The question that now concerned Compton—as it already had aboard ship (Problem 1. II)—was whether or not monochromatic *X-rays* could excite a similar soft "fluorescent" *X-radiation,* which, especially at the larger angles, predominated in the secondary beam.

To answer this question experimentally, Compton placed absorbers of increasing thickness first in the primary X-ray beam and then in the secondary beam—the same technique which he had used at the Cavendish. Using his Bragg spectrometer as outlined above, he selected two different primary wavelengths, 0.12 Å and 0.50 Å, and measured the relative amounts of "truly scattered" and "fluorescent" X-radiation each produced in the secondary beam at two different scattering angles, 90° and 30°. He found "an effect identical in character with that observed with gamma rays, but not so prominent." Specifically, he found that for "$\lambda = 0.12$ [Å] . . . the scattered rays form about 15 per cent. at 90° with the primary beam and 70 per cent. of the total radiation at 30°. For $\lambda = 0.50$ [Å] the scattering is relatively much more important, but the fluorescent radiation is still a large part of the total radiation."[89] In other words, he found that for the short-wavelength radiation (0.12 Å), the amount of longer-wavelength "fluorescent" radiation in the secondary beam increased from 30% to 85% of the total in going from 30° to 90°. For the longer-wavelength radiation (0.50 Å), he found a similar increase with increasing scattering angle, but the actual amount of "fluorescent" radiation at each angle was less.

Several conclusions followed. In the first place, a general criticism could be leveled at the results of all previous X-ray and γ-ray scattering experiments, for example those of Barkla, Crowther, and Owen, because none of these experimentalists had recognized that not all of the secondary radiation "represents truly scattered rays." The error was less severe at small scattering angles, but at large scattering angles the "truly scattered" radiation was "much less" than that predicted by Thomson's theory, "especially for the shorter wave-lengths, being less than 1/1,000th of the theoretical value in the case of gamma rays."[90] This great reduction, Compton noted, could of course be satisfactorily explained by "the hypothesis that the electron is of sufficient size (about 3×10^{-10} cm.) to give rise to interference effects. . . ." He specified neither the shape of the electron, nor any reason for changing his estimate of its radius.

Compton's most significant conclusion by far, however, was that he had now also discovered a new type of X-ray "fluorescent" radiation which was entirely analogous to the new γ-ray "fluorescent" radiation he had discovered at the Cavendish. "These results," he wrote, "both in the case of the gamma rays and of the x-rays, seem to be independent of the material used as radiator. . . ." Therefore, they "indicate the existence of a type of fluorescent high frequency radiation whose wave-length is independent of the particular substance used as radiator, depending only on the frequency of the exciting primary rays."[91]

Because of its close analogy to the γ-ray fluorescent radiation, Compton's first thought was to interpret the X-ray fluorescent radiation as before, as the radiation "excited by the impact of the secondary cathode rays liberated in the radiator by the action of the primary radiation." A difficulty arose, however: "While such an explanation accounts for many of the characteristics of this radiation, quantitative measurements showed that at least in some cases as much as 50 per cent. of the energy of the primary rays appeared as general fluorescent radiation. Since the efficiency of production of x-rays by cathode rays is less than 1 per cent., it does not seem that this can be the real origin of the secondary rays." This difficulty prompted Compton, who was assisted by Charles F. Hagenow, to plan further experiments: "It occurred to us that some information on this point might be obtained from a careful measurement of the polarization of the fluorescent radiation."[92] The results of these experiments were sent to *The Physical Review* in April, 1921, along with those discussed above.

Compton and Hagenow's experiments were modeled after Barkla's classic 1906 double-scattering experiments, except that Compton and Hagenow "employed much more sensitive methods" of detecting the radiation emerging from the second scatterer, labeled R_2 in Fig. 10.[93] "The first radiator [R_1] was placed about 12.5 cm. directly above the second radiator [R_2], which was in turn placed on the crystal table of a Bragg spectrometer. The degree of polarization could then be measured by noting the scattering from the second radiator parallel and at right angles with the primary beam." By introducing suitable absorbers into the primary beam, Compton and Hagenow insured that "under the conditions of our experiment approximately 70 per cent. of the total secondary radiation at 90° . . . [was] of the general fluorescent type. . . ."[94] They then compared the intensities recorded by the ionization chambers I_2 and I_1. They found that the ratio I_2/I_1 was 0.05 for paper and sulfur scatterers, and 0.04 for an aluminum scatterer. When these results were corrected for the "imperfection of the geometrical conditions," it meant that the polarization of the secondary beam was "complete within experimental error." Since the "scattered" radiation by itself *had* to be

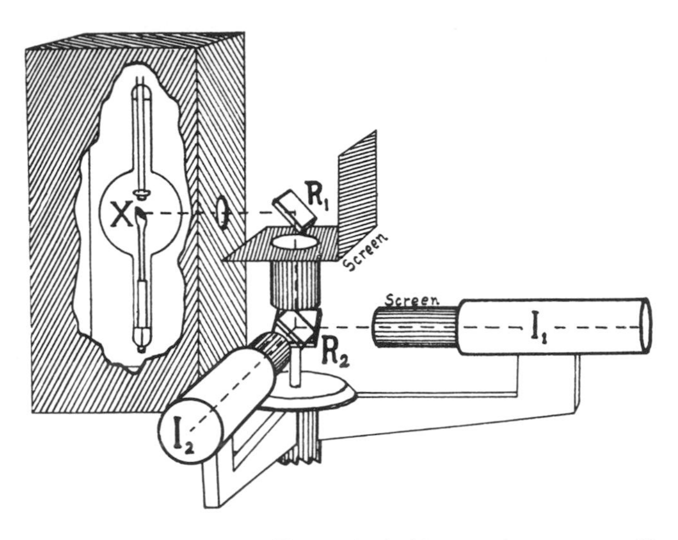

FIG. 10. Diagram of Compton and Hagenow's double-scattering apparatus. The scatterers are labeled R_1 and R_2; I_1 and I_2 are ionization chambers.

completely polarized, the unavoidable conclusion was "that the fluorescent as well as the scattered rays are completely polarized."[95]

This was a startling conclusion. All X-ray characteristic fluorescent radiations were known to be completely *un*polarized. Only *scattered* radiation was known to be polarizable, the degree of polarization depending upon the experimental conditions. Compton and Hagenow's observation therefore represented one more fact to be added to the growing list of facts that set the "general fluorescent radiation" completely apart from all other known fluorescent radiations. Instead of clarifying the "efficiency difficulty" discussed above, Compton and Hagenow had uncovered a further, "polarization difficulty."

This difficulty was particularly acute, because it was already known since Barkla's experiments that when cathode rays are rapidly stopped in a metal plate, they produce *partially* polarized, not *completely* polarized, X-rays. Hence, it was "difficult to support the view" that Compton's "fluorescent" X-rays were emitted by rapidly moving secondary β-particles—the interpretation he had offered for the origin of his "fluorescent" γ-rays. At the very least, this interpretation required modification. Eventually, Compton offered the following suggestion: "The hypothesis that this [fluorescent] radiation is emitted at the instant of liberation of the secondary cathode rays from the atoms . . . seems to offer a possible explanation of both the degree of polarization and the efficiency of production of these fluorescent rays."[96]

Compton apparently reasoned as follows: Concerning the degree of polarization, any departure from complete polarization could be attributed to the effects of multiple scattering of the secondary electron. Hence, by insisting that the "general fluorescent radiation" was emitted "at the instant of liberation" of the electron, these effects should be reduced to zero, and the emitted radiation should be completely polarized. Concerning the efficiency of production, the low efficiency of X-ray tubes could be attributed to the electron's kinetic energy being transformed into heat losses as the electron moved within the struck plate. Hence, by insisting that the "general fluorescent radiation" was emitted "at the instant of liberation" of the electron, these heat losses should be much reduced, and much higher efficiencies might be expected.

In sum, with these modifications, Compton's secondary β-ray hypothesis survived as a viable explanation of the origin of the unusual "fluorescent" radiation—for X-rays as well as for γ-rays. It is noteworthy, however, that Compton's attention had now been focused on "the instant of liberation" of the secondary β-rays.

4. A Brief Theoretical Excursion

In mid-1921 Compton reported further work, this time theoretical in nature.[97] While this work—an analysis of X-ray spectral line broadening—appears to be quite remote from Compton's current research interests, and in fact represents a contribution to the Bohr-Sommerfeld atomic theory,[98] it also had clear implications for his studies on scattering and absorption. Compton proved,[99] in effect, that the inherent inhomogeneity of characteristic X-ray spectral lines could not account for the production of his longer-wavelength secondary radiation, should such characteristic lines be used as the primary radiation—as they had been, for example, in Sadler and Mesham's 1912 experiments. Compton found a maximum relative inhomogeneity $d\lambda/\lambda$ of less than 1% for 1.2 Å X-rays, which (since $d\mu/\mu = 3d\lambda/\lambda$) corresponds to an observed relative change in absorption coefficient of less than 3%—less than that typically observed. Never, in fact, would Compton in the future suggest that the inherent inhomogeneity of the primary radiation could account for the longer-wavelength secondary radiation.

F. Compton's *Experimentum Crucis* (1921)

1. Gray's Post-War X-Ray Experiments Raise the Issue

Arthur Compton spent the summer of 1921 as a Visiting Lecturer at the University of California at Berkeley.[100] In the fall, almost immediately on his return to Washington University, he resumed his researches. Although they were closely related to his earlier work, their precise nature was determined by a more powerful consideration. They were carried out in direct response to a serious challenge which suddenly confronted his "fluorescent radiation" interpretaton of the softened secondary X-radiation. In order to understand that challenge, and Compton's response to it, it is necessary to first examine certain experiments performed after the war in 1920 by J. A. Gray of McGill University.

Gray's post-war experiments[101] grew naturally out of his remarkable pre-war γ-ray work, which we discussed in Chapter 1. He already knew, both from his own work and from that of Florance, that the scattered rays were of "quite a different type to that of the primary rays, the quality of the rays depending . . . on the angle of scattering." What troubled Gray now was that "most writers" disagreed with his and Florance's final interpretation of these results, holding with Kaye that the scattered rays "are identical with the primary rays in quality, and can in fact be conveniently regarded as so many unchanged primary rays, which have been merely scattered or deviated by the substance." This view led naturally to the "usually accepted explanation" of the change in quality, namely, that since "the primary rays are heterogeneous, [the] softer rays are scattered relatively more strongly at larger angles than [the] harder rays." Gray (and finally Florance as well) disagreed with this "selective scattering" interpretation. Hence, it seemed to Gray to be "of great theoretical importance" to begin by proving conclusively, if possible, that there actually "is a change in the scattered rays, depending on the angle of scattering. . . ."[102]

First, by essentially repeating some of his earlier experiments, Gray measured the angular variation in the quality (absorption coefficient) of secondary γ-rays scattered by several different substances. He proved by explicit calculations that the results of these experiments agreed with his earlier ones. He summarized: "This variation of quality with the angle of scattering, and the fact that it is approximately independent of the nature of the radiator, show that the effect we have observed is not due to the formation of characteristic radiations in the radiator. We conclude, therefore, that in some way an ordinary beam of γ-rays is changed in quality when scattered, the change depending on the angle of scattering."[103]

A fundamental question—the same question that Compton had asked—was whether or not the same phenomenon could be observed with X-rays. Referring to himself in the third person, Gray stated that he "had intended to use, as primary rays, X-rays reflected from a crystal, but it was not until after demobilization last year, that he was able, through the kindness of Sir William Bragg and the British Scientific and Industrial Research Council, to carry out further experiments at the University College, London. In the time at his disposal he was not able to obtain a reflected beam of sufficient intensity to work with satisfactorily. However, one that was approximately homogeneous . . . was obtained by filtering X-rays through a sheet of tin."[104]

To compare the quality of the primary rays with that of the secondary rays, Gray argued, one could compare their spectra, but that was "often impracticable," and "such a comparison has invariably been made by absorption measurements." Gray's present experiments were no exception. He reported his results in the following unambiguous language: "The scattered rays were again found to be more absorbable than the primary rays. For example, . . . rays scattered through 35° have an absorption coefficient in aluminum 6 per cent. greater than that of the primary rays; rays scattered through 70°, an absorption coefficient 12 per cent. greater; and rays scattered through 110°, one 18 per cent. greater. The increase is thus approximately proportional to the angle of scattering. It also appeared to be independent of the thickness of aluminum passed through." In sum, these "experiments . . . verify the previous results obtained with γ-rays, although, in the latter case, the comparative softening of the scattered rays is much more marked."[105]

Gray therefore had reached precisely the same general conclusion that Compton had reached—and at least five months before Compton. The similarity between Compton's and Gray's work extended to their general experimental method, except for the fact that Compton had scattered crystal-reflected X-rays, while Gray had scattered the characteristic lines of tin. But that was where the similarity ended. Compton and Gray differed, once again, in the most fundamental way in the interpretation of their observations: Compton was always careful to distinguish between the "truly scattered" and "fluorescent" secondary radiations, while Gray never made such a distinction. Gray always referred to the secondary beam as the "scattered" beam, thereby expressing his conviction—which he had held already in 1913—that the change in quality was produced in the scattering process. In 1913, however, Gray had postulated no mechanism to produce this change in quality. In 1920, he formulated such a mechanism.

Gray introduced his ideas by considering the phenomenon of "excess scattering," choosing as a particularly simple model[106] the interference of two rays scattered by two electrons, e_1 and e_2, separated by a distance $2p$, which he assumed to be small compared to the wavelength λ of the incident radiation (see Fig. 11). From the sketch it is easy to see that the phase difference δ between the two scattered rays is $\delta = (2\pi/\lambda)[4p\sin(\theta/2)]$. Therefore, the intensity of the scattered radiation—the square of the vector sum of the individual amplitudes—is proportional to $(1 + \cos\delta) = \{1 + \cos[(8\pi p/\lambda)\cdot$

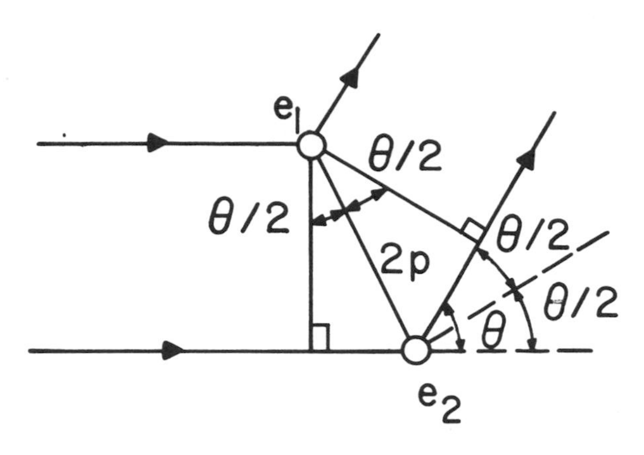

FIG. 11. Author's sketch of Gray's "excess scattering" model. The two electrons, e_1 and e_2, are separated by a distance $2p << \lambda$.

$\sin(\theta/2)]\}$. Gray assumed that this expression modulated the Thomson $(1 + \cos^2\theta)$ factor, and he demonstrated that in this way he could account at least qualitatively for several of the well-known features of "excess scattering." Of most concern to him, however, was the change in quality of the radiation. His explanation for it was based on the above model, but, he argued, before it could be understood, it was necessary to understand certain general aspects of the scattering process.

In the first place, it was crucial to carefully distinguish between "ordinary scattering"—the scattering of X-rays coming directly from an X-ray tube —and the scattering of monochromatic, crystal-reflected X-rays. While the former X-rays might be changed in wavelength during the scattering process, Gray took "an experiment of Moseley's and Darwin's,"[107] as well as "further experiments at University College, London," to prove "that a wave train of definite wave-length is *not* altered in wave-length by scattering."[108] This "fact" determined Gray's interpretation of the change in quality. He emphatically denied that the production of secondary β-rays (Compton's interpretation) had anything to do with it.[109] Of course, if "rays of a definite frequency were altered in wave-length" during the process of scattering, the change in quality could be readily explained. But, Gray asserted, "we may not always be justified in looking on a beam of X-rays in this way. . . . We have very little idea of the actual mechanism by which X-rays are produced from β-rays, but it seems probable that X-rays consist rather of a series of thin pulses than of regular waves occurring in trains of great length. . . ."

Since typical currents in X-ray tubes were equivalent to about 10^{16} electrons per second, and since typical efficiencies were roughly 1 percent, about 10^{14} pulses per second left an X-ray tube with the speed of light (3×10^{10} cm/sec), implying that they were spatially separated by about 3×10^{-4} cm.

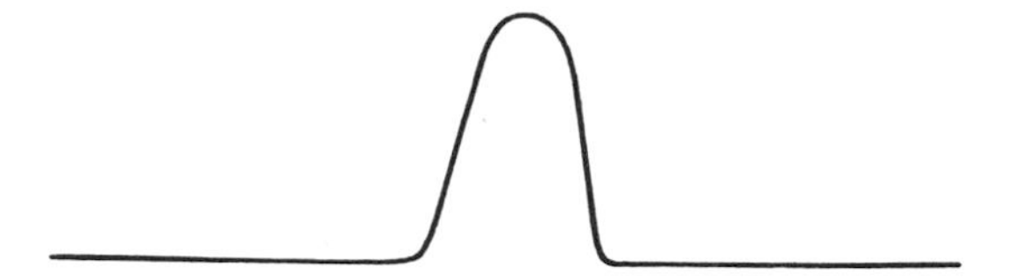

FIG. 12. Gray's illustration of an X-ray pulse.

This distance was to be compared with a typical atomic dimension, 10^{-8} cm, which Gray took to be on the order of magnitude of the pulse thickness. "The average distance between pulses is, therefore, much greater than the average thickness of a pulse." We do not know the "actual form of a pulse. . . , but for purposes of illustration, we can imagine it to be of the shape indicated by . . . [Fig. 12], in which, at some definite time, the electric displacement in the pulse is supposed plotted against the distance from its origin, the pulse traveling on unchanged in form when not passing through matter."[110]

Therefore, Gray resurrected the pulse theory of X-rays in 1920. Moreover, he argued that this theory could be combined with the two-electron model discussed above to account for the change in quality. For imagine that a single incident pulse strikes each electron, after which they combine by interference. If the two are exactly in step, "the thickness of the combined pulses will be the same as before, but its intensity will be four times that of a single pulse just as in the case when two waves of the same length and in the same phase reinforce each other. If the two pulses are slightly out of step, the resultant pulse will be slightly thicker and less penetrating and the resultant intensity will be less than four times that due to a single pulse." Clearly, "the thickness of the scattered pulse will depend on [the scattering angle] θ and will be larger the greater the angle of scattering."[111] While the situation with γ-rays was not so clear, theoretically "it appears possible to explain the results of experiment on the hypothesis that X-rays consist of a series of pulses. . . . When the primary rays are rays of definite wave-length reflected from a crystal, the scattered rays should be of the same wave-length."[112]

2. *Plimpton's Experiments Support Gray*

Everything indicates that if Arthur Compton actually knew of Gray's pulse theoretical interpretation of the softening when it appeared in November 1920, he placed no stock in it whatsoever. If Compton actually rejected it, he had good reason for doing so. He knew that Laue's discovery of the crystal diffraction of X-rays was opposed to it; and he knew that D. L. Webster had proven that the pulse theory was inconsistent with the sharp frequency cutoff associated with the Duane-Hunt law.[113] In fact, almost a year passed before

Compton suddenly paid a great deal of attention to Gray's theory. The reason for Compton's sharp reversal in attitude may be directly traced to the appearance, in September, 1921, of a paper entitled "On the Scattering of Rays in X-ray Diffraction"[114] by S. J. Plimpton, who at the moment was working in the laboratory of William H. Bragg at the University of London.

Plimpton may or may not have been familiar with any of Compton's or Gray's work—there are no references in his three-page paper—but he believed, as Gray had, that if the rays incident on a scatterer "could be made strictly monochromatic, there should be no difference in the measured absorption when the [absorbing] screen is changed from the incident to the scattered beam." Only if the primary rays were inhomogeneous could the secondary beam appear softer, provided that the "longer waves . . . [were] more efficiently scattered." Contrary to what Gray believed, however, Plimpton seemed to think that there was evidence that even monochromatic primary rays underwent a change in quality, but he regarded this effect as probably spurious. It "would seem to suggest," he wrote, "that possibly the mechanism of scattering is not so simple as supposed but involves a more indirect process such as absorption and subsequent re-emission of energy. Because of its very great theoretical importance it has seemed worth while to test this matter with considerable care." He continued: "The best known method for obtaining monochromatic rays is by crystal reflexion, but . . . it has hitherto been thought impracticable to attempt accurate measurements of the scattered . . . radiation produced by these weak reflected beams."[115]

In spite of the dim prospects for success, Plimpton set up an apparatus (Fig. 13[116]) which actually utilized the beam emerging from a reflecting crystal of mica "bent into an equiangular spiral" (2 in the diagram). The scatterer (3) was "usually paraffin or water." The ionization chamber (4) was "filled with ethyl bromide and connected with a Wilson filled electrometer."[117] To compare the qualities of the primary and secondary X-rays, an aluminum absorbing screen was alternately placed in positions 10 and 11, before and after the scatterer. After a number of measurements, Plimpton concluded: "the ionization produced by the scattered rays did not appear to be measurably affected by changing the absorbing screen from position (10) to position (11). The readings were reproducible to within about one per cent."[118] Plimpton had found, therefore, what he had expected to find.

3. *Compton's Response and First Calculations of a Change in Wavelength*

Plimpton's conclusion evoked an immediate and sharp response from Compton. The reason for Compton's agitation was given in a short note which he

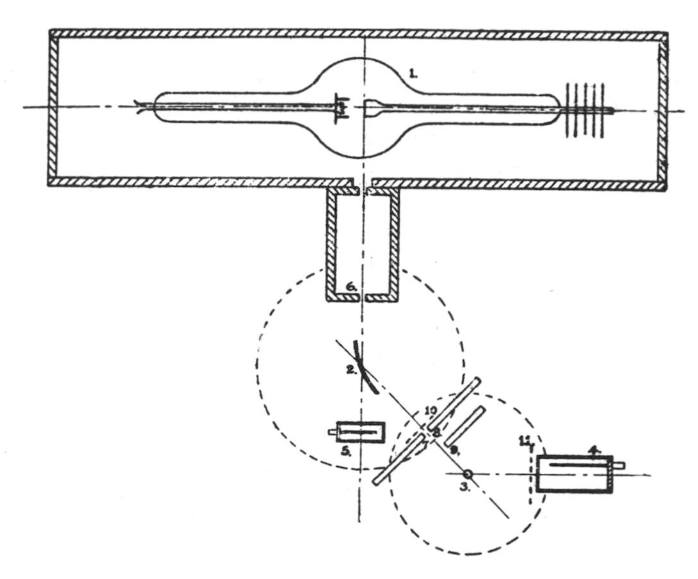

FIG. 13. Plimpton's X-ray scattering apparatus.

later published in *Nature*.[119] He outlined the problem as follows: "A number of experimenters have noticed that when a beam of X-rays or γ-rays traverses any substance, the secondary rays excited are less penetrating than the primary rays. Prof. J. A. Gray . . . and the present writer . . . have shown that the greater part of this softening is not due, as was at first supposed, to a greater scattering of the softer components of the primary beam, but rather to a real change in the character of the radiation." There was, however, a clear difference of opinion regarding the interpretation of these facts: "My conclusion was that this transformation consisted in the excitation of some fluorescent rays of wave-length slightly greater than that of the primary rays. Prof. Gray, on the other hand, showed that if the primary rays came in thin pulses, as suggested by Stokes's theory of X-rays, and if these rays are scattered by atoms or electrons of dimensions comparable with the thickness of the pulse, the thickness of the scattered pulse will be greater than that of the incident pulse. He accordingly suggests that the observed softening of the secondary rays may be due to the process of scattering."[120]

This difference of opinion suggested an experiment to Compton, but before its principle could be understood it was absolutely essential to understand, Compton wrote, *"that if the X-rays are made to come in long trains, as by reflection from a crystal, the scattering process can effect no change in wave-length* [italics added]." The principle of the experiment could now be described: "On Gray's view, therefore, if X-rays reflected from a crystal are allowed to traverse a radiator, the incident and the excited rays should both have the same wave-length and the same absorption coefficient. If, on the other hand, the softening is due to the excitation of fluorescent rays, as I had suggested, reflected X-rays should presumably be softened by scattering in the same manner as unreflected rays. An examination of the absorption coefficient of reflected X-rays before and after they have been scattered should therefore afford a crucial test of the two hypotheses."[121]

This therefore was no ordinary experiment. It was a "crucial test," an *experimentum crucis,* in Newton's venerable terminology. The blunt truth of the matter was, however, that the experiment had already been carried out! Plimpton's experiment was precisely the experiment Compton had in mind, and Plimpton's results were totally disastrous to Compton's "fluorescent" radiation hypothesis. Judging from Compton's laboratory notebooks, which cover this period of time, as soon as Compton had read Plimpton's paper he put everything aside and reacted to the challenge. His first entry, as if time were unimportant, is dated "About Oct. 10, 1921."[122] Let us follow Compton's response in detail.

He decided, in the first place, to use as primary X-rays one of the characteristic K-lines of tungsten after its reflection from a rock salt (NaCl) crystal. He had already constructed an X-ray tube with a tungsten target. (Compton was an expert glass blower and consistently manufactured most of his own apparatus.[123]) He therefore looked up the wavelengths of the tungsten K-lines and K-absorption edge,[124] and for each K-line computed the angle of reflection from the NaCl crystal. He then set up his spectrometer and began observing these reflected rays directly—before he was forced to interrupt his work for two days, until October 12.

When he resumed his experiments, he immediately set out to answer the question foremost in his mind: Was Plimpton's conclusion correct? In the reproduction in Fig. 14[125] we see (1) Compton's apparatus—note that the X-rays traveled from right to left, and that the crystal was used as a "wavelength selector"; (2) his experimental data; and (3) his answer to that question: "Secondary rays softer than Primary." Therefore, Compton proved to his own satisfaction that Plimpton's experiments must somewhere contain an error, and at the same time he reassured himself that his own experiments carried out the preceding spring must have been substantially correct.

Compton next began a calculation—the first steps of which are on the bottom of the page shown in Fig. 14—to determine whether or not *crystal-reflected* X-rays undergo the same change in quality as *unreflected* X-rays when passing through an absorber. In Compton's notation (Fig. 14), I_A and I_B are the intensities recorded with the absorber (a copper screen 1 mm thick) in position A or B, i.e., in the primary or secondary beams (R de-

Oct. 12, 1921. 3

Intensity of Scattered Rays reflected from crystals

B R A X

C = Na Cl

R = Graphite

Abs. screen = 1 mm. Cu.

θ = c.83°

Position A.	B. Count	B. Count	Position B
29.0	6.0	4.5	16.5
28.3	4.8	3.0	13.0
27.0	3.8	2.9	15.0
26.0	2.6	1.6	14.8
110.3	17.2	12.0	59.3
17.2			12.0
93.1			47.3

$\frac{I_B}{I_A} = \frac{47.3}{93.1} = 0.51$

i.e. ~~Scattered~~ Secondary rays softer than Primary.

$\Delta\mu = \frac{1}{x} \log_e \frac{I_A}{I_B} = 6.74$; $\frac{\Delta\mu}{\rho} = .753$

8.8

15.5

FIG. 14. Compton's apparatus, initial data, and first steps of a calculation used to check Plimpton's experimental results.

notes the radiator or scatterer—note that the scattering angle is "c.83°"). From the absorption equations $I_A = I_0 e^{-\mu_A x}$ and $I_B = I_0 e^{-\mu_B x}$, it is easy to see that

$$\Delta\mu = \mu_B - \mu_A = \frac{1}{x} \log_e \frac{I_A}{I_B} \tag{4.12}$$

the equation at the bottom of the page in Fig. 14. Dividing $\Delta\mu$ by the density of copper ($\rho = 8.9 \text{g/cm}^3$), and substituting the correct values for the other quantities, one obtains Compton's result, namely, that $\Delta\mu/\rho = 0.753 \text{ cm}^2/\text{g}$.

But μ_A (and hence μ_A/ρ) may be determined separately from intensity measurements with copper absorbers of various thicknesses at position A. Compton made these measurements, finding $\mu_A/\rho = 0.985$ cm^2/g. Hence, by simply adding, Compton also found $\mu_B/\rho = (0.985 + 0.753)$ cm^2/g $= 1.738$ cm^2/g. In other words, he found that the ratio μ_B/μ_A was $1.738/.985 = 1.762$. Then, by apparently locating the values of μ_B/ρ and μ_A/ρ on a graph of μ/ρ *vs* wavelength λ, Compton determined the corresponding wavelengths $\lambda_B = 0.222$ Å and $\lambda_A = 0.178$ Å. He therefore could conclude, finally, that "$\lambda_B/\lambda_A = 1.25$; $\lambda_B - \lambda_A = .044$ [Å]."[126]

Therefore, as far as we are able to tell from existing materials, October 12, 1921, was the date on which Arthur Compton first explicitly computed from experimental data a change in wavelength of the proper order of magnitude between primary and secondary X-rays.

But Compton's *motive* for making this calculation was relatively prosaic: *he simply wanted to determine whether or not crystal-reflected X-rays and unreflected X-rays undergo the same change in wavelength when passing through an absorber.* His subsequent calculations leave no doubt on this point. Immediately below his conclusion that $\lambda_B - \lambda_A = 0.044$ Å, Compton listed three data points (λ, μ/ρ) which he previously determined for "unreflected rays," as follows: (0.218, 1.64), (0.153, 0.75), and (0.119, 0.424). These three points define an "*un*reflected ray" plot of μ/ρ *vs* λ, and Compton saw at a glance (as may be easily verified) that by striking off the "reflected ray" interval (0.178 − 0.222) Å on it, he would obtain an "*un*reflected ray" interval for $\Delta\mu/\rho$ very close to 0.753 cm^2/g, the calculated "reflected ray" value. The conclusion was obvious. Unreflected and reflected X-rays, for the same scattering angle, undergo the same change in quality when passing through matter.

By now Compton was completely satisfied that he was on the right track. On October 13 he listed the following "Proposed expts. 1. Repeat at some ∠ [angle] with greater care, using Paraffine [*sic*]. 2. Perform Plimpton's expt, using Mo target."[127] Actually, Compton found time to repeat his experiment not at one, but at two scattering angles (30° and 90°) using a paraffin scatterer (recall that Plimpton had used paraffin). Finally, precisely as outlined above, and for precisely the same reasons, Compton calculated the ratio μ_B/μ_A, and the corresponding difference in wavelength $\Delta\lambda = \lambda_B - \lambda_A$ between the primary and secondary X-rays at both scattering angles. His results[128] are summarized in Table 4.

Continuing his work on October 15, Compton changed from a tungsten to a molybdenum target, looked up the wavelengths of molybdenum's K-lines and K-absorption edge, and calculated the angles of reflection of the K-lines from rock salt and calcite crystals. Scattering these rays from paraffin

Table 4. Summary of Compton's results at two scattering angles (θ)

θ	$(\mu_B/\mu_A)_{Cu}$	$\Delta\lambda$	=	λ_B	—	λ_A
30°	1.225	0.017 Å	=	0.203 Å	—	0.186 Å
90°	1.520	0.034 Å	=	0.220 Å	—	0.186 Å

and proceeding exactly as before, Compton obtained the results[129] summarized in Table 5.

Table 5. Summary of Compton's results for μ_B/μ_A (paraffin scatterer, aluminum absorber) at two scattering angles

θ	$(\mu_B/\mu_A)_{Al}$
20°	1.06
90°	1.29

Note that this time Compton computed no changes in wavelength corresponding to each scattering angle. He did not have to. He had already demonstrated to his own satisfaction that reflected and unreflected rays undergo the same change in wavelength when passing through matter.

When Compton published the results of his experiments in *Nature* in mid-November of 1921, one of the first things he did was to pinpoint the sources of Plimpton's experimental inaccuracies: "Apparently his measurements were made on the secondary rays at comparatively small angles, and this, together with the relatively long wave-lengths employed, form the conditions under which the least change in hardness occurs. . . ." Compton then outlined his own experiments, concluding: "The softening thus observed when reflected X-rays are scattered is substantially the same as that found when unreflected rays of the same hardness are employed. . . . The conclusion seems necessary, therefore, that *the softening of secondary X-rays is due, not to the process of scattering, but to the excitation of a fluorescent radiation in the radiator* [italics added]."[130]

4. *Conclusions*

We see that Gray had argued that only if X-rays were pulses could they give rise to a longer-wavelength secondary radiation: monochromatic X-rays could undergo no change in wavelength in the scattering process. Compton wholeheartedly *agreed* with Gray that monochromatic X-rays could undergo no change in wavelength in the *scattering* process, but at the same time he believed that they necessarily liberated secondary β-rays in the scatterer, which, at the instant of liberation, emitted the unusual, longer-wavelength

(Doppler-shifted) "fluorescent" radiation. His present experiments—which contradicted Plimpton's—proved that even monochromatic X-rays can excite the longer-wavelength secondary radiation. This conclusively disproved Gray's pulse interpretation. However, Compton fell into the trap of believing that the *only possible alternative* to Gray's pulse interpretation of the softened radiation was his own "fluorescent radiation" interpretation. He therefore concluded that his "crucial test"—his *experimentum crucis*—had decided in favor of his own "fluorescent radiation" interpretation of the softening. Monochromatic radiation could not be, and was not, changed in wavelength in the scattering process. Or so Compton still believed in late 1921.

G. References

[1] Marjorie Johnston (ed.), *The Cosmos of Arthur Holly Compton* (New York: Knopf, 1967), p. 26.

[2] *Ibid.*, p. 27.

[3] For some of Compton's lecture notes, see Compton Notebook A.1. (Wash. U., A.I.P.), pp. 76–79, 81, and 83; for books he purchased, see p. 103.

[4] Johnston (ed.), *Cosmos*, pp. 30 and 26; also see E. N. da C. Andrade, *Rutherford and the Nature of the Atom* (New York: Doubleday, 1964), pp. 156–213.

[5] Johnston (ed.), *Cosmos*, pp. 26 and 30.

[6] Compton Notebook A.1. (Wash. U., A.I.P.), p. 99.

[7] *Nature* 104 (1919):412. A full report of this work was later published under Compton's name alone. See "Radioactivity and the Gravitational Field," *Phil. Mag.* 39 (1920):659–662.

[8] Quoted in Johnston (ed.), *Cosmos*, p. 29.

[9] *Ibid.*, p. 34.

[10] The five most important of Compton's papers that are closely tied to his work at the Cavendish are: (1) "Possible Magnetic Polarity of Free Electrons," *Phil. Mag.* 41 (1921):279–281—hereafter cited as "Polarity"; (2) "The Absorption of γ-Rays by Magnetised Iron" *Phys. Rev.* 17 (1921):38–41—hereafter cited as "Absorption"; (3) "The Degradation of Gamma-Ray Energy," *Phil. Mag.* 41 (1912):749–769—hereafter cited as "Degradation"; (4) "The Wave-length of Hard Gamma Rays," *Phil. Mag.* 41 (1921):770–777—hereafter cited as "Wave-Length"; (5) "Classical Electrodynamics and the Dissipation of X-ray Energy," *Wash. U. Stud.* 8 (1921): 93–129—hereafter cited as "Electrodynamics." From internal evidence, it appears that these papers were governed by the following chain of events: *(a)* Compton wrote (3) first of all and submitted it to *The Philosophical Magazine; (b)* Just before leaving the Cavendish and England, he wrote the two short papers, (1) and (2), and submitted the former to *The Philosophical Magazine* (signed at the Cavendish on August 20, 1920); *(c)* Shortly after leaving the Cavendish and arriving at Washington University, St. Louis, he apparently received proofs of (3), which he revised somewhat and returned to *The Philosophical Magazine,*

signing it September 24, 1920, at St. Louis; (*d*) Two months later, he analyzed the data for (4), wrote the paper and submitted it to *The Philosophical Magazine,* signing it December 1, 1920, at St. Louis; *(e)* During the course of some of the work cited here he wrote (5), which was published in January, 1921.

[11] See section C.3.

[12] Compton, "Degradation," pp. 750–751.

[13] *Ibid.*, p. 749.

[14] *Ibid.*, pp. 751–752.

[15] *Ibid.*, p. 752.

[16] *Ibid.*, p. 753. I have changed Compton's notation I'/I *to* I_B/I_A throughout for greater clarity.

[17] *Ibid.*, p. 754.

[18] *Ibid.*, pp. 754, 757.

[19] *Ibid.*, p. 757.

[20] *Ibid.*, p. 758.

[21] *Ibid.*, p. 751; see also p. 759.

[22] *Ibid.*, p. 760.

[23] *Ibid.*, pp. 761–762.

[24] *Ibid.*, p. 762.

[25] "The Emission of Röntgen Rays from Thin Metallic Sheets," *Proc. Cambridge Phil. Soc.* 15 (1909):262–272.

[26] "Primary and Secondary Gamma Rays," *Phil. Mag.* 18 (1909):275–291, especially p. 289. Also see "The Secondary γ Rays due to the γ Rays of Radium C." *Phil. Mag.* 16 (1908):224–234.

[27] *Radioactive Substances and their Radiations* (Cambridge: Cambridge University Press, 1913), pp. 273–276, for example.

[28] For Stark's work, see Chapter 1, section D.4.

[29] "Origin of the General Radiation Spectrum of X-rays," *Phys. Rev.* 13 (1919):303–305.

[30] Compton, "Degradation," pp. 762, 764.

[31] *Ibid.*, p. 765; O. W. Richardson, *The Electron Theory of Matter,* 2d ed., (Cambridge: Cambridge University Press, 1916), p. 256. While this equation correctly reflects the forward peaking through its $1/(1 - \beta\cos\theta)$ factor raised to a high power, it does not seem to agree with that given in R. B. Leighton, *Principles of Modern Physics* (New York: McGraw-Hill, 1959), p. 412, where one finds that

$$R = \frac{I_1}{I_2} = \left(\frac{1 - \beta\cos\theta_2}{1 - \beta\cos\theta_1}\right)^5 \left(\frac{\sin^2\theta_1}{\sin^2\theta_2}\right).$$

[32] Compton, "Degradation," (reference 10) p. 766.

[33] *Ibid.*, pp. 765–766.

[34] "Die Streuung der γ-Strahlen," *Phys. Z.* 21 (1920):193–198. Kohlrausch carried out his experiments in an attempt to check the 1915 scattering theory of Peter Debye (see

Chapter 1, section C.2—and concluded by criticizing it for its limitations, as follows (pp. 197–198):

> [Even if it were possible to solve other difficulties,] there remains unexplained a phenomenon closely connected with the occurance of secondary or scattered waves, namely, the secondary excited β-radiation. In this respect experiment shows: That the atomic radiation coefficient of the secondary β-radiation increases with the atomic number of the radiator. . . ; that the 'incident' radiation is softer than the 'emergent,' that is hardness is independent of the nature of the radiator, and that the asymmetry increases with decreasing atomic number. . . . For the scattered γ-radiation . . . experiments show: That the atomic scattering coefficient . . . increases with atomic number, that the [total] 'incident' radiation is softer, that the hardness is independent of the material of the radiator, and that the asymmetry shows no distinct dependence on Z [i.e., the atomic number]. Thus, up to the last point we have on all counts qualitatively similar results. Even if the two series of observations do not allow us to set up quantitative relationships, they nevertheless seem to me to point the way toward which the theoretical treatment of the problem will have to move.
>
> In the first place, it has to be explained why electrons which are accelerated to the point of ejection by the transverse electric field of the γ-waves do not move only, or at least predominantly, perpendicular to the γ-direction [of propagation], but rather prefer to move in that direction.

Compton may or may not have pondered these insightful remarks.

[35] Compton, "Degradation," (reference 10) p. 768.

[36] *Ibid.;* Compton also noted (p. 763) that if one assumed that in heavier elements more secondary β-ray oscillators move in a direction opposite to the primary beam than in light elements, one could also qualitatively understand why heavier elements emit a larger amount of as well as more penetrating "fluorescent radiation" at large scattering angles than light elements.

[37] *Ibid.* See also Chapter 3, section D.7.

[38] Compton, "Absorption," p. 38.

[39] See Chapter 3, end of section D.1.

[40] Compton, "Absorption," p. 40.

[41] *Ibid.*, pp. 40–41.

[42] Compton, "Wavelength," (reference 10) pp. 770–777.

[43] See Chapter 3, section D.7.

[44] See Chapter 1, section C.2, and Debye's paper "Zerstreuung von Röntgenstrahlung," *Ann. Physik* 46 (1915):809–823; translated in *Collected Papers* (New York: Interscience, 1954); for relevant equation, see p. 45.

[45] "Note on the Scattering of X-rays and Atomic Structure," *Phil. Mag.* 31 (1916):222–232.

[46] Compton, "Wave-Length," p. 773. I have modified Compton's notation to $I(\theta,\lambda)/I_1N$ throughout for clarity.

[47] *Ibid.*, p. 775.

[48] *Ibid.*, p. 777.

[49] Compton, "Electrodynamics," (reference 10) p. 93.

[50] *Ibid.*, pp. 93–94.

[51] See Chapter 1, section C.2.

[52] "The Scattering of X- and γ-Rays by Rings of Electrons—A Crucial Test of the Electron Ring Theory of Atoms," *Proc. Roy. Soc. (London)* [A] 96 (1920):395–423. See also "On the Scattering of "X-Rays By Hydrogen," *Phys. Rev.* 23 (1924): pp. 9ff, and A. W. Conway, "Professor G. A. Schott," *Roy. Soc. Obit. Not. (London)* 2 (1936–1938):451–454. It is noteworthy that Schott also provided a list of Thomson's basic assumptions; however, Schott did not list the validity of "the usual electrodynamics," while Compton put it at the head of his list, and Schott did not cite Thomson's point charge assumption, while Compton knew very well the advantages gained by disregarding this assumption. In this connection it is also interesting to note that R. A. Houstoun also did not recognize the possible fruitfulness of giving up Thomson's point charge assumption. In his paper, "Note on the Scattering of X-rays," *Proc. Roy. Soc. (Edinburgh)* 40 (1919–1921):43–50 (received April 19, 1919; read June 2, 1919; revised October 14, 1919; issued separately March 25, 1920), we find the following statement (pp. 47–48): "We have now to consider the low value of the scattering coefficient for γ-rays. This is a great difficulty. . . . J. J. Thomson's formula offers an irreducible minimum. There appears to be no way of getting below it except by reducing the number of scattering electrons, *i.e.*, by making this less than the atomic number. So the question must be left open at present."

[53] Compton, "Electrodynamics," (reference 10) p. 99.

[54] Schott, "Scattering by Rings," (reference 52) p. 407.

[55] Compton promised to publish his derivation but never did. He seems, however, to have started from Debye's 1915 expression, in which he first rewrote the electron separation distance in terms of the polar coordinates of the individual electrons, and then averaged over all angles.

[56] Compton, "Electrodynamics," p. 100.

[57] Schott, "Scattering by Rings," (reference 52) p. 416.

[58] Compton, "Electrodynamics," p. 103.

[59] "The Magnetic Electron," *J. Franklin Inst.* 192 (1921):147–148.

[60] "The Elementary Particle of Positive Electricity," *Nature* 106 (1921):828.

[61] Compton, "Electrodynamics," p. 105.

[62] *Ibid.*, p. 111. He also compared his work with Owen's 1911 data.

[63] *Ibid.*, p. 114.

[64] *Ibid.*, pp. 112–113.

[65] See Chapter 2.

[66] Compton, "Electrodynamics," pp. 113–114.

[67] C. W. Hewlett, however, still regarded the ring electron theory as viable later in 1921. See "The Mass Absorption and Mass Scattering Coefficients for Homogeneous X-Rays of Wavelength between 0.13 and 1.05 Angstrom units in Water, Lithium, Carbon, Nitrogen, Oxygen, Aluminum, and Iron," *Phys. Rev.* 17 (1921):287; 298–299. H. S. Allen also

published a paper in 1921 on "The Angular Momentum and some Related Properties of the Ring Electron," *Phil. Mag.* 41 (1921):113–120, which exploited the model in atomic theory; see also his "Optical Rotation, Optical Isomerism, and the Ring-Electron," *Phil. Mag.* 40 (1920):426–439.

[68] Compton, "Electrodynamics," (reference 10) p. 114.

[69] *Ibid.*, p. 116.

[70] See Chapter 3, section D.5.

[71] For underdamped motion ($l << q$) and for $x = \dot{x} = 0$ at $t = 0$, one finds $x = A_1 e^{-kt}\cos(p_1 t + \Delta) + Be^{-lt}\cos(q_1 t - \beta)$, where A_1 and B are new amplitudes and Δ and β are new phase factors. The energy dissipated by the electron is then $E_a = (m/2)\,(\dot{x})^2 \doteq A^2 e^2 R'/2mp^2$, where

$$R' \equiv \frac{q^2}{p^2}\left[1 + \left(\frac{p^2}{q^2} - 1\right)\sin^2\Delta\right]$$

Since the incident energy is

$$E_0 = \frac{c}{4\pi}\ A^2 e^{i(p_1 t+\delta)-kt} = \frac{cA^2}{8\pi}\int_0^\infty e^{-2kt}dt = \frac{cA^2}{16\pi k}$$

if $k/r = e^2p^2/3mc^2$, which Compton probably derived from the Larmor formula $dE/dt = (2/3)(e^2/c^3)a^2$ (a = acceleration), then the new absorption coefficient a/ρ is given by $a/\rho = NE_a\phi/E_0\rho = \sigma_0\ \phi\ R'r/\rho$, where the "acceleration factor" ϕ has been taken into account.

[72] This follows from the equation for a/ρ in reference 71 because (i) r is on the order of unity, and (ii) $< R' > = \frac{2}{\pi}\int_0^{\pi/2} R'd\Delta = \frac{2}{\pi}\int_0^{\pi/2}\left[\frac{q^2}{p^2} + \left(1 - \frac{q^2}{p^2}\right)\sin^2\Delta\right]d\Delta$

$$= \tfrac{1}{2}(1 + q^2/p^2) \doteq \tfrac{1}{2}$$

(i.e., on the order of unity and independent of λ), since typical X-ray frequencies p are considerably larger than electron frequencies q.

[73] The experimental data had not been enlarged since 1919. See Compton, "Electrodynamics," pp. 125–126.

[74] *Ibid.*, pp. 125–126.

[75] *Ibid.*, pp. 126–127.

[76] *Ibid.*, p. 129.

[77] "Arthur Holly Compton, Research Physicist," *Science* 138 (1962):796. Copyright © 1962 by the American Association for the Advancement of Science.

[78] Johnston (ed.), *Cosmos*, p. 34.

[79] *Ibid.*, p. 31.

[80] Johnston (ed.), *"Cosmos,"* p. 31.

[81] These are included as loose sheets in Compton Notebook B. 1. (Wash. U., A.I.P.). The page that is reproduced (by permission of Mrs. A. H. Compton) is p. i.

[82] *Ibid.*, pp. i–ii.

[83] *Ibid.*, p. iii.

[84] This is stated on an undated description written by A.L. Hughes, which is on deposit in the Washington University Archives; its most important lines are quoted in Chapter 5, section D.5.

[85] "A Recording X-Ray Spectrometer, and the High Frequency Spectrum of Tungsten," *Phys. Rev.* 7 (1916):646–659; 8 (1916):753.

[86] Compton Notebook B.1. (Wash. U., A.I.P.), loose sheets iv–v.

[87] *Ibid.*, pp. vi–vii.

[88] "Secondary High Frequency Radiation," *Phys. Rev.* 18 (1921):96.

[89] *Ibid.*, pp. 96–97.

[91] *Ibid.*, p. 96.

[92] A. H. Compton and C. F. Hagenow, "The Polarization of Secondary X-Rays," *Phys. Rev.* 18 (1921):97.

[93] This diagram was published three years later in "A Measurement of the Polarization of Secondary X-Rays," *J. Opt. Soc. Am.* 8 (1924):488.

[94] Compton and Hagenow, "Polarization," pp. 97–98.

[95] *Ibid.*

[96] *Ibid.*, p. 98.

[97] "The Width of X-Ray Spectrum Lines," *Phys. Rev.* 18 (1921):322; 19 (1922):68–72. This work represented an analysis of certain observations Compton had already made as a graduate student at Princeton University.

[98] Related work is discussed in "A Possible Origin of the Defect of the Combination Principle in X-Rays,' *Phys. Rev.* 18 (1921):336–338.

[99] By differentiation of the Bragg equation, $n\lambda = 2d\sin\phi$, we have that $d\lambda/\lambda = d\phi/\tan\phi$, or for two different angles, ϕ and ϕ', $d\lambda/\lambda = (d\phi - d\phi')/(\tan\phi - \tan\phi')$. Using data on the second and fourth orders, Compton found $d\lambda/\lambda = 0.0007$ radian, which corresponded to $d\phi_1 = (d\lambda/\lambda)\tan 11.8° = 0.50'$ for the first order. This was the very small broadening (less than 1%) representing the relative inhomogeneity for roughly 1.2 = Å X-rays. This is not to be confused with line *broadening*, for which Jauncey in his paper "The Effect of Damping on the Width of X-Ray Spectrum Lines," *Phys. Rev.* 19 (1922):64–67, Fourier analyzed the damped oscillation

$$e^{-kt}\sin(pt + A)$$

and proved that a finite line width is predicted, the exact amount depending on the damping constant k. It is worth noting that his calculation is the classical analogue of line broadening due to the Uncertainty Principle, a principle that neither Jauncey, nor Compton, of course had any knowledge of. See R. B. Leighton, *Principles of Modern Physics,* (reference 31) pp. 283–284, especially Exercise 8–11.

[100] For a complete chronology of Compton's life, see Johnston (ed.), *Cosmos,* pp. 449–457.

[101] "The Scattering of X- and γ-Rays," *J. Franklin Inst.* 190 (1920):633–655.

[102] *Ibid.,* pp. 635–636. In discussing the "usually accepted" selective scattering hypothesis, Gray cites, among other works, the 1919 work of Compton.

[103] *Ibid.,* p. 642.

[104] *Ibid.,* pp. 642–643.

[105] *Ibid.,* pp. 640, 643.

[106] Gray did not make his model nearly as explicit as I have described it, but is clear that the one given is what he had in mind.

[107] See "On Reflexion of the X-rays," *Phil. Mag.* 26 (1913):210–232.

[108] Gray, "X- and γ-Rays," (reference 101) p. 645; italics added.

[109] *Ibid.,* p. 655.

[110] *Ibid.,* pp. 644–645.

[111] *Ibid.,* p. 645, 654.

[112] *Ibid.,* p. 655.

[113] See Chapter 2, section D.4.

[114] *Phil. Mag.* 42 (1921):302–304.

[115] *Ibid.,* pp. 302–303.

[116] *Ibid.,* p. 303.

[117] *Ibid.,* pp. 303–304.

[118] *Ibid.,* p. 305.

[119] "The softening of Secondary X-rays," *Nature* 108 (1921):366–367.

[120] *Ibid.,* p. 366. The word "number" in the first quotation is in small capital letters in the original.

[121] *Ibid.*

[122] Compton Notebook B.2. (Wash. U., A.I.P.), p. 1.

[123] Johnston (ed.), *Cosmos,* p. 427.

[124] Compton obtained this information from William Duane's report, "Data Relating to X-Ray Spectra," *Bull. Natl. Res. Counc.* 1 (1920):392.

[125] Compton Notebook B.2. (Wash. U., A.I.P.), p. 3 (reproduced by permission of Mrs. A. H. Compton). "Blank" obviously is the ionization chamber reading without the radiator *R* in position.

[126] Compton Notebook B.2. (Wash. U., A.I.P.), p. 4.

[127] *Ibid.,* p. 5.

[128] *Ibid.,* pp. 5–6.

[129] *Ibid.,* pp. 7–9.

[130] Compton, "Softening," (reference 119) pp. 366–367.

CHAPTER 5

The Classical-Quantum Compromise

A. Introduction

Arthur Compton, as we have just seen, interpreted his crucial experiment as establishing unambiguously the composition of the secondary radiation described in Chapter 4. He asserted explicitly that "in addition to scattered radiation there appeared in the secondary rays a type of fluorescent radiation, whose wave-length was nearly independent of the substance used as radiator, depending only upon the wave-length of the incident rays and the angle at which the secondary rays were examined."[1] He reached that conclusion—the culmination of years of research—in October 1921. During the succeeding year, however, he gradually began to question its validity. He gradually began to suspect that there might be a more fundamental, even revolutionary, interpretation of the evidence. He gradually began to doubt, in a word, the universal validity of classical electrodynamics. In this chapter we shall examine his reasons for doing so.

B. Compton's First Spectroscopic Experiments and the First Stage of His Classical-Quantum Compromise (1921)

On October 18, 1921, three days after Compton had repeated Plimpton's experiment, he made the following entry in his notebooks: "Proposed expts. 1. Help Jauncey finish work on scattering by crystal. 2. Get ionization spectrum of scattered x-rays using a. Mo tube. b. W. tube."[2] The first experiment does not concern us—its object was to determine certain properties of the crystal itself.[3] The second experiment, in contrast, represents the most fundamental change in Compton's research program since he turned from γ-ray to X-ray experiments. He had just decided, in order to obtain "more definite information with regard to the characteristics of the secondary x-radiation,"[4]

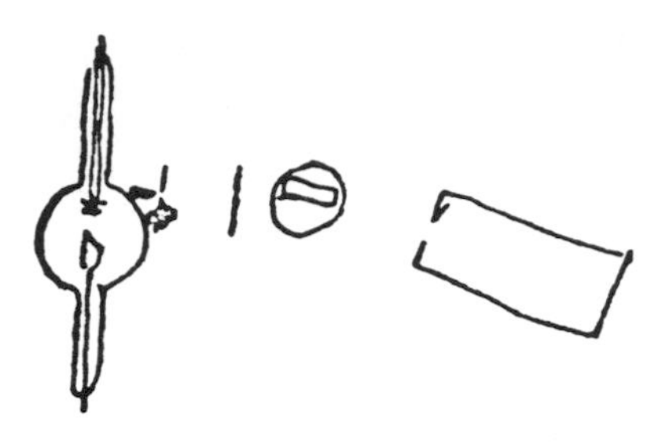

FIG. 1. Unlabeled sketch showing (from left to right) X-ray tube, crystal, and ionization chamber.

to use his spectrometer *as a spectrometer* rather than as a "wavelength selector." This change is clearly indicated by the position of the crystal midway between the X-ray tube and ionization chamber, as shown in the small, unlabeled, but self-explanatory sketch in Fig. 1, which has been reproduced from Compton's notebooks,[5] and which should be compared with Fig. 14 of the last chapter. Compton spent just over one month (October 19–November 20, 1921) taking spectra of the secondary and primary radiations. Most of his raw data have survived, a particularly fortunate circumstance, since it enables us to delineate rather precisely the next stages in the development of his thought.

Compton set up his initial experimental arrangement (X-ray tube with molybdenum target, celluloid scatterer, Bragg spectrometer with NaCl crystal, and ionization chamber) on October 19, and immediately took a number of preliminary measurements with it to determine, for example, the zero position of the ionization chamber. The following day, still orienting himself, he examined the primary and secondary molybdenum spectra in some detail (probably using a scattering angle of 90°), and found that the resolution and alignment of his spectrometer left a great deal to be desired. He evidently traced some of his difficulties to the NaCl crystal, since he decided to permanently replace it with one of calcite.

One week later, between October 27th and November 4th,[6] Compton once again took a number of preliminary spectra, using a zirconium filter to obtain "a primary beam consisting principally of the $K\alpha$ line from molybdenum [λ = 0.708 Å] together with some fluorescent K rays from zirconium . . . ."[7] His data indicated the puzzling result that the secondary wavelength was actually *shorter* than the primary. Removing the zirconium filter left this puzzling situation unimproved. Other experimental problems also appeared. For instance, he found that decreasing the potential difference in his X-ray tube caused the primary peak to shift, and reinserting filters into the primary and secondary beams caused the secondary peak to shift—rather spectacularly, in fact. The latter, obviously spurious shifts were too much to accept, and

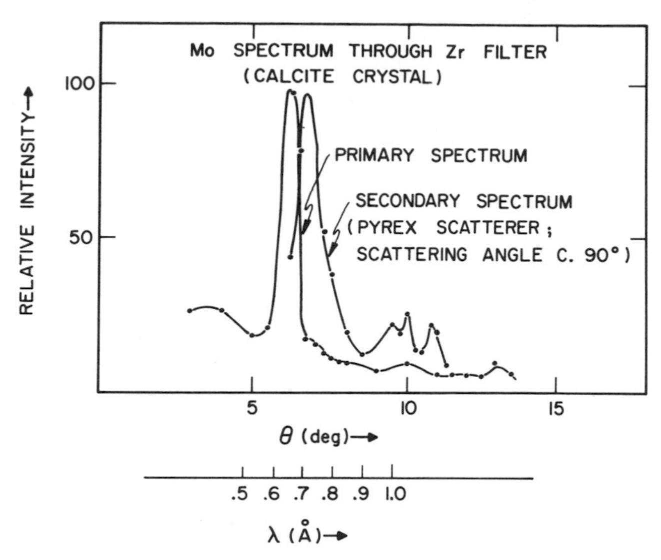

FIG. 2a. Author's plot of Compton's data for MoKα rays (Zr filter) scattered by Pyrex through approximately 90°. The two abscissae are related by the Bragg equation $\lambda = 2d\sin\phi$, where $d = 3.028$ Å for calcite.

Compton closed down his experiment for fifteen days, until November 19. The reason became apparent from the caption of his notebook entry on that date: "With new tube, . . . New Motor."[8]

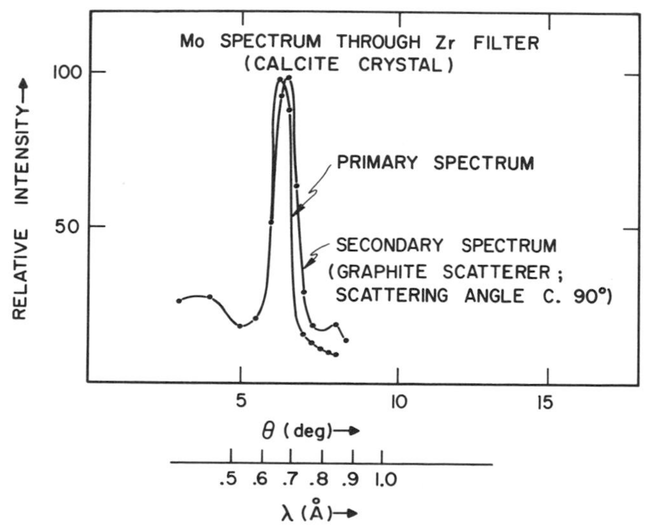

FIG. 2b. Author's plot of Compton's data for MoKα rays (Zr filter) scattered by graphite through approximately 90°.

Using this new equipment (which he no doubt constructed himself), Compton took nine new spectra on November 19–20. At first, he was apparently troubled by some of his old difficulties, but then things improved. His best data, which I have plotted in Figs. 2a and 2b, were obtained using Pyrex and graphite (carbon) scatterers, and a scattering angle of "c. 90°."[9] All of his initial statements and conclusions were based on his Pyrex spectra, which were taken over a much wider range of wavelengths than his graphite spectra. He reported his results (unaccompanied by his spectra) in an abstract which he sent to the editor of *The Physical Review,* probably in early December 1921. He explained:

> The spectra obtained show lines identical in wave-length with the primary *K* lines from molybdenum, thus proving that a part of the secondary radiation is truly scattered and unchanged in wave-length. In addition to these lines, a general radiation is observed which is more prominent in the secondary than in the primary beam. When the x-rays incident upon the radiator [scatterer] were unfiltered, the general secondary radiation had a broad intensity maximum at a wave-length slightly under 1 [Å.] On introducing a zirconium filter between the x-ray tube and the radiator . . . a much sharper maximum in the secondary fluorescent radiation was observed. This result has been verified by means of photographic spectra, which show a maximum of the general radiation at about 0.95 [Å.], which is about 35 per cent. greater than the wave-length of the exciting ray.[10]

It is obvious from this passage, first, that Compton took more data than those contained in his surviving laboratory notebooks—he specifically mentions "photographic spectra," and all of the spectra recorded in his notebooks are ionization spectra. But by far the most important observation to make is that Compton says explicitly that the maximum of the "secondary fluorescent radiation" is "at about 0.95 [Å.], which is about 35 per cent. greater than the wave-length of the exciting ray." Since Compton knew that the wavelength of the MoK$_\alpha$ line is 0.708 Å, and since, indeed, 1.35 times 0.708 Å is about 0.95 Å, only one conclusion is possible: *Compton completely ignored the small shift in the MoK$_\alpha$ peak and focused his entire attention on the small peaks seen in Fig. 2a around 9.5°*. We now realize that these peaks simply represent the second-order spectrum of the scattered rays,[11] but Compton did not realize that. To Compton the wavelength of the primary radiation, as well as that of the secondary "truly scattered" radiation, was 0.708 Å; the wavelength of the "secondary fluorescent radiation" was 0.95 Å; and the "shift in wavelength" was therefore 0.24 Å!

Why did Compton at this time attach no significance to the *small* displacement of the MoK$_\alpha$ peak? Two plausible and related answers are suggested by his data taking and data analysis techniques. Thus, in taking his ionization chamber readings in the vicinity of the MoK$_\alpha$ peak, Compton chose angular intervals of ¼°, and it is easy to show from the Bragg equation that this corresponds to wavelength intervals of 0.026 Å. Consequently, it would be natural, even essential, for Compton to ignore wavelength shifts of less than about 0.026 Å. Furthermore, when he reduced his data for analysis, he converted his raw intensity measurements into relative units by simply assigning the value 100 to the largest absolute intensity, and scaling all others accordingly. Since, however, he took his readings at each ¼° position, he may have tended to regard one such position as the precise position of the MoK_α peak, the position to be assigned the relative intensity 100. But the precise position of the MoK_α peak could of course be anywhere; and it would certainly be fortuitous if it were to fall on a ¼° mark. Perhaps both of these factors led Compton to ignore any shifts in wavelength on the order of about 0.026 Å, if they did not simply escape his attention.

In any event, believing that he had established the location of the "secondary fluorescent radiation" peak at 0.95 Å, Compton felt obligated to account for its existence at that position. He explained:

> The energy in this general radiation is roughly 30 per cent. as great as the energy of the scattered *K* rays. Previous experiments have shown that when shorter wave-lengths are employed, the energy in the fluorescent rays may be even more prominent than the truly scattered rays. If we suppose that the incident x-ray beam ejects electrons moving forward with a kinetic energy hc/λ, where λ is the wave-length of the exciting ray, and if the ejected electron is oscillating at such a frequency that as observed in the direction of motion the wave-length is λ, on account of the Doppler effect the wave-length of the radiation at right angles with the primary beam will be very close to that of the fluorescent rays observed in these experiments.[12]

This explanation must be analyzed piecemeal.

In the first place, by once again referring to Fig. 2a, we observe that the height of the small peak at about 0.95 Å is indeed roughly 30% as high as that of the MoK_α peak, which simply removes all doubt as to which peak Compton associated with the secondary "fluorescent" radiation. Secondly, and far more significantly, we note that Compton—for the very first time—explicitly used the "quantum relation" in an explanation of the "fluorescent" radiation. To see precisely how he used the quantum relation, however, it is essential to carry out in detail the calculation sketched in the above passage.

Compton first suggests that each "ejected electron" is supplied with kinetic energy hc/λ by the "incident x-ray beam." Note that he refrains from using the term "quanta"—and he does *not* say that each electron takes up the energy of a *single* quantum. All we are justified in concluding from Compton's statement is that

$$\frac{hc}{\lambda} = mc^2 \left(\frac{1}{\sqrt{1 - v^2/c^2}} \right) - 1 \doteq \tfrac{1}{2} mv^2, \tag{5.1}$$

where v is the velocity, and m the rest mass, of the "ejected electron." (The last step is justified because, as may easily be verified, $(hc/\lambda)/(mc^2) \ll 1$.) Next, Compton suggests that the "ejected electron is oscillating," and therefore that the Doppler effect must be introduced into the argument. The exact relativistic expression for the Doppler effect may be written as

$$\frac{\lambda_1}{\lambda_2} = \frac{1 - (v/c) \cos \theta_1}{1 - (v/c) \cos \theta_2}, \tag{5.2}$$

where λ_1 and λ_2 are the wavelengths observed in the directions θ_1 and θ_2, respectively. (There is absolutely no question that this is the expression Compton had in mind. We have seen that he used it earlier, and he would also use it again in the near future.) Now, in Compton's particular experimental arrangement, $\theta_1 = 0°$ and $\theta_2 = \pi/2$, so that equation (5.2) reduces to

$$\frac{\lambda_1}{\lambda_2} = 1 - \frac{v}{c}. \tag{5.3}$$

One may now eliminate the velocity v by using equation (5.1), realizing that $\lambda = \lambda_1$, to obtain

$$\frac{\lambda_1}{\lambda_2} = 1 - \frac{1}{c}\sqrt{\frac{2hc}{m\lambda_1}} = 1 - \sqrt{\frac{2hc/\lambda_1}{mc^2}}. \tag{5.4}$$

Substituting $\lambda_1 = 0.708$ Å $\doteq 0.71$ Å, and $mc^2 = 0.51$ Mev, we find

$$\frac{\lambda_1}{\lambda_2} = 1 - \sqrt{\frac{2(0.017 \text{ Mev})}{0.51 \text{ Mev}}} = 1 - 0.26 = 0.74.$$

This is the ratio that Compton, in effect, asks us to compare with the "observed" ratio $\lambda_1/\lambda_2 = 0.71$ Å$/0.95$ Å $= 0.75$. Who could ask for better agreement?

C. Compton's Discovery of the Total Internal Reflection of X-Rays (March 1922)

The above train of argument represents the first stage in what might be termed Compton's "classical-quantum compromise" to account for his unusual secondary "fluorescent" radiation. A second stage followed, but only after Compton had been diverted momentarily from his studies in scattering and absorption. In March 1922, certain work came into his hands which suggested that X-rays might be totally reflected from substances,[13] and since

Compton was completely committed to understanding X-rays in all of their aspects, he decided to pursue this idea. It developed that Compton's subsequent experiments were of the greatest importance intrinsically—and they also had a marked influence on his developing ideas on scattering and absorption.

The stimulus for Compton to interrupt his work on scattering and absorption was apparently the appearance of a short paper by Manne Siegbahn of the University of Lund which was presented to the French Academy of Sciences on December 19, 1921.[14] In it Siegbahn drew attention to the thesis research of W. Stenström,[15] who in the course of taking X-ray spectra had observed very slight departures from Bragg's law, $n\lambda = 2d\sin\phi$. Stenström had found, in effect, that for observed values of ϕ_1, ϕ_2, and ϕ_3, corresponding to the first, second, and third orders, the calculated wavelength λ did not remain constant, but decreased slightly in going from the first to higher orders. Stenström interpreted this systematic decrease in wavelength as arising from a slight refraction of the incident X-rays at the air–crystal interface. A short time later, however, William Duane and R. A. Patterson of Harvard University carried out similar experiments and concluded that these very slight departures from Bragg's law were within experimental error.[16] Almost simultaneously, P. Knipping of the University of Munich maintained that even if they existed, they should be attributed not to refraction, but to a slight increase in the lattice constant of the surface atoms.[17] Knipping's analysis, in turn, was challenged by his colleague, P. P. Ewald, who demonstrated the quantitative correctness of Stenström's original refraction hypothesis,[18] a fact that Siegbahn, in a second paper, corroborated.[19]

Compton did not enter into this debate. Rather, with unerring insight, he concluded that *if* Stenström's *observations* were correct, that is, *if* the wavelength decreased slightly with increasing order, the index of refraction μ for X-rays must be slightly *less* than unity. Consequently, if X-rays were incident on a substance at a very small glancing angle, they should be totally reflected. By Snell's law the glancing angle ϕ is given by

$$\cos\phi = \mu = \sqrt{1 - \sin^2\phi}, \text{ so that } \sin\phi = \sqrt{(1 + \mu)(1 - \mu)} \doteq \sqrt{2\delta},$$

where μ is the index of refraction, and where $\delta = 1 - \mu$ represents the departure from unity. But δ could also be evaluated from the Drude–Lorentz dispersion formula,[20] which predicts that

$$\delta = 1 - \mu = \frac{Ne^2}{2\pi m\nu^2} = \frac{Ne^2\lambda^2}{2\pi mc^2}, \tag{5.5}$$

where the symbols have their usual meaning, N being the number of electrons per unit volume in the refractive medium. Compton's laboratory notebook shows[21] that on March 23, 1922, he arbitrarily set λ = 1 Å, estimated N

for a density of 3 g/cm^3, calculated δ, and concluded that the corresponding glancing angle ϕ is 0.16° or 9.6′—a very small, but readily measurable angle.

Compton verified this prediction on March 23–25; on April 5 he sent an abstract of his results to the editor of *The Physical Review;* and on April 22 he gave a more complete report at the Washington meeting of the American Physical Society. He concluded:

> Experiments have been performed which show that regular reflection of x-rays does occur at angles of a few minutes of arc from polished surfaces of glass and silver, and that except near the critical angle the greater part of the incident energy is reflected. The critical angles for the reflection of $\lambda = 1.279$ [Å] from plate glass and silver mirrors have been measured, giving values of $1 - \mu = 5.0 \times 10^{-6}$ and 20.9×10^{-6} respectively. Taking the number of electrons per atom to be the atomic number, the corresponding theoretical values are 5.2 and 19.8×10^{-6}. The agreement is thus quite satisfactory.[22]

Judged solely on their intrinsic scientific merits, these results were extremely significant. First, they represented the first direct determination of the index of refraction of X-rays, a measurement that had eluded physicists since the time of Röntgen's discovery.[23] Second, they represented the first experimental confirmation of the Drude–Lorentz dispersion formula. All earlier attempts to test this formula using optical frequencies—for which the formula assumes a more complicated and structure-dependent form—had failed.[24] Finally, by assuming the correctness of the dispersion formula, the number of electrons per atom could be calculated. This was one of the first such calculations that was independent of atomic structure assumptions, or even of the assumption that the number of electrons per atom is equal to the atomic number.[25] It is not difficult, therefore, to agree with S. K. Allison that "This work alone . . . would have established . . . [Compton] in the first rank of experimental physicists.[26]

What influence, however, did Compton's reflection experiments exert on his developing ideas on the nature of high-frequency radiation? Did they influence his ideas on scattering and absorption? The answer to the first question seems obvious, since not many phenomena are more characteristic of classical electrodynamics than total reflection. To answer the second question, we need only observe that for Thomson's theory of scattering to hold even approximately, the incident X-rays must be able to set a large number of electrons into oscillation; Thomson's expression, equation (1.3), contains a factor N. The interaction of the incident rays with a *single* electron would yield a result only $1/N$ as large as the experimental value. Now, since Compton had just confirmed the Drude–Lorentz dispersion formula, equation (5.5), *which contains the same factor N,* had he not just strikingly demonstrated precisely the same point, that the incident rays do not interact with a *single* electron? Moreover, within a month, Compton found further evidence for this point from X-ray diffraction studies on powdered crystals.[27] Nothing in any of these experiments suggested that X-rays interact with *single* electrons, as demanded by the quantum theory.

D. Compton's National Research Council Report (October 1922)

1. *Its Origin and Purpose*

Compton spent the summer of 1922, as he had one year earlier, at the University of California at Berkeley as a Visiting Lecturer.[28] Throughout the summer, and probably even earlier, he began gathering information for a National Research Council report on the "Secondary Radiations Produced by X-Rays, and some of their Applications to Physical Problems."[29] This report had the following origin. In 1920 a committee of the Division of Physical Sciences of the National Research Council was formed to report on the current status of X-ray research. As originally constituted, the members of this committee were William Duane (Harvard University), Chairman; Bergen Davis (Columbia University), A. W. Hull (General Electric Research Laboratory), and D. L. Webster (Leland Stanford Junior University). By the end of 1920, reports had already been published by Duane,[30] Davis,[31] and Webster [32] (Hull apparently did not submit a report). In 1922 Arthur Compton was asked to become a fifth member of this committee and to prepare its final report, which, as it turned out, was published in October 1922.

The primary purpose of these National Research Council reports was to present a critical review of the literature pertaining to the subject in question —in Compton's case, the literature on the secondary radiations produced by X-rays. It is hardly necessary to emphasize that such a review—Compton's ran to 56 pages—always stimulates new insights into the problem at hand, and acquaints the author with older work with which he had not been previously familiar. Compton was no exception. Most significantly, in preparing his report, Compton was led to appraise the relative merits of the classical and quantum theories of radiation.

2. *Compton Begins to Seriously Doubt His Large Electron Model*

Compton's subject demanded that he investigate both types of secondary radiation, X-rays and β-rays, in detail. He began with the former, immediately subdividing them further into scattered X-rays and fluorescent X-rays —thereby maintaining his old dichotomy: "If the primary ray is homogeneous, and if the electrons traversed are at rest, so that no Doppler effect occurs, the scattered beam will be homogeneous and of the same wave-length; whereas the fluorescent rays will differ in wave-length from the primary."[33] With this

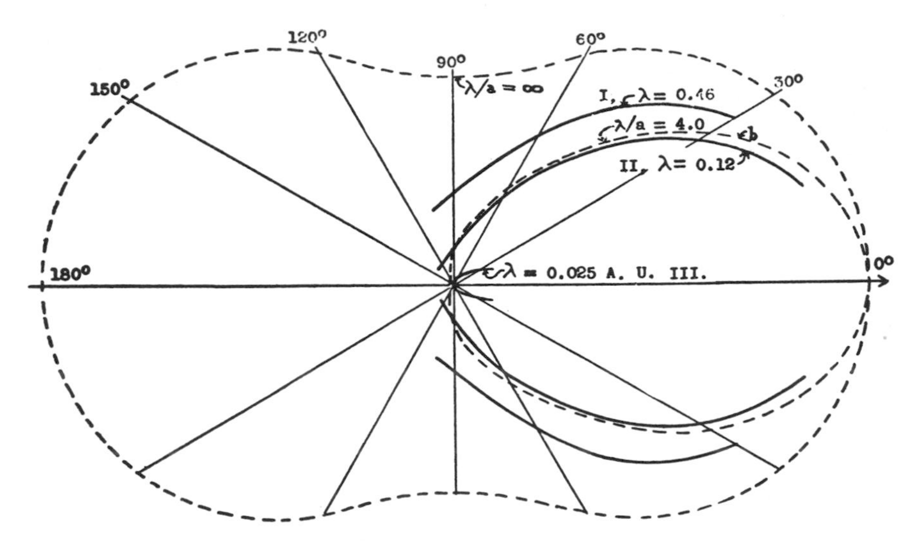

FIG. 3. Compton's plots of $I(\lambda/a,\ \theta)$ *vs* θ for comparison between theory (dashed curves) and experiment (solid curves) for his large electron scattering theory.

sharp dichotomy still in the forefront of his mind, it is not surprising that in treating the "scattered" X-rays he followed a well-traveled route, discussing the basic assumptions and predictions of Thomson's theory, the phenomenon of "excess scattering,"[34] and finally, his large electron hypothesis. In the course of his discussion of his large electron hypothesis, however, it became apparent that once again he had achieved new insights into its validity.

To compare his large electron hypothesis with experiment, Compton first plotted (Fig. 3[35]), as he had on several occasions in the past, the angular distribution $I(\lambda/a,\ \theta)$ *vs* scattering angle θ. He displayed theoretical (dashed) curves for scattering both by a point charge ($\lambda/a = \infty$), and by a large, flexible, spherical shell electron with a wavelength to radius ratio of 4 ($\lambda/a = 4.0$). He plotted experimental (solid) curves for three different primary wavelengths, 0.46 Å (I), 0.12 Å (II), and 0.025 Å (III). These experimental curves were based on his own scattering data: Curves I and II represented his Washington University X-ray data, while curve III represented his Cavendish γ-ray data. It is seen that while his X-ray data fit his theoretical curve for $\lambda/a = 4.0$ reasonably well, his γ-ray data missed the curve completely.

This discrepancy obviously demanded scrutiny, and Compton therefore examined the same data from another, also familiar, point of view. He plotted (Fig. 4[36]), the "finite electron factor" Φ *vs* a/x or $\lambda/4\pi\sin(\theta/2)$. Theoretical (solid) curves are shown for (a) a spherical shell electron of radius 0.032 Å,

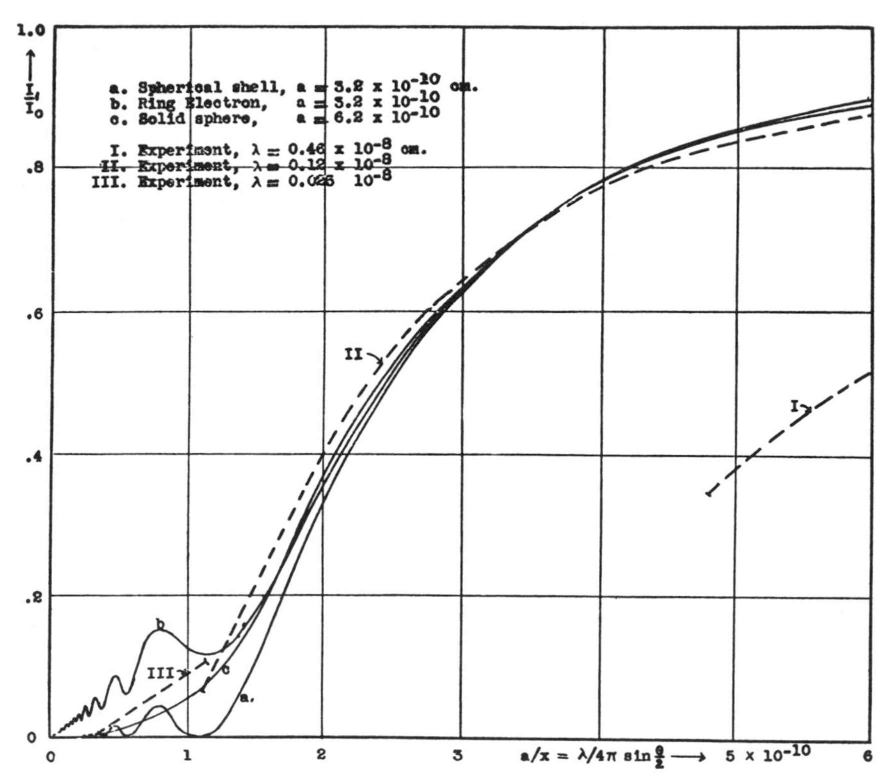

FIG. 4. Compton's experimental data (dashed curves) and theoretical plots (solid curves) of Φ *vs* a/x for three different electron models. One symbol (Φ) has been changed for consistency in notation.

for which $\Phi = \sin^2 x/x^2$; (*b*) a ring electron of radius 0.032 Å, for which

$$\Phi = \frac{1}{x}\sum_{n=0}^{\infty} J_{2n+1}(2x);$$

and (*c*) a solid sphere electron of radius 0.062 Å, for which

$$\Phi = \left[\sum_{n=0}^{\infty} \frac{(-1)^n x^{2n}}{(2n+1)(2n+1)!}\right]^2.$$

Experimental (dashed) curves are shown for the data that were used in Fig. 3. It is obvious that the long-wavelength X-ray data (curve I) are in marked disagreement with *all three* theoretical curves.

One possible explanation of this difficulty was that the electron, irrespective of its shape, was not perfectly flexible. While this possibility was "doubtful," Compton noted: "This is an important matter for further theoretical and experimental investigation, for if this discrepancy is real it would not seem possible to reconcile the results with the classical electrodynamics." His final conclusion, however, was dictated by the fact that a point charge electron would scatter according to the upper Thomson line, where $\Phi = 1$: "the fact that experiment gives consistently lower values when short wave-lengths are used, indicates that the electron is not sensibly a point charge of electricity. We find, indeed, that this reduced scattering for small values of a/x can be accounted for by interference between the rays from different parts of the electron, if its radius is of the order of 4×10^{-10} cm."[37]

Notwithstanding his new estimate of the electron's radius, and notwithstanding his apparent continuing confidence in his large electron hypothesis, Compton fully realized, in general, the disturbing nature of the plots in Fig. 4. For these plots represented the first evidence indicating that *no* large electron model could encompass *all* of the scattering data. Indeed, inspection shows that for close agreement between theory and experiment for *all* incident wavelengths, the size of the electron would somehow have to increase with increasing wavelength. Compton knew that it would be extremely difficult, even if it were plausible, to devise a physical mechanism capable of bringing about such a change.

3. Compton Achieves New Insights from the Behavior of Secondary Beta-Rays

Setting aside for the moment Compton's discussion of the secondary "fluorscent" X-radiation and turning to his discussion of the secondary β-rays, it is not difficult to find further indications here that Compton was seriously questioning his old ideas. Never before had Compton examined in detail the properties of the secondary β-rays. Three appeared to be most important: (1) their forward-backward asymmetry when emitted; (2) their number; and (3) their velocity of emission.

Compton had known for a long time that many more secondary β-rays are emitted in the forward direction than in the backward direction. He had even intended to investigate this asymmetry himself (recall the fifth of his Problems to be tackled at Saint Louis), but had never undertaken the experiments. He had, however, learned a great deal from the results of others, especially perhaps in the course of his studies as a graduate student[38] and at the Cavendish. It was this knowledge that guided his literature search—and led him to rediscover the Bragg–Barkla controversy over the nature of X-rays and γ-rays, which we discussed in Chapter 1. Compton now became thoroughly familiar with the experiments and theories of Bragg and Barkla (1908–1910),

Madsen (1909), Cooksey (1908), Stuhlmann (1910–1911), and Kleeman (1910).[39] Collectively, this work indicated that the forward-backward asymmetry remained fairly constant for primary wavelengths between 3000 Å and 0.5 Å, but increased sharply for wavelengths somewhere between 0.5 Å and 0.05 Å.[40]

Several explanations occurred to Compton. First, he remarked that Bragg's "neutral pair" hypothesis, while "difficult to defend" after Stuhlmann's ultraviolet light experiments and even "untenable" after Laue's discovery, nevertheless provided an obvious qualitative explanation for the asymmetry, namely, "that these 'neutrons,' on colliding with electrons, transferred to them their energy and momentum," hurling them forward. A second possible explanation was O. W. Richardson's (1912), "that as the electron absorbs a quantum $h\nu$ of energy, the momentum of the absorbed radiation is also transferred to the electron, causing a resultant motion in the forward direction" at an average velocity of $v^2/2c$. Compton felt that Richardson's result qualitatively explained the increase in asymmetry with decreasing wavelength but that it predicted a "greater and more uniform" increase than actually was observed. Compton finally concluded that the key to its understanding lay in the fact that the asymmetry increased sharply in exactly the same primary wavelength region (0.5–0.05) Å as the asymmetry in the secondary *X-radiation:* "This suggests that both phenomena may have a common origin, and tends to support the view that the secondary x-rays are emitted by electrons which are moving forward at high speed."[41]

When Compton turned to the second property of the secondary β-rays, the number ejected by the primary radiation, the increasing importance of quantum considerations in his thought became apparent. He noted that on the quantum view the number liberated per unit length should be given by $E_i\tau/h\nu$, where E_i is the energy of the incident radiation and τ is the fluorescent absorption coefficient. In other words, the incident energy E_i and the number liberated should be proportional to each other, which was apparently the case for both X-rays and γ-rays. Moreover, even though "no direct experimental determination of the factor of proportionality has been made," Compton asserted, "there seems no reason to doubt that this factor is the energy quantum $h\nu$."[42]

The general trend of Compton's thought became even clearer when he discussed the third property of the secondary β-rays, their velocities of emission. These velocities had been studied recently by C. D. Ellis[43] and M. de Broglie[44] who used magnetic spectrum analysis techniques. The former had made one observation that struck Compton as particularly unusual: "It is remarkable that in Ellis's paper no magnetic spectrum lines are recorded which are due to the expulsion of electrons from the outer rings . . . . We should . . . have expected these outer electrons . . . to give rise to a strong line for which [their kinetic energy] $T = h\nu$." Compton reconciled this difficulty by arguing, in effect, that the outer electrons are ejected with a *spread* in energies, which would be the case if "an electron scatters a whole quantum of energy at a time, and receives the momentum of the incident quantum, [so that] the average momentum in the forward direction of an electron which has scattered rays of wave-length λ"[45] is given by

$$M = \frac{h\nu}{c} = \frac{m\nu}{\sqrt{1-\beta^2}}, \tag{5.6}$$

where $\beta = v/c$. Those electrons ejected in different directions should have different momenta, and hence different energies.

This entire picture involved an important implication: "Ellis's failure to observe photoelectrons with the maximum kinetic energy $h\nu$ seems to support the hypothesis that most of the secondary β-rays excited by γ-rays are not a result of fluorescent absorption, but are rather a by-product of the scattering process." Compton speculated even further: it "is possible," he wrote, "that there may be two types of photoelectrons, those whose liberation excites the characteristic fluorescent radiation, and those which recoil after scattering a quantum of energy . . . . The energy of the second type will ordinarily be small compared with that of the first type. The evidence is thus consistent with the view that each photoelectron represents the removal of one quantum of energy from the primary beam, and that no other energy is lost except perhaps through true scattering."[46] This was the first time that Compton recognized the possible existence of *recoil* electrons—electrons which were apparently entirely different from those emitting the Doppler shifted "fluorescent" radiation.

4. Barkla's J-Radiation Hypothesis and Compton's Reaction to It

The origin of the unusual "fluorescent" radiation therefore again became the focus of Compton's attention. Owing to the nature of his report, however, Compton could not simply confine himself to an exposition of his own interpretation, but was obliged to examine all candidates. One in particular had attracted a great deal of attention, although Compton himself had never seriously considered it before. This was the hypothesis that the incident X-radiation excites in the scatterer a *characteristic* fluorescent X-radiation of longer wavelength than the primary radiation, but of shorter wavelength than the characteristic radiations of the K-series. For this reason the series was known as the J-series. By 1922 it had had a long and tangled history, which it is now appropriate to sketch.

The existence of the J-series, "a new series of characteristic radiations,"[47] was announced in a Bakerian Lecture of 1916 by the discoverer of the other characteristic radiations, C. G. Barkla, who was then at the University of Edinburgh. The following year Barkla and White published plots of μ/ρ versus λ^3 for X-rays in paraffin, water, aluminum, and copper,[48] each of which exhibited a discontinuity at the same point—the wavelength at which the J-series was presumably excited. While its position for copper was questionable, Barkla and White declared without hesitation that the J-line had a wavelength of 0.42 Å when emitted by carbon, 0.39 Å when emitted by oxygen, and 0.37 Å when emitted by aluminum. E. A. Owen (University College, London) even observed "splitting," declaring that the J_β line had a wavelength of 0.559 Å when emitted by carbon, 0.519 Å when emitted by oxygen, and 0.493 Å when emitted by silicon.[49] Both Barkla's and Owen's experiments were supported in 1918 by C. M. Williams.[50] In general, it is not difficult to see why all of this work generated a great deal of excitement. As we pointed out in Chapter 2, there was, of course, no room for the J-series in Bohr's theory of the atom.

For this reason, Barkla's discovery also generated a great deal of skepticism between 1919 and 1921. The first adverse evidence was brought forward by William Duane and Takeo Shimizu of Harvard University, who tried to locate the J-series in the *emission* (rather than the absorption) spectrum of aluminum. They neatly bracketed the work of Barkla and Owen and concluded that "aluminum has no characteristic lines in its emission spectrum, between the wavelengths $.1820 \times 10^{-8}$ cm. and 1.259×10^{-8} cm. that amount to as much as 2 per cent. of the general radiation in the neighborhood. . . ."[51] This evidence directly conflicted with Barkla's, and prompted the French physicist A. Dauvillier to attempt new *absorption* measurements in 1920. Dauvillier actually found evidence, so he claimed, that supported Barkla's original experiments. He therefore concluded that the aluminum J-series was anomalous—it did not show up in the *emission* spectrum, but appeared only in the *absorption* spectrum.[52]

This point was challenged, still in 1920, by the American physicists F. K. Richtmyer and Kerr Grant. Using incident wavelengths between 0.08 Å and 48.0 Å, Richtmyer and Grant carried out absorption experiments not only on aluminum, but on water, copper, and molybdenum as well. They found "no evidence of a 'J' series of lines within this region." They suggested "that the apparent discontinuities observed by Barkla and White may really have been due to a combination of a rather large experimental error with such an accidental grouping of the observations as would suggest a discontinuity."[53] Richtmyer and Grant's results were confirmed in 1921 by very precise absorption measurements by C. W. Hewlett,[54] as well as by further work by Richtmyer himself.[55]

In spite of this almost overwhelming evidence against the J-series, however, papers continued to appear on the subject. In late 1921, J. A. Crowther —in a paper communicated to the *Philosophical Magazine* by Ernest Rutherford—declared that:

> The existence of this 'J' radiation does not appear to have been generally admitted, partly because such radiation had not been directly detected, and partly because radiation of this type is not indicated by the current theory of the structure of the atom. The latter objection, though probably the more influential of the two in producing this suspension of judgment, is not one which can be maintained in the face of experimental facts. The former, and more vital, objection it is hoped that the results described in the present paper may do something to remove.[56]

Crowther then went on to conclude from his absorption measurements that he had actually observed the excitation of various lines in the J-series.

No proponent of the J-series, however, was more adamant than C. G. Barkla, who had meanwhile won the Nobel Prize of 1917— for his discovery of the characteristic X-radiations! In his Nobel lecture (which was delivered after the war in 1920) Barkla discussed rather thoroughly the evidence in favor of the J-series, although he was forced to admit that these radiations had "proved more elusive than early experiments led us to believe."[57] Barkla in fact pursued his hypothesis unrelentingly well into the 1930's. The spirit in which he pursued it is illustrated by the following remarks he made in 1926 in a letter to H. S. Allen: "Yes! the J-phenomenon is very interesting and is so fundamentally new, but it may take a generation to work it out thoroughly. I have said a great deal about it in my papers, so will not inflict any more upon you! I am convinced that it is of the very greatest importance. And control of the phenomenon, to which most people seem to attach the highest importance, I am not greatly concerned about. However, with suitable arrangements, we can get control."[58]

Arthur Compton, citing evidence from his own experiments and from those of others, assessed the state of affairs unambiguously in his October 1922 report: "the evidence is . . . strongly against the existence of a characteristic J radiation."[59]

5. Compton Reexamines His Spectroscopic Data and Progresses to the Second Stage of His Classical-Quantum Compromise

Compton also dismissed Gray's pulse theoretical interpretation as well as the old "selective scattering" interpretation of the softened radiation. In the latter case he cited Sadler and Mesham's 1912 experiments, which proved that it was *not* the *softer,* but actually the harder rays which were more

FIG. 5. Actual photograph of Compton's X-ray scattering apparatus which he used in Eads Hall at Washington University, St. Louis. Its components have been described by A. L. Hughes, as quoted in the text.

strongly scattered. There was in fact now no doubt in Compton's mind that a "real change in the character of the radiation"[60] was occurring. His recent spectroscopic measurements, he felt, had shown that conclusively. And it is important to note at this point that these measurements, in turn, had been possible only through his use of a superior apparatus, which he himself had designed and built.

An actual photograph of Compton's apparatus which he used in Eads Hall is shown in Fig. 5, and a schematic diagram of it is shown in Fig. 6.[61] Note that the radiator (or scatterer) *R* has been mounted directly onto the outer wall of the X-ray tube. Note also that the geometry is such as to yield a scattering angle of 90°.[62] A. L. Hughes much later described Compton's apparatus in an illuminating note written out in longhand:

The lead box, made by AHC, housed the x-ray tube, also made by A.H.C . . . . The 'Bragg spectrometer' was brought from Cambridge, England in 1920. The calcite crystal on the spectrometer table had a ground and polished face, a technique developed by A.H.C. which roughly doubled the intensity of the reflected x-ray

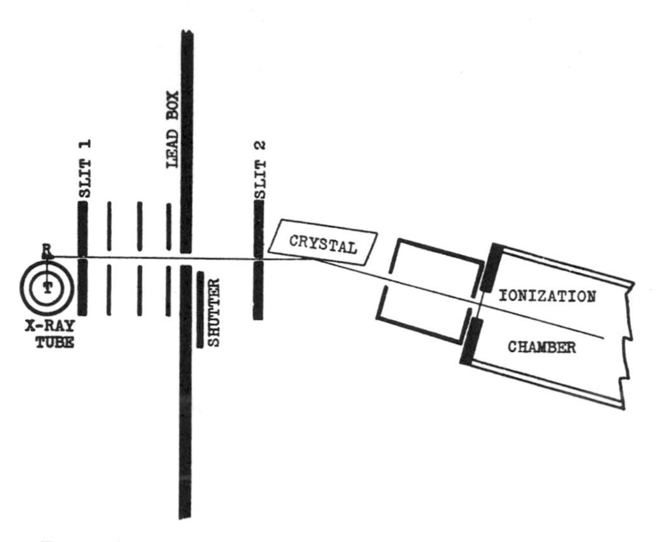

FIG. 6. Schematic diagram of Compton's scattering apparatus.

beam . . . . [The] Compton electrometer with which the measurements were made . . . was the joint invention of Karl T Compton and A.H.C., and was at this time the most sensitive electric current measuring instrument made in the U.S.[63]

Professor Hughes' description points out the crucial role played by Compton's apparatus in securing for him key experimental advantages over his contemporaries.

Compton first published his spectra in his October 1922 report, and many years later he himself recalled the circumstances: ". . . I found myself engaged, as a member of a committee of which William Duane of Harvard was the chairman, in preparing a report for the National Research Council on secondary radiations produced by x rays. When it came to publication of the report, Duane objected to including my revolutionary conclusion that the wavelength of the rays was increased in the scattering process . . . because he felt that the evidence was inconclusive. At the insistence of A. W. Hull, however, this portion of my report was included in the publication."[64] R. S. Shankland recalls that Compton always regarded Hull's attitude as stemming from his conception of a proper approach to science, rather than as constituting momentary support for Compton as an individual scientist.[65]

The exact wording of the portion of Compton's report that was included was as follows: "Recent spectroscopic measurements by the writer show that the secondary rays [solid curve] have suffered a distinct change in wavelength. Thus in [Fig. 7] the spectrum of the molybdenum rays after being scattered at 90° by graphite show the K lines at angles distinctly greater than

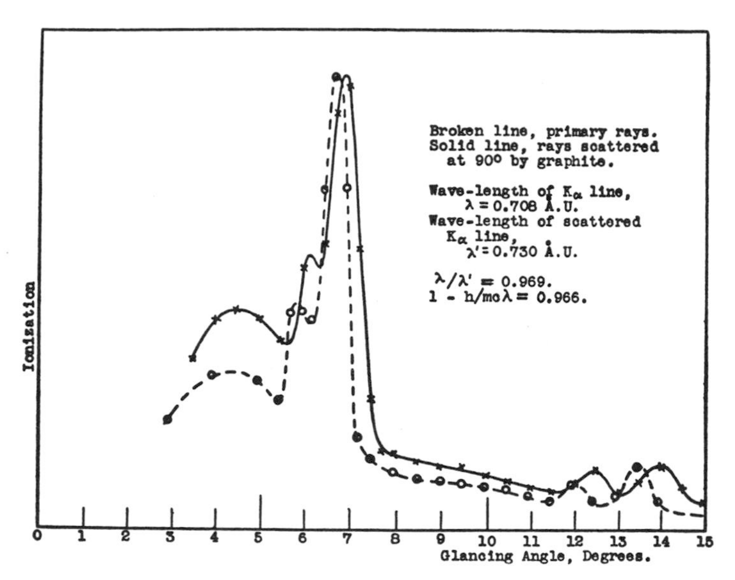

FIG. 7. Compton's first publication of the primary and secondary MoKα spectra.

those at which they occur for the primary beams." He noted that these spectra seemed "to support the absorption measurements in showing the existence of both the unchanged and the longer wave-lengths in the secondary beam. In any case the spectrum shown in [Fig. 7] leaves no doubt but that a large part of the secondary X-rays have suffered a real change in wave-length. According to the writer's *absorption* measurements, over the range of primary rays from .7 to .025 [Å], the wave-length of the secondary X-rays at 90° with the incident beam is roughly 0.03 [Å] greater than that of the primary ray which excites it."[66]

I have italicized the word "absorption" in the foregoing passage for the following reason. Figure 7 shows very clearly that Compton took measurements in the vicinity of the peak at angular intervals of ¼°, which as we have seen corresponds to wavelength intervals of about 0.026 Å. These spectra, therefore, may have been taken as early as those which he had taken in March (Fig. 2b). Yet, Compton has now concluded that the change in wavelength is "roughly 0.03 [Å]"—and not an order of magnitude larger, as he had maintained earlier. He now realized, therefore, that the relative change in wavelength for X-rays is *small,* not large—and it seems quite possible that he may have been led to this most significant conclusion by reexamining his old *absorption* measurements.

Given that the wavelength of the secondary X-rays was only about 0.03 Å greater than that of the primary X-rays, however, the most fundamental question of all still remained: how could one account for this change in wavelength? Compton explained:

> If the incident X-rays are homogeneous, as in the writer's experiment, the scattered rays must be homogeneous and of the same wave-length unless a Doppler effect is present. But in order to account for the observed softening of the secondary rays as due to a Doppler effect, the scattering particles would have to be moving in the direction of the primary beam at a speed comparable with that of light. This is not possible on the classical theory, which supposes that all the electrons in the radiator are effective in scattering. Thus the classical electrical theory appears irreconcilable with the view that the part of the secondary rays that are of greater wave-length than the primary beam are truly scattered. *The assumption of a general fluorescent radiation is the obvious and apparently the only alternative.* On this view, only that part of the secondary radiation whose wave-length is identical with that of the primary beam is truly scattered, and that of greater wave-length is fluorescent.[67]

The issue was apparently closed—once again a "general fluorescent radiation" of the proper wavelength had been excited in the scatterer! No hint of a quantum interpretation was present. Yet, immediately following the remarks quoted above, Compton included a section significantly entitled *The Doppler Effect in Secondary X-rays.* It almost seems—and this may indeed have been the case—as if Compton had written his entire report when, at the last moment, he decided to insert an alternative explanation of the change in wavelength. He wrote:

> On the basis of the quantum theory a different hypothesis may be formed. Let us suppose that each electron when it scatters X-rays receives a whole quantum of energy and reradiates the whole quantum in a definite direction. The momentum which the scattering electron receives from the radiation will then be $h\nu/c$ . . . . This will result in a velocity in the forward direction which will produce a Doppler effect as the scattered rays are observed at different directions. In addition, as the electron radiates a quantum of energy toward the observer, the conservation of momentum principle demands that the electron shall recoil with a momentum $h\nu'/c$, where ν' is the average frequency of the scattered radiation. For cases in which the resulting velocity of the electron is small compared with the speed of light, it can be shown on this basis that the ratio of the average frequency of the rays scattered at 90° to that of the incident rays should be $1 - h/mc\lambda$. In the case of the molybdenum $K\alpha$ line ($\lambda = .708$ Å) this calculated ratio is 0.966, while the value of the ratio taken from the experiments shown in [Fig. 7] is 0.969. This close numerical agreement would suggest that we should consider scattering as a quantum phenomenon instead of obeying the classical laws of electricity. . . .[68]

After years of research on the scattering of high-frequency radiation, Compton finally saw in the quantum theory of radiation the key to the explanation of the observed softening. Or did he? Let us examine this remarkable paragraph in detail and carry out the calculation indicated in it.

Compton first suggests that each electron receives from the radiation a whole quantum of energy, which carries with it momentum $h\nu/c$. As a result, the electron moves in the direction of the incident radiation with a certain velocity, say v, which may be determined from conservation of momentum:

$$\frac{h\nu}{c} = mv, \qquad \text{or} \quad v = \frac{h\nu}{mc} = \frac{h}{m\lambda}. \tag{5.7}$$

Next, *and precisely as he had stated earlier in the year,* Compton suggests that the secondary electron—evidently some kind of oscillator—reradiates this energy as it travels forward, so that the Doppler effect may be invoked to determine its wavelength or frequency. In the paragraph immediately following the one above, Compton gave an explicit expression for the Doppler effect:

$$\frac{\lambda_1}{\lambda_2} = \frac{1 - \beta \cos \theta_1}{1 - \beta \cos \theta_2} = \frac{\nu_2}{\nu_1}, \tag{5.8}$$

where $\beta = v/c$. This is precisely the expression he used earlier, and substituting $\theta_1 = 0°$ and $\theta_2 = \pi/2$ to agree with Compton's experimental configuration, it once again reduces to

$$\frac{\nu_{\pi/2}}{\nu_0} = 1 - \frac{v}{c}. \tag{5.9}$$

One may now eliminate v by using equation (5.7) to obtain

$$\frac{\nu_{\pi/2}}{\nu_0} = 1 - \frac{h}{mc\lambda} = \frac{\lambda_0}{\lambda_{\pi/2}}, \tag{5.10}$$

which is precisely the expression Compton cited, and which is also shown along side of his spectra in Fig. 7. (Note of course that $\lambda_0 = \lambda$ and $\lambda_{\pi/2} = \lambda'$.)

In his earlier paper Compton had eliminated the electron velocity v by using conservation of *energy*, $h\nu = \frac{1}{2}mv^2$—and had thereby accounted for the large (35%) change in wavelength. Now, Compton eliminated the electron velocity v by using conservation of *momentum*, to obtain

$$\lambda_0/\lambda_{\pi/2} = 1 - h/mc\lambda = 1 - 0.034 = 0.966.$$

This is the ratio that Compton asks us to compare with the experimentally observed ratio $\lambda_0/\lambda_{\pi/2}$ = 0.708 Å/0.730 Å = 0.969. Once again, who could ask for better agreement?

In retrospect, no doubt the most dramatic aspect of Compton's present derivation of the change in wavelength—the second stage in what might be called his "classical-quantum compromise"—is that it rested *not* on a wedding of conservation of momentum with conservation of energy, but on conserva-

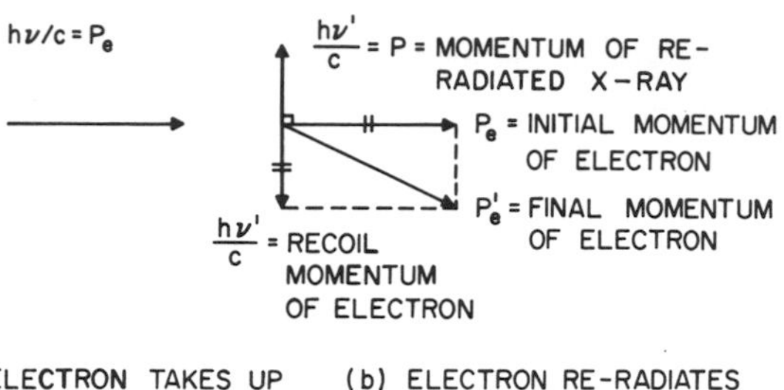

Fig. 8. Vector diagrams illustrating Compton's two-step scattering process.

tion of momentum *alone*. Compton has *not* described a *single-step* "billiard ball" collision process. Rather, he has described a *two-step* process, in which the electron *first* takes up momentum $h\nu/c$, and *then*—note Compton's word—"'reradiates" a quantum of radiation "in a definite direction." This two-step process which Compton envisioned may be indicated schematically in the *two* vector diagrams shown in Fig. 8 (note the key importance of a scattering angle of 90°). This two-step process seems to be the only interpretation consistent both with conservation of momentum, in the sense described above, and with Compton's description of his present "quantum conception of scattering."

6. *Other Evidence and a Critique of Compton's Compromise*

In addition to his spectra, Compton cited two other pieces of evidence for his present theory of scattering. First, he stated that the "view that these softened secondary X-rays are really scattered is apparently confirmed by a study of their polarization."[69] That is, Compton now *reversed* his earlier opinion, and took his and Hagenow's experiments to prove that the softened secondary X-rays are *not* fluorescent, but scattered X-rays—although, as we have seen, he had a very special interpretation of the word "scattered." Second, Compton stated that the "view that much of the secondary radiation comes from electrons moving at high speed is supported by the apparent Doppler effect observed in the case of secondary γ-rays."[70] Thus, Compton substituted into equation (5.8) the two sets of values $(\lambda_1, \theta_1) = (0.03\ \text{Å}, 20°)$ and $(\lambda_2, \theta_2) = (0.08\ \text{Å}, 135°)$—which he himself had previously determined from γ-ray *absorption* measurements—and calculated β to be 0.52. This value of β, Compton argued, agreed both with estimates based on the observed secondary γ-ray asymmetry, and with A.S. Eve's old (1904) estimate of $\beta = ½$.

In fact, this train of argument suggests a plausible explanation of how Compton actually arrived at his present "quantum conception of scattering."

Thus, the simple Doppler expression, $\lambda_0/\lambda_{\pi/2} = 1 - v/c$, shows that the observed *large* change in *absorption coefficient for γ-rays* scattered through 90° could obviously be accounted for if the secondary β-ray oscillators were traveling forward at about one-half the velocity of light ($\beta = v/c = ½$ implies $\lambda_0/\lambda_{\pi/2} = ½$). Moreover, for typical γ-ray wavelengths, Compton could easily have seen that this velocity is roughly the velocity to be *expected* if the β-rays receive *momentum* $mv = h\nu_0/c$ from the incident γ-rays. This would have led him to suspect that conservation of momentum was the key, and not conservation of energy, as he had believed earlier when attempting to account for a presumed *large* wavelength shift for *X-rays*. Then, applying conservation of momentum *alone* to the scattering of *X-rays* through 90°, Compton would have immediately found that it predicts a much *smaller* change in wavelength, and hence in *absorption coefficient*. By then closely reexamining his X-ray *spectroscopic* observations, he would also have found, indeed, that the MoK_α *X-rays* had actually undergone a *small* change in wavelength, thereby vindicating his conservation of momentum picture.

It was worth noting, parenthetically, that Compton seems to have been unaware of certain experimental evidence published in 1922 by J. A. Gray[71] which argued against the very foundation of his interpretation. Gray had designed experiments which indicated that the intensity of secondary X-radiation produced is not directly related to the number of secondary β-particles liberated, which he took to mean "that the observed change in frequency and consequent softening of secondary X-rays is not due to their being formed by secondary β-rays. . . ."[72] Nowhere did Compton hint that he was aware of Gray's conclusion.

All of which puts the objection that William Duane raised against Compton's work in a new light. In the first place, it seems to be *true* that Compton's spectroscopic measurements could in fact be considered "inconclusive"—as we have seen, the maximum uncertainty in each measurement was about 0.026 Å, or roughly as large as the change in wavelength (0.03 Å) which he reported. Secondly, as we have also seen, it would be very misleading to conclude that *because* Compton had published evidence for the above *small* change in wavelength, he must have been in complete possession of his correct quantum theory of scattering—which was presumably what Duane was fundamentally objecting to. We have seen exactly what Compton meant by his "quantum conception of scattering" in October 1922. Perhaps Duane was not completely familiar with Compton's past work, but if he was, he would certainly have recognized Compton's present theoretical ideas as a natural outgrowth of his earlier ones. It was indeed, as Compton wrote, a "revolutionary conclusion that the wavelength of the rays was increased in the scattering process"—but his October 1922 explanation of that increase in wavelength was not.

Compton's thought at this time was clearly vacillating between classical and quantum ideas, between the old and the new. It is therefore natural to ask how Compton, while in this state of conceptual flux, viewed the most fundamental issue of all, the ultimate nature of radiation. His uncertainty on this issue, it developed, was actually increased unexpectedly by what he had meanwhile learned about a related, but seemingly prosaic subject—the separate contributions to the mass-absorption coefficient μ/ρ. He regarded these as three in number, which he symbolized by σ/ρ, τ/ρ, and ω/ρ. The latter two require brief comment.

Compton found, in considering the subject further, that he could obtain a "formal derivation" of the $Z^4\lambda^3$ variation of the fluorescent-absorption coefficient τ/ρ by assuming that the incident ("attenuated") electromagnetic radiation sets into motion "slightly damped electronic oscillators . . . of natural frequencies distributed between the critical absorption frequency and infinity." This was not a very "plausible" assumption, as Compton himself admitted, but it led to the correct result. Compton stated: "the fact that without introducing the quantum concept we can thus obtain a formula which is in many ways so satisfactory, strongly suggests [that] the X-ray absorption is not a quantum phenomenon. The results support rather the view that X-ray absorption is a continuous process, obeying the usual laws of electrodynamics."[73]

That would have been Compton's final conclusion if, after the above discussion had been written, L. de Broglie had not published a paper[74] in which he had derived the $Z^4\lambda^3$ dependence on entirely different assumptions. According to Compton, de Broglie had assumed "1. that Wien's energy distribution law holds for black body radiation for X-ray frequencies, 2. that Kirchhoff's law relating the emission and absorption coefficients of a body is valid, and 3. that Bohr's hypothesis connecting the energy of the electron's stationary states with the corresponding critical absorption frequencies is correct . . . ." On these assumptions de Broglie had obtained an expression which differed from Compton's only by a factor of ½. In a letter to Compton, however, de Broglie had pointed out "that his theory does not definitely determine the value of this constant, so that there is no positive contradiction between the two results." "It is certainly remarkable," concluded Compton, "that results so nearly identical should be obtained on the basis of wholly different assumptions."[75]

Equally remarkable were the competing ways in which the third contribution, symbolized by ω/ρ, could be understood. Compton termed this contribution, which had "almost escaped detection," the "momentum absorption coefficient." This coefficient was not to be confused with the coefficient a/ρ, which he had earlier rejected as unimportant for his purposes. Rather, he explained that: "Two alternative hypotheses of the origin of this . . . absorption . . . may be presented. 1. We may think of this energy as spent in exciting a form of general fluorescent radiation through the medium of the secondary β-rays. Or 2. we may suppose, on the quantum idea of scattering, that when an electron receives a quantum of γ-ray energy, the momentum of the quantum gives to the scattering electron a considerable momentum, so that the scattered energy is equal to the energy absorbed less the kinetic energy of recoil of the scattering electron."[76]

Therefore, not only was there a classical and a quantum interpretation of τ/ρ, there was also a classical and a quantum interpretation of ω/ρ. Which interpretation was correct? Were X-rays and γ-rays electromagnetic waves, or quanta of radiation? Could X-rays interact only with groups of electrons, or did they interact with single electrons? Was scattering a classical, or a quantum phenomenon? Compton discussed these questions and came to definite answers in his report in a long concluding section entitled *Nature of Radiation.* This section, written on the eve of discovery, constitutes one of the most fascinating discussions in the literature of physics. It merits quotation *in extenso*:

Let us recall that it has been found possible to account satisfactorily for many of the experiments on the scattering of X-rays on the basis of the classical electrodynamics . . . . The *inadequacy* of the quantum conception as applied to the scattering of radiation may be shown in the following manner. Experiment shows that a cathode electron at impact may give rise to one quantum of radiant energy of frequency $\nu = Ve/h$, where V is the potential applied to the tube, or to several quanta of lower frequency. Let us consider the case where $\nu > Ve/2h$, so that not more than one quantum of energy of this frequency can be radiated at the impact of each electron. We find that the resulting X-ray when scattered by matter shows the phenomenon of excess scattering. But this phenomenon is to be accounted for by the fact that the incident X-ray excites secondary radiation of the *same frequency* from *several* electrons in the same atom, with the result that these radiations cooperate in the forward direction and partially interfere with each other at larger angles with the incident beam.

If, however, it is assumed that each electron both absorbs and emits radiation in quanta, the primary ray can excite radiation from only *one* electron, and these interference effects become inexplicable. . . . The same difficulty arises in explaining the reflection of X-rays by crystals . . . . Thus the interference phenomenon occurring in the scattering and reflection of X-rays are *inconsistent* with the view that an electron always emits scattered radiation in quanta.

There remains the possibility that the radiation itself always occurs in quanta, but that when scattering occurs the quantum of energy is radiated not by a single electron but by some group of electrons, affected by the incident wave. On this view it is a matter of probability whether or not an incident ray shall pass through a scattering body. If it passes through, it remains undiminished as a whole quantum;

if it is scattered, some group of electrons in the body are effective, and the whole quantum is scattered, none passing through. This view also requires radiation of energy in discrete quanta, though now the energy quantum may be radiated by any number instead of by a single electron.

It is important therefore to point out that a quantum of radiant energy cannot always retain its integrity, and that its parts may be separately scattered and absorbed. Hence scattered energy is not necessarily radiated in quanta, nor is radiation necessarily absorbed in integral quanta.

Perhaps the most convincing example of the division of radiation into parts of less than a quantum is that of the Michelson interferometer. We may suppose that one quantum of energy of wave-length λ strikes the semi-reflecting mirror, half being reflected to the movable mirror and half transmitted to the fixed one. The returning waves recombine at the semi-silvered surface, and in order that the energy quantum may remain complete, we may suppose that both components proceed together, either toward the eye or in the direction of the source. If the mirror is exactly half-silvered, the probability will be equal for the two directions. It is essential, however, that the initial division of the quantum occur at the half mirror, for if all the radiation during any given period went to either the movable or the fixed mirror, there would be nothing with which it could interface on returning to the half mirror. It seems that the only possible interpretation of this interference is that while part of the energy quantum goes to the movable mirror, another part is simultaneously going to the fixed one. Thus neither part of the divided beam can carry the whole quantum of energy.

It remains to show that such a *part* of the original quantum can *by itself* be absorbed or scattered. Let us suppose the silvering of the mirror is accurate, so that the interference of the rays at the eye is complete. Then imagine a thin absorption screen placed in the path of one of the divided beams. This absorption might for example be due to a slight tarnish on one of the mirrors. That absorption actually occurs is made evident by the incomplete interference of the recombined beam. But we have seen that *neither part* of the divided light beam possesses a whole quantum of energy. *There accordingly appears to be no escape from the conclusion that radiation may be absorbed in amounts less than a quantum.*

In a similar manner the light may be reflected or scattered from the surfaces of an interferometer mirror. It seems that in this scattering process we have an example of radiation of energy in smaller units than a quantum. The only escape from this conclusion would be to imagine a gradual absorption of energy by the electrons on the mirror surface, until a whole quantum is accumulated, and then a simultaneous emission of the radiation by the electrons each in its proper phase to produce interference effects. Such a view presents very grave difficulties.

A consideration of these and similar interference phenomena, whose importance must not be minimized, seems to lead *with certainty to the conclusion that under certain conditions radiation does not occur in a definite direction, nor in definite quanta; that radiation may be absorbed in fractions of a quantum; and that, in the process of scattering at least, radiation may be emitted in fractions of a quantum.* Most of the experiments on X-ray scattering and reflection, on the other hand, receive satisfactory explanation if the X-rays spread over a wide enough solid angle to excite oscillations of a large number of electrons in their proper phases. *This study therefore supports the view that radiation occurs in waves spreading throughout space in accord with the usual electrical theory.*

The experiments described above . . . , showing that the wave-length of the scattered X-ray is greater than that of the incident X-ray, present, however, a serious difficulty to this conclusion. This change in wave-length was found to receive quantitative explanation on the view that the radiation was received and emitted by each scattering electron in discrete quanta. No alternative explanation has as yet suggested itself. *Nevertheless, the cogency of the argument based on interference phenomena is so great that it seems to me questionable whether the quantum interpretation of this experiment is the correct one.*

If then radiation may under certain conditions be emitted in infinitesimal fractions of a quantum, and if absorption is a continuous process, the question arises, has the quantum any real physical significance? To this our study of X-rays gives a definite affirmative answer. Thus we have seen that de Broglie's and Ellis' experiments . . . are in accord with the view that each photoelectron leaves its normal position in the atom with a kinetic energy $h\nu$, where ν is the frequency of the incident rays. Conversely, Duane and Hunt's experiments indicate that if the whole kinetic energy of a cathode ray is transformed to radiation at a single impact, then $E_{\text{kinetic}} = h\nu$. A similar relation also expresses quantitatively the frequency of the rays emitted as an electron falls from one energy level to another within the atom. It thus appears that the quantum law may describe a reversible mechanism whereby energy may be interchanged between radiation and the kinetic energy of an electron.

This mechanism is presumably that which is responsible for the fluorescent absorption of X-rays, since it is the energy thus absorbed which appears again as the kinetic energy of the photoelectrons. But thc energy dissipated in scattering is not thus transformed, and need not therefore have any dependence upon the quantum mechanism. *On this view there is no reason to question the application of the classical electrodynamics to the problem of scattering; so calculations on this basis may be used in studying the structure of matter. Such calculations may be employed with the greater confidence since we have found them capable of explaining the principal phenomena of X-ray scattering.*[77]

In the final analysis, in October 1922, Compton found the quantum theory of radiation wanting.

E. References

[1] "The Spectrum of Secondary X-rays," *Phys. Rev.* 19 (1922):267.

[2] Compton Notebook B.1. (Wash. U., A.I.P.), p. 10.

[3] See G. E. M. Jauncey, "Secondary X-rays from Crystals," *Phys. Rev.* 19 (1922):435–436; "The Scattering of X-Rays by Crystals," *Phys. Rev.* 20 (1922):405–420.

[4] Compton, "Spectrum," p. 267.

[5] Compton Notebook B.1. (Wash U., A.I.P.), p. 11 (reproduced by permission of Mrs. A. H. Compton).

[6] *Ibid.*, pp. 20–32; Compton Notebook B.2. (Wash. U., A.I.P.), pp. 1–2.

[7] Compton, "Spectrum," p. 267.

[8] Compton Notebook B.2. (Wash. U., A.I.P.), p. 3.

[9] *Ibid.*, pp. 6–10. Compton actually took some data at "C. 125°" also, but evidently never used them. The *absolute* intensities of the primary Mo spectrum were of course much larger than those of the secondary spectrum. In plotting, I simply assigned the $K\alpha$ peaks in both Mo spectra the same *relative* value of 100. The graphs are plotted as Compton probably would have plotted them at the time. Later, it became evident that there are only two, not three, small peaks in the secondary spectrum. See, for example, Compton's paper "The Scattering of X-Rays," *J. Franklin Inst.* 198 (1924):60–61.

[10] "Spectrum," pp. 267–268.

[11] See A. H. Compton, "The Scattering of X-Rays," *J. Franklin Inst.* 198 (1924):60–61.

[12] Compton, "Spectrum," (reference 1) p. 268. For consistency of notation I have replaced Compton's capital C with a lower-case c.

[13] I previously summarized this work in "Arthur Holly Compton and the Discovery of the Total Reflexion of X-rays," *Acts XIIth International Congress in the History of Science* (Paris, 1968), pp. 101–105. See also R. S. Shankland's introduction to Compton's *Scientific Papers* (Chapter 3, reference 7) for information on where this work eventually led; for example, it led to the experiments of Compton, J. A. Bearden, and R. L. Doan.

[14] "Nouvelles measures de précision dans le spectre de rayons X," *Compt. Rend.* 173 (1921):1350–1352, especially p. 1352.

[15] University of Lund Dissertation, 1919.

[16] "On the X-Ray Spectra of Tungsten," *Phys. Rev.* 16 (1920):532.

[17] "Zur Frage der Brechung der Röntgenstrahlen," *Z. Phys.* 1 (1920):40–41.

[18] "Abweichung vom Braggschen Reflexionsgesetz der Röntegenstrahlen," *Phys. Z.* 21 (1920):617–619.

[19] "Sur le degré d' exactitude de la loi de Bragg pour les rayons X," *Compt. Rend.* 174 (1922):745–746.

[20] See H. A. Lorentz, *The Theory of Electrons,* 2d ed. (New York: Dover, 1952), p. 149.

[21] Compton Notebook B.2. (Wash. U., A.I.P.), p. 12.

[22] "Total Reflection of X-rays from Glass and Silver," *Phys. Rev.* 20 (1922):84. For the index of refraction, I have replaced Compton's n with μ.

[23] Upper limits on δ had, however, been already determined by C. G. Barkla, "Experiments to Detect Refraction of X-Rays," *Phil. Mag.* 31 (1916):257–260, as well as D. L. Webster and H. Clark, "A Test for X-Ray Refraction Made with Monochromatic Rays," *Phys. Rev.* 8 (1916):528–533. Only in 1924 were direct measurements made with a prism. See A. Larson, M. Siegbahn, and I. Waller, "Der Experimentelle Nachweis der Brechung von Röntgestrahlen," *Naturwiss.* 12 (1924):1212–1213.

[24] See A. H. Compton, "Secondary Radiations Produced by X-Rays, and some of their Applications to Physical Problems," *Bull. Natl. Res. Counc.* 4 (1922):48.

[25] *Ibid.*, p. 51.

[26] "Arthur Holly Compton," *Natl. Acad. Sci. Biog. Mem.* 37 (1965):86.

[27] "The Intensity of X-ray Reflection from Powdered Crystals," *Nature* 10 (1922):38 (with N. L. Freeman).

[28] Marjorie Johnston (ed.), *The Cosmos of Arthur Holly Compton* (New York: Knopf, 1967), p. 450.

[29] *Bull. Natl. Res. Counc.* 4 (1922):1–56.

[30] "Data Relating to X-Ray Spectra," *Bull. Natl. Res. Counc.* 1 (1920):383–408.

[31] "Intensity of Emission of X-Rays and their Reflection from Crystals," *Ibid.*, pp. 410–426.

[32] "Problems of X-Ray Emission," *Ibid.*, pp. 427–455.

[33] Compton, "Secondary Radiations," (reference 24) p. 3.

[34] However, Compton brought the work on this subject up to date. In addition to citing all of the physicists (including himself) whose work on "excess scattering" has already been discussed in this book, Compton discussed R. Glockner and M. Kaupp's "Atomstruktur and Streustrahlung," *Ann. Physik* 64 (1921):514–565, in which the scattering electrons were assumed to be in coplanar rings rotating at different speeds, as well as R. Glockner's "Die Streuwirkung des räumlichen Kohlenstoffatoms," *Z. Phys.* 5 (1921):54–60, in which the atom was assumed to be "Landés pulsating tetrahedronal carbon atom."

[35] Compton, "Secondary Radiations," p. 8.

[36] *Ibid.*, p. 10.

[37] *Ibid.*, pp. 10–11.

[38] Recall that his brother Karl and he himself (briefly) had been students of O. W. Richardson, who was completely abreast of current research, including of course Bragg's.

[39] See Chapter 1, sections B and C.

[40] Compton, "Secondary Radiations," (references 24) p. 23.

[41] *Ibid.*, pp. 24–25.

[42] *Ibid.*, p. 29.

[43] "The Magnetic Spectrum of the β-Rays Excited by γ-Rays," *Proc. Roy. Soc. (London)* [A] 99 (1921):261–271; "β-Ray Spectra and Their Meaning," *Proc. Roy. Soc. (London)* [A] 101 (1922):1–17.

[44] "Sur les spectres corpusculaires et leur utilisation pour l'étude des spectres de rayons X," *Compt. Rend.* 173 (1921):1157–1160; "Sur les spectres corpusculaires des éléments," *Compt. Rend.* 174 (1922):939–941; "Sur le modèle d'atome de Bohr et les spectres corpusculaires," *Comp. Rend.* 172 (1921):746–748 (with L. de Broglie). See also M. de Broglie, *Les Rayons X* (Paris: A. Blanchard, 1922); *Introduction à la physique des rayons X et gamma* (Paris: Gauthier-Villars, 1928).

[45] Compton, "Secondary Radiations," p. 27; three misprints have been corrected.

[46] *Ibid.*, pp. 27–29.

[47] "On X-rays and the Theory of Radiation," *Phil. Trans. Roy. Soc. London* [A] 217 (1917):357.

[48] "Notes on the Absorption and Scattering of X-Rays, and the Characteristic Radiations of J-series," *Phil. Mag.* 34 (1917):284.

[49] "The Absorption of the Radiation emitted by a Palladium Anticathode in Rhodium, Palladium and Silver," *Proc. Roy. Soc. (London)* 94 (1918):341.

[50] "On the Absorption of X-Rays in Copper and Aluminum," *Proc. Roy. Soc. (London)* [A] 94 (1918):573.

[51] "On the Spectrum of X-Rays from an Aluminum Target," *Phys. Rev.* 14 (1919):393. See also their joint paper "Are the Frequencies in the K Series of X-rays the Highest Frequencies Characteristic of Chemical Elements?" *Phys. Rev.* 13 (1919):289–291.

[52] "Recherches spectrométriques sur les rayons X," *Ann. Phys. (Paris)* 14 (1920):82.

[53] "The Mass-Absorption Coefficient of Water, Aluminium [*sic*], Copper and Molybdenum for X-Rays of Short Wave-length," *Phys. Rev.* 15 (1920):548.

[54] "The Mass Absorption and Mass Scattering Coefficients for Homogeneous X-Rays of Wave-length between 0.13 and 1.05 Angström units in Water, Lithium, Carbon, Nitrogen, Oxygen, Aluminum, and Iron," *Phys. Rev.* 17 (1921):294.

[55] "Absorption of X-Rays," *Phys. Rev.* 18 (1921):29.

[56] " 'J' Radiation," *Phil. Mag.* 42 (1921):720–721.

[57] See reprinted lecture in *Nobel Lectures: Physics*, Vol. 1 (Amsterdam: Elsevier, 1967), p. 398.

[58] Quoted in H. S. Allen, "Charles Glover Barkla," *Roy. Soc. Obit. Not. (London)* 5 (1945–1948):356.

[59] Compton, "Secondary Radiations," (reference 24) p. 36.

[60] *Ibid.*, p. 16.

[61] I am very grateful to Professor R. E. Norberg, Chairman of the Physics Department of Washington University, for sending me this photograph, and for permission to reproduce it here. The diagram was first published in Compton's paper "The Spectrum of Scattered X-rays," *Phys. Rev.* 22 (1923):410.

[62] For complete information on the tube voltage, current, etc., see S. K. Allison, "Arthur Holly Compton," (reference 26) p. 84.

[63] Undated description on deposit in the Washington University Archives, St. Louis. In "The Scattering of X-Rays," *J. Franklin Inst.* 198 (1924):59–60, Compton notes explicitly the crucial importance of obtaining high intensities for successful spectroscopic work.

[64] "The Scattering of X Rays as Particles," *Am. J. Phys.* 29 (1961):817.

[65] Private communication to author, September 8, 1971.

[66] Compton, "Secondary Radiations," pp. 16–17; italics added.

[67] *Ibid.*, p. 18; italics added.

[68] *Ibid.*, pp. 18–19.

[69] *Ibid.*, p. 20.

[70] *Ibid.*, p. 19.

[71] "Primary and Secondary β-Rays," *Trans. Roy. Soc. Can.* 16 (1922):125–128; "The Softening Exhibited by Secondary X-Rays," *Trans. Roy. Soc. Can.* 16 (1922):129–134.

[72] Gray, "Softening," p. 134.

[73] Compton, "Secondary Radiations," p. 42.

[74] "Sur la théorie de l'absorption des rayons X par la matière et la principe de correspondance," *Compt. Rend.* 173 (1921):1456–1458; see also "Rayons X et equilibre thermodynamique," *J. d Phys. e Radium* 3 (1922):33–45.

[75] Compton, "Secondary Radiations," (reference 24) pp. 43–44.

[76] *Ibid.*, p. 45.

[77] *Ibid.*, pp. 52–55; italics (with the exception of the second set) have been added, subsection numbers have been omitted, and two misprints have been corrected.

CHAPTER 6

The Compton Effect

A. Introduction: The Relative Autonomy of Compton's Research Program (1917–1922)

One of the most striking aspects of Compton's research program, when viewed in its entirety, was its relative autonomy. His major theoretical insights depended in large measure on his own expertly conceived experiments at the University of Minnesota, at the Cavendish Laboratory, and at Washington University. He experienced no sudden conceptual breakthroughs stimulated by a forceful theoretical achievement of another physicist. Nothing could be further from the truth, for instance, than the impression often gained from textbook accounts,[1] that Compton, after somehow learning of Einstein's light quantum hypothesis, quickly (in a flash of insight) solved the billiard-ball collision problem and designed a clever X-ray scattering experiment to search for the predicted change in wavelength.

Quite the contrary. Compton only very gradually came to take Einstein's light quantum hypothesis and its associated energy and momentum concepts more and more seriously. In fact, when Compton finally cited Einstein's 1905 paper in his October 1922 report, he did so in connection with the photoelectric effect equation and still remarked in a skeptical vein that this "equation was first proposed by Einstein . . . , but was shown by Richardson to be a direct consequence of Planck's radiation formula."[2] That Compton was not influenced in any important way by Einstein's work, and quite likely never even read Einstein's 1905 paper, is evident not only from internal considerations, but also by the way in which Compton again and again repeated his belief, actually a historical fiction as we saw in Chapter 1, that: "It was to account for this [photoelectric] effect that Einstein . . . suggested the reversion to Newton's corpuscular theory of radiation."[3] Compton may have discussed the quantum relation with any number of physicists over the years, but his confidence in it grew only as rapidly as his own experimental program progressed.

This in fact constituted the decisive advantage that Compton's program possessed over those of his contemporaries—his theoretical insights were derived from, and anchored in, his own precise experiments. Compton, and no one else, eventually had the conclusive spectroscopic evidence in hand—the obstacle he had to overcome was to discover its correct interpretation. By contrast, at least two other physicists, Albert Einstein and Erwin Schrödinger, were at this very same time (1921-1922) convinced of the validity and fruitfulness of the light quantum hypothesis, but lacked definitive experimental evidence for it.[4] (The particularly striking case of a third physicist, Peter Debye, will be treated later.) Because they display the overwhelming importance of Compton's experimental researches, it is very instructive to examine Einstein's and Schrödinger's achievements before turning to the final phase in the development of Compton's thought.

B. Einstein's and Schrödinger's Researches (1921–1922)

1. Einstein's Crucial Experiment for Light Quanta

The central importance of locating conclusive experimental evidence for light quanta was apparent to Einstein, who in December 1921 presented a paper to the Berlin Academy in which he proposed a means of doing just that.[5] Einstein suggested observing a beam of excited canal rays (positive ions) moving perpendicularly to the common optical axis of two lenses, L_1 and L_2, positioned in such a way that L_1 would pick up the light emitted by the canal rays and refocus it at the focal point of L_2, which would therefore collimate it. Now, on the one hand, Einstein argued, according to the wave theory the emitted light should be Doppler shifted in such a way that the wavelength of the light leaving L_2 should be greater at the top of the beam than at the bottom. The surfaces of constant phase should therefore fan out from bottom to top, and hence if a dispersive medium were placed beyond L_2, the light should be deviated a few degrees under reasonable conditions. On the other hand, according to the quantum theory of emission, one had reason to believe that the emitted light was monochromatic, and hence it should neither be Doppler shifted nor deviated. On Einstein's urging, this experiment was actually carried out in Berlin by Hans Geiger and Walter Bothe—who found no deviation whatsoever. According to Einstein's analysis, this result did not necessarily constitute positive evidence for quanta, but it definitely constituted negative evidence for waves.

This was a startling conclusion, and one which attracted widespread attention. On the one hand, Arnold Sommerfeld, while admitting that he did not fully understand Einstein's experiment "in spite of Geiger's explanations in Jena," wrote Einstein from Munich on January 11, 1922, that he was happy to take Einstein's word for its validity, and to accept the fact that Einstein had finally discovered a way to puncture the wave theory at some point, and thereby end the dualism in viewpoints.[6] On the other hand, Einstein's Berlin colleague Max von Laue challenged Einstein's conclusion in a physics colloquium, and Einstein's friend Paul Ehrenfest challenged it in correspondence.[7] Ehrenfest argued that Einstein had erred in treating the light traveling in the dispersive medium as consisting of infinitely long plane waves propagating with a certain phase velocity. Rather, it should be treated as consisting of wave *groups* of finite length traveling with a certain *group* velocity. Ehrenfest's key point was that while the individual composite wave-normals do in fact fan out, the wave-normal of the group as a whole does not, and hence one can expect no deviation of the light even on classical theory. Geiger and Bothe's results were therefore inconclusive. Einstein at first disputed Ehrenfest's argument, but at the end of January 1922, less than two months after he had proposed his experiment, Einstein saw that Ehrenfest was "absolutely right."[8] The fundamental conclusion was indisputable. Einstein's experiment was not the decisive experiment Einstein had sought for quanta.

2. *Schrödinger's Quantum Interpretation of the Doppler Effect*

Schrödinger's achievement, which according to A. E. Ruark[9] attracted considerable attention in certain East Coast circles after it became known in mid-1922, was to prove that the Doppler effect could be completely understood by combining Einstein's light quantum hypothesis with Bohr's frequency condition for atomic transitions.[10] This was a very significant achievement, since previously the Doppler effect had been considered by everyone—including Compton and Einstein, as we have just seen—to be characteristic only of wave motion.

Schrödinger motivated his treatment by noting that if one supposes with Schwarzschild, Heurlinger, and Lenz that band spectral lines arise from an optical electron term, a nuclear term, and a molecular rotational term, one could hardly avoid asking what possible meaning might be attached to a molecular *translational* term. The answer, Schrödinger believed, was that such a term would correspond to a Doppler broadening of the spectral lines. In fact, already in 1920 Försterling had attempted to derive the Doppler shift using Bohr's frequency condition for atomic transitions, but his result was unsatisfactory because it predicted only the transverse effect—it implied, as Pauli noted, that the molecule had emitted no linear momentum during its transition, which contradicted what Einstein had termed the "main result" of his 1917 paper.[11] In Schrödinger's words, "according to Einstein's arguments . . . , the emitted quantum $h\nu$ always—and in every co-ordinate system—

must possess linear momentum $h\nu/c$, which is the largest possible momentum that can be associated with this energy." Schrödinger therefore intended to show "that the resulting 'velocity jump' [of the molecule], according to Bohr's frequency condition, yields the Doppler shift precisely, and indeed in detail, as demanded by the theory of relativity."[12]

To illustrate his fundamental ideas, Schrödinger first analyzed the simplest conceivable version of the process under consideration, the case in which a molecule of mass m and translational velocity v_1 suddenly emits a quantum of radiation in its direction of motion, and thereby experiences a reduction in its velocity to v_2. By conservation of energy, using Bohr's frequency condition,

$$h d\nu = (m/2)(v_1^2 - v_2^2); \tag{6.1}$$

and by conservation of momentum,

$$h\nu/c = m(v_1 - v_2). \tag{6.2}$$

Dividing equation (6.1) by equation (6.2) and rearranging, one obtains

$$\frac{d\nu}{\nu} = \left(\frac{v_1 + v_2}{2}\right)\left(\frac{1}{c}\right). \tag{6.3}$$

This is precisely the expression for the simple Doppler shift, provided that the molecule's velocity is given neither by its initial velocity v_1, nor by its final velocity v_2, but by the arithmetic mean of the two.

The same process could now be analyzed more generally by introducing relativistic considerations. Imagine a system of coordinates in which the molecule's center of mass is at rest, and further imagine that the molecule is capable of possessing two different energies, E_1 and E_2, corresponding to two different rest masses, E_1/c^2 and E_2/c^2, in that system of coordinates. Assume now that the molecule is *moving* and that it suddenly emits a quantum of radiation in its direction of motion, thereby reducing its velocity from v_1 to v_2. In the moving system, both the initial and final energies of the molecule are modified by the relativistic factor $1/\sqrt{1 - v^2/c^2}$, so that now by conservation of energy,

$$h\nu = \frac{E_1}{\sqrt{1 - v_1^2/c^2}} - \frac{E_2}{\sqrt{1 - v_2^2/c^2}}; \tag{6.4}$$

and by conservation of momentum,

$$\frac{h\nu}{c} = \frac{E_1 v_1}{c^2\sqrt{1 - v_1^2/c^2}} - \frac{E_2 v_2}{c^2\sqrt{1 - v_2^2/c^2}}. \tag{6.5}$$

Combining these two equations, one obtains

$$E_1\phi_1 = E_2\phi_2, \tag{6.6}$$

where $\phi_1 = \sqrt{(c - v_1)/(c + v_1)}$ and $\phi_2 = \sqrt{(c - v_2)/(c + v_2)}$ have been introduced as convenient abbreviations. Equation (6.4) may now be rewritten either in terms of ϕ_1 or ϕ_2:

$$h\nu = \frac{E_1^2 - E_2^2}{2\,E_1\phi_1} = \frac{E_1^2 - E_2^2}{2\,E_1\phi_2}, \tag{6.7}$$

which also implies that

$$h\nu = \frac{1}{\sqrt{\phi_1\phi_2}}\,\frac{E_1^2 - E_2^2}{2\sqrt{E_1E_2}}. \tag{6.8}$$

If one now defines a frequency $\nu^* = (E_1^2 - E_2^2)/2h\sqrt{E_1E_2}$, the meaning of which will become apparent, one has from equation (6.8) that

$$\nu^* = \nu\sqrt{\phi_1\phi_2} = \nu\sqrt{\frac{c - v_1}{\sqrt{c^2 - v_1^2}}\,\frac{c - v_2}{\sqrt{c^2 - v_2^2}}}. \tag{6.9}$$

Since the usual (relativistic) expression for the longitudinal Doppler shift is given by

$$\nu' = \nu\frac{1 - v/c}{\sqrt{1 - v^2/c^2}} = \nu\frac{c - v}{\sqrt{c^2 - v^2}}, \tag{6.10}$$

one sees that the frequency ν^* plays the role of a frequency observed in the molecule's rest system, where $v_1 = v_2 = 0$. Consequently, equation (6.9) is the correct expression for the longitudinal Doppler shift, provided that the factor multiplying ν is associated neither with the molecule's initial velocity v_1, nor with its final velocity v_2, but with the geometric mean of the two.

It is possible to go even one final step further by removing the restriction that the quantum of radiation is emitted in the direction of motion of the molecule. Thus if the molecule's direction of motion before and after its transition makes the angles θ_1 and θ_2, respectively, with the direction in which the quantum is emitted, the energy equation (6.4) remains the same, but the momentum equation (6.5) must be replaced by *two* equations, that is

$$\begin{aligned} \frac{h\nu}{c} &= \frac{E_1 v_1 \cos\theta_1}{c^2\sqrt{1 - v_1^2/c^2}} - \frac{E_2 v_2 \cos\theta_2}{c^2\sqrt{1 - v_2^2/c^2}} \\ 0 &= \frac{E_1 v_1 \sin\theta_1}{c^2\sqrt{1 - v_1^2/c^2}} - \frac{E_2 v_2 \sin\theta_2}{c^2\sqrt{1 - v_2^2/c^2}}. \end{aligned} \tag{6.11}$$

By now carrying through algebraic manipulations analogous to those above, Schrödinger proved that

$$\nu^* = \nu\sqrt{\frac{c - v_1\cos\theta_1}{\sqrt{c^2 - v_1^2}}\cdot\frac{c - v_2\cos\theta_2}{\sqrt{c^2 - v_2^2}}}. \tag{6.12}$$

Once again, providing that the geometric mean replaces the usual factor multiplying ν, equation (6.12) is *precisely* the expression for the relativistic Doppler shift, incorporating as it does both the longitudinal and transverse effects.[13] The key point is that Schrödinger's entire analysis was possible *only* because he assigned momentum $h\nu/c$ to the emitted quantum of radiation—an assumption which he knew was completely in accord with Einstein's 1917 conclusion.

3. *Conclusion: Bohr's Continuing Skepticism*

Notwithstanding Einstein's experiment and Schrödinger's derivation, it would be very misleading to conclude that by 1922 a substantial number of physicists took Einstein's light quantum hypothesis seriously. The most prominent skeptic was Niels Bohr. From 1913 to 1922 Bohr had steadfastly pursued his researches on atomic structure, but had published nothing substantial on the nature of radiation, in spite of the fact that his atomic model entailed obvious non-classical radiative conceptions.[14] Rather, as Bohr's thought progressed he became more and more convinced that to understand radiative processes, or more generally to understand the link between the quantum and classical realms, he would have to rely on his correspondence principle, the roots of which were already present in Part I of his 1913 trilogy. This principle served as the major guide for Bohr's own researches, and in general it served as one of the two major guides for the old quantum theory, the other being Ehrenfest's adiabatic principle. Concerning radiation, Bohr felt that his correspondence principle provided a touchstone with classical electrodynamics, that it made possible the search for a quantum theory of spectra which would be a "natural generalization of our ordinary ideas of radiation." He refused to admit light quanta into his physics, and at least as early as 1919[15] was entertaining the possibility of abandoning the conservation laws, if necessary, to avoid doing so—a theme that became more and more prominent in his thought over the next several years, as we shall see.

The essence of Bohr's convictions on light quanta was captured in certain remarks he made in his Nobel Lecture in Stockholm on December 11, 1922. At one point he turned to "The Origin of the Quantum Theory" and in this connection discussed the significance of the photoelectric effect. He described it as a phenomenon which was "entirely unexplainable on the classical theory," and he admitted that "the predictions of Einstein's theory have received such exact experimental confirmation in recent years, that perhaps the most exact determination of Planck's constant is afforded by measurements on [it]." He felt compelled to add, however, that: "In spite of its heuristic value . . . , [Einstein's] hypothesis of light-quanta, which is quite irreconcilable with so-called interference phenomena, is not able to throw light on the nature of radiation."[16] Bohr had not yet learned that Arthur Holly Compton—just ten days earlier—had publicly presented his classical paper, "A Quantum Theory of the Scattering of X-rays by Light Elements,"[17] to the American Physical Society in Chicago.

C. The Compton Effect (1922)

1. *Compton's Final Insight and His Discovery of the Quantum Theory of Scattering*

Word of Compton's discovery spread like a shock wave emanating from Chicago. Eventually it would completely and radically change the situation described by Bohr, for it provided precisely the kind of experimental proof for quanta that Einstein had so recently sought. As such, Compton's discovery was beyond doubt one of the most fundamental and far-reaching discoveries of this century.

It was also one of the most unanticipated—even, to a large degree, by Compton himself. One is reminded of Compton's remark that when a scientist attempts to understand a phenomenon, he "tries all the possible answers that he can think of to see which one of them works best";[18] or of Einstein's remark that the route to a discovery involves a scientist's struggling with his problems, his trying everything to find a solution which comes at last often by very indirect means;[19] or of R. P. Feynman's remark which he made when he recalled how he grappled with a certain problem: "I guess I knew every way to modify quantum electrodynamics known to man at the time."[20] An analogous statement could have been made by Arthur Compton. By the time Compton finally enunciated his quantum theory of scattering, he had explored every modification of *classical* electrodynamics known to man at the time. He had, in fact, spent over five years in painstaking theoretical and experimental research before discovering the Compton effect.

Compton's final insight—after having rejected a quantum interpretation of his scattering experiments only a few months earlier—was apparently achieved while scrutinizing his Doppler interpretation of the secondary "fluorescent" radiation; in particular, while scrutinizing its implications for the motion of the radiating secondary β-rays. Compton knew that to produce the large change in wavelength observed for γ-rays scattered through $90°$, the simple Doppler effect, as expressed by $\lambda/\lambda' = 1 - v/c$, demanded that the velocity of the secondary β-particles be roughly one-half the velocity of light ($v/c \sim \frac{1}{2}$)—and this, he wrote, was "an assumption obviously contrary to fact."

It was clear that if any electrons were moving in this manner, it was only a very small fraction of the whole number in the scattering material, and that it must be this small fraction which was responsible for the scattering. The idea thus . . . [presented] itself that an electron, if it scatters at all, scatters a complete quantum of the incident radiation; for thus the number of electrons which move forward would just be equal to the number of scattered quanta.[21]

It was this picture that Compton developed in detail for his audience at Chicago on December first or second, 1922.

Less than two weeks later, after returning to St. Louis, Compton sent the full text of his paper to the editor of *The Physical Review*. Its opening sentence, in complete contrast to the tone of his October National Research Council report, was clear, decisive, and unmistakably challenging: "J. J. Thomson's classical theory of the scattering of X-rays, though supported by the early experiments of Barkla and others, has been found incapable of explaining many of the more recent experiments." The major deviations from Thomson's theory—the low scattering coefficient at short wavelengths, and the forward-backward asymmetry in the angular distribution—now stood out clearly in Compton's mind. In blunt words he spoke of his own efforts—extending up to very recent times—which he had directed toward an explanation of these deviations, particularly the former:

Several years ago the writer suggested that this reduced scattering of the very short wave-length X-rays might be the result of interference between the rays scattered by different parts of the electron, if the electron's diameter is comparable with the wave-length of the radiation. By assuming the proper radius for the electron, this hypothesis supplied a quantitative explanation of the scattering for any particular wave-length. But recent experiments have shown that the size of the electron which must thus be assumed increases with the wave-length of the X-rays employed, and the conception of an electron whose size varies with the wave-length of the incident rays is difficult to defend.[22]

A far more serious difficulty, however, was that the wavelength of the secondary radiation was actually greater than that of the primary.

Such a change in wave-length is directly counter to Thomson's theory of scattering, for this demands that the scattering electrons . . . shall give rise to radiation of exactly the same frequency as that of the radiation falling upon them. Nor does any modification of the theory such as the hypothesis of a large electron suggest a way out of the difficulty. This failure makes it appear improbable that a satisfactory explanation of the scattering of X-rays can be reached on the basis of the classical electrodynamics.[23]

In a footnote Compton acknowledged that the very early view of Gray and Florance, "that the softening is in some way an accompaniment of the

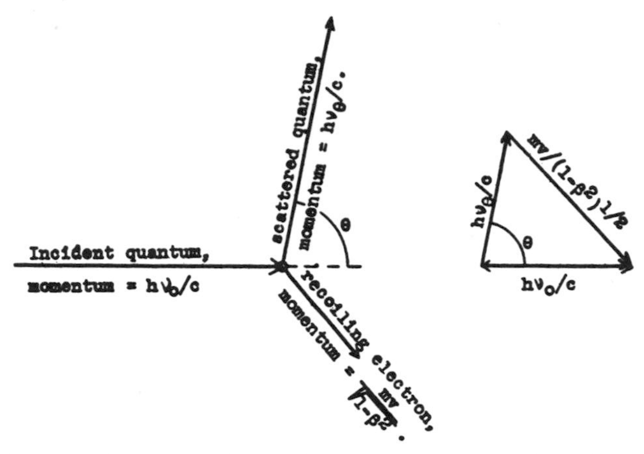

FIG. 1. Quantum-electron collision and momentum conservation diagram for the Compton effect.

scattering process," was indicated by the "considerations brought forward in the present paper . . . [to be] the correct one."[24]

Specifically, these considerations were as follows:

> According to the classical theory, each X-ray affects every electron in the matter traversed, and the scattering observed is that due to the combined effects of all the electrons. From the point of view of the quantum theory, we may suppose that any particular quantum of X-rays is not scattered by all the electrons in the radiator, but spends all of its energy upon some particular electron. This electron will in turn scatter the ray in some definite direction, at an angle with the incident beam. This bending of the path of the quantum of radiation results in a change in its momentum. As a consequence, the scattering electron will recoil with a momentum equal to the change in momentum of the X-ray. The energy in the scattered ray will be equal to that in the incident ray minus the kinetic energy of the recoil of the scattering electron; and since the scattered ray must be a complete quantum, the frequency will be reduced in the same ratio as is the energy. Thus on the quantum theory we should expect the wave-length of the scattered X-rays to be greater than that of the incident rays.
>
> The effect of the momentum of the X-ray quantum is to set the scattering electron in motion at an angle of less than 90° with the primary beam. But it is well known that the energy radiated by a moving body is greater in the direction of its motion. We should therefore expect, as is experimentally observed, that the intensity of the scattered radiation should be greater in the general direction of the primary X-rays than in the reverse direction.[25]

To describe this scattering process quantitatively, Compton drew his classical vector diagrams (Fig. 1[26]) and set up the corresponding equations for conservation of energy and conservation of momentum:

$$h\nu_\theta = h\nu_0 - mc^2\left(\frac{1}{\sqrt{1-\beta^2}} - 1\right) \qquad (6.13)$$

$$\left(\frac{mc}{1-\beta^2}\right)^2 = \left(\frac{h\nu_0}{c}\right)^2 + \left(\frac{h\nu_\theta}{c}\right)^2 + 2\,\frac{h\nu_0}{c}\cdot\frac{h\nu_\theta}{c}\cos\theta, \qquad (6.14)$$

where m is the rest mass of the electron, $\beta = v/c$, and ν_θ is the frequency observed in the direction θ. Solving these two equations simultaneously for ν_θ and β, Compton obtained:

$$\frac{\nu_\theta}{\nu_0} = \frac{1}{1 + 2\alpha\sin^2(\theta/2)}, \qquad (6.15)$$

and

$$\beta = 2\alpha\sin(\theta/2)\,\frac{\sqrt{1 + (2\alpha + \alpha^2)\sin^2(\theta/2)}}{\sqrt{1 + 2(\alpha + \alpha^2)\sin^2(\theta/2)}}, \qquad (6.16)$$

where $\alpha = h\nu_0/mc^2 = h/mc\lambda_0$. Equation (6.15) describes the angular variation of the frequency of the scattered quantum, while equation (6.16) describes the angular variation of the velocity of the recoil electron.

2. *Compton's Major Supporting Evidence—a Critique*

A number of significant points may be brought out by examining in some detail the experimental evidence Compton cited in support of the foregoing equations. Let us first transform equation (6.15) into a more familiar form by inverting it, as follows:

$$\frac{\nu_0}{\nu_\theta} = \frac{\lambda_\theta}{\lambda_0} = 1 + 2\alpha\sin^2\frac{\theta}{2} = 1 + \alpha(1 - \cos\theta). \qquad (6.17)$$

The scattered radiation therefore undergoes a discrete change in wavelength, $\Delta\lambda$, given by

$$\Delta\lambda = \lambda_\theta - \lambda_0 = (h/mc)(1 - \cos\theta). \qquad (6.18)$$

To check this equation, Compton relied on his X-ray spectroscopic measurements—the same measurements, precisely, which he had published a few months earlier in his National Research Council report. The spectra are identical (see Fig. 2[27]), but note that the equation to the right of the MoK_α peaks has been changed. The present equation obviously agrees with equation (6.18) for $\theta = 90°$, which yields $\Delta\lambda = h/mc$. The earlier equation (5.10), however, is given by $\lambda_0/\lambda_{\pi/2} = 1 - h/mc\lambda_0$, which is equivalent to

$$\Delta\lambda = \frac{h}{mc}\left(\frac{\lambda_{\pi/2}}{\lambda_0}\right). \qquad (6.19)$$

These two expressions differ—but not by much. To see precisely how much, let us begin from more general considerations, by assuming an arbitrary scattering angle θ.

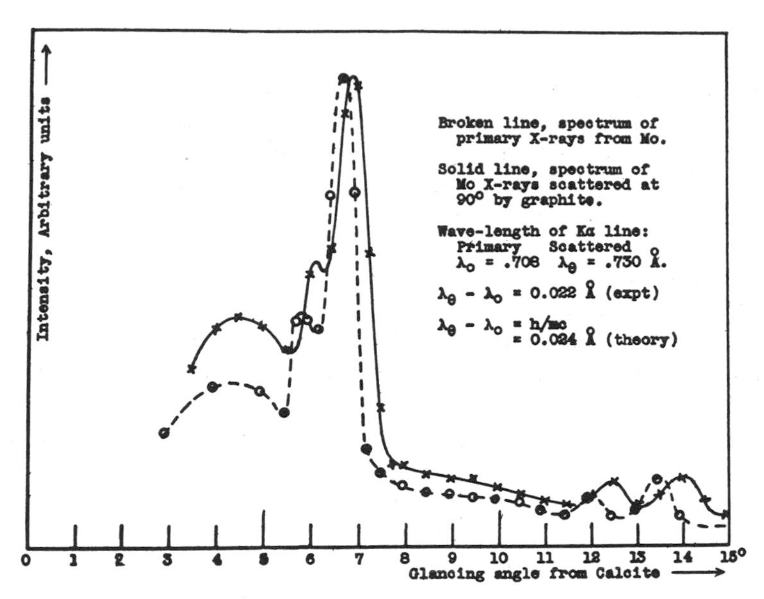

FIG. 2. Primary (dashed) and secondary (solid) spectra for MoKα X-rays scattered through 90° by graphite, shown with revised captions.

Compton's earlier Doppler expression, equation (5.8), may then be rewritten in terms of α and transformed by expanding its denominator, as follows:

$$\frac{\lambda_\theta}{\lambda_0} = \frac{1 - (v/c)\cos\theta}{1 - v/c} = \frac{1 - (h/mc\lambda_0)\cos\theta}{1 - h/mc\lambda_0} = \frac{1 - \alpha\cos\theta}{1 - \alpha}$$

$$= (1 - \alpha\cos\theta)(1 - \alpha)^{-1} = (1 - \alpha\cos\theta)(1 + \alpha + \alpha^2 \ldots). \qquad (6.20)$$

By combining terms of the same order in α, one obtains

$$\frac{\lambda_\theta}{\lambda_0} = 1 + \alpha(1 - \cos\theta)(1 + \alpha + \alpha^2 + \ldots), \qquad \text{or}$$

$$\Delta\lambda = \lambda_\theta - \lambda_0 = \frac{h}{mc}(1 - \cos\theta)(1 + \alpha + \alpha^2 + \ldots). \qquad (6.21)$$

We see, therefore, that Compton's new *and exact* expression, equation (6.18), is simply the zero-order term of his old expression for arbitrary θ, equation (6.21).[28] The derivations of these two expressions, however, rest on entirely different assumptions. The old expression was obtained by combining momentum conservation with the Doppler principle; the new expression was obtained by combining momentum conservation with energy conservation.

The question therefore arises whether or not Compton could have distinguished between these two interpretations on the basis of his spectroscopic data. To do so, we see that Compton would have had to distinguish between terms on the order of $(\Delta\lambda)(\alpha)$, and hence we also see that the answer is clearly negative: $\Delta\lambda$ is about 0.02 Å, and for MoK_α X-rays, $\alpha = h\nu/mc^2 = (0.017 \text{ Mev})/(0.51 \text{ Mev}) \approx 0.03$, so that Compton would have had to distinguish between changes in wavelength of about (0.02 Å) (.03) = 0.0006 Å—which in fact he was unable to do. Therefore, while Compton was perfectly justified in stating, as he did, that the observed change in wavelength (0.022 Å) was in "very satisfactory agreement" with the theoretical value (0.024 Å), he could *not* have used his X-ray spectroscopic data to eliminate his *old* interpretation of this change in wavelength in favor of his *new* interpretation.

The neutrality of Compton's X-ray spectroscopic data in another respect may be seen by analyzing a second theoretical claim, namely, "that the increase [in wavelength] should be the same for all [incident] wavelengths."[29] Compton had already noted in his National Research Council report that the increase in wavelength appeared to be roughly constant at 0.03 Å for incident wavelengths between 0.025 and 0.7 Å, which was of course completely consistent with his present theory. It actually contradicted his old theory, however, since that theory predicts that $\Delta\lambda$ depends on α, and hence on the incident wavelength λ_0. The actual dependence may be estimated by summing the series in equation (6.21), calculating the derivative $d(\Delta\lambda)/d\lambda_0$, and dividing the result by the original equation to obtain

$$\left| \frac{d(\Delta\lambda)}{\Delta\lambda} \right| = \frac{\alpha}{1-\alpha} \frac{d\lambda_0}{\lambda_0}. \tag{6.22}$$

Since $\alpha = 0.03$ for MoK_α X-rays, $\left| \frac{d(\Delta\lambda)}{\Delta\lambda} \right| \approx .03 \frac{d\lambda_0}{\lambda_0}$ for these rays. That is, a relatively large change in λ_0 is translated into a relatively small change in $\Delta\lambda$. For example, a 10% change in λ_0 is translated roughly into a (.03) (.1) = 0.3% change in $\Delta\lambda$, or into an absolute change of about (0.02 Å) (.003) = 0.00006 Å, which to Compton was again completely undetectable. We are therefore forced to conclude, once again, that Compton could not have used his spectroscopic data in this respect to distinguish between his old and new interpretations.

He could, however, have used his *γ-ray absorption* data. It is easy to see for example that for 0.025 Å γ-rays, α is roughly 0.97, and hence a relatively large change in λ_0 is translated into a much larger change in $\Delta\lambda$—so large in fact that it is wholly inappropriate to use the methods of differential calculus to estimate it. Furthermore, the predicted change in wavelength at some angle θ should be $1/(1-\alpha) = 33$ times larger on his old theory than on his new one. These differences make it apparent that Compton's new interpretation was to be preferred over his old one.

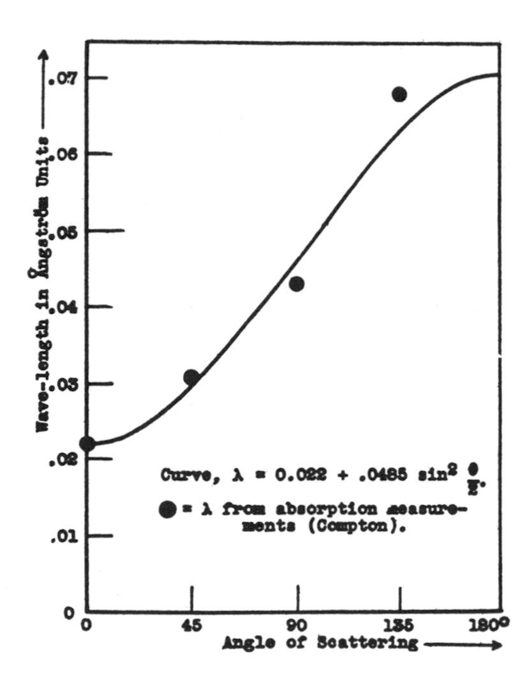

FIG. 3. Plot of angular variation of wavelength of scattered radiation.

It might be anticipated that the most conclusive evidence Compton offered in favor of his new interpretation rested on equation (6.18), the predicted variation in the change in wavelength $\Delta\lambda$ with *scattering angle* θ. Unfortunately, however, Compton had not yet measured this variation *spectroscopically* with *X-rays*—his X-ray data had been taken only at a single scattering angle (90°)—and hence Compton was here once again forced to rely on his old γ-ray *absorption* measurements. We recall that Compton had obtained values for the mass-absorption coefficient μ/ρ (in lead) for RaC γ-rays scattered through 0°, 45°, 90°, and 135° by three different scatterers, iron, aluminum, and paraffin. By taking an average at each scattering angle θ, and by converting the above variation of μ/ρ with θ into a variation of λ with θ, Compton first calculated the wavelength λ_0 of the primary radiation (= 0.022 Å) and then plotted its angular variation (Fig. 3[30]). The result, he concluded, was "in satisfactory accord" with the experimental data. Compton was therefore fully justified in rejecting his old interpretation in favor of his new one, but, as we have just seen, it is significant that to do so he could not have relied *solely* on his X-ray *spectroscopic* experiments, which at key points were neutral to this question.

3. *Recoil Electrons and Secondary Quanta. Compton's Conclusions*

The confirmation of a given theory, in general, does not rest on only one of its predictions, but rather on *all* of its predictions. In the case of Compton's theory, one additional prediction was particularly striking, bold, and suggestive: the existence of recoil electrons. There was a complete lack of recognized experimental evidence on recoil electrons at the time Compton advanced his theory, and Compton himself did not have such evidence in hand, but this did not prevent him from seeing that he could use their *presumed* existence to actually calculate the angular distribution $I(\theta, \alpha)$, and the "scattering absorption coefficient" $\sigma(\alpha)$, of the *secondary quanta.* Noting the striking formal similarity between equation (6.15) and the Doppler expression,

$$\frac{\lambda_0}{\lambda_\theta} = \frac{1-\beta}{1-\beta\cos\theta} = \frac{1}{1+[2\beta/(1-\beta)]\,[\sin^2(\theta/2)]}, \tag{6.23}$$

Compton saw that one could be translated into the other by simply replacing α by $\beta/(1-\beta)$, or β by $\alpha/(1+\alpha)$. Hence, he wrote: "It is clear . . . that so far as the effect on the wave-length is concerned, we may replace the recoiling electron by a scattering electron moving in the direction of the incident beam at a velocity such that $\bar{\beta} = \alpha/(1+\alpha)$. We shall call $\bar{\beta}c$ the 'effective velocity' of the scattering electrons."[31] For RaC γ-rays (λ_0 = 0.022 Å, α = 1.09), Compton found $\bar{\beta}$ to be equal to 0.52, which, he noted, was "in accord" with Rutherford's estimate of roughly ½.[32]

This analogy could be pursued further, however. Compton asserted without proof that it "seems obvious that since these two methods of calculation result in the same change in wave-length, they must also result in the same intensity of the scattered beam."[33] Therefore (in a long calculation which will not be repeated here), he first obtained an expression for the angular distribution produced by an *oscillator* moving with a relativistic velocity $v = \beta c$, and then made use of the correspondence $\beta/(1-\beta) \leftrightarrow \alpha$ to convert it into an expression for the angular distribution of the scattered *quanta.* His final result was equivalent to

$$I(\theta,\alpha) = \frac{Ne^4}{m^2c^4}\left\{\frac{1+\cos^2\theta+2\alpha(1+\alpha)\,(1-\cos\theta)^2}{2[1+\alpha(1-\cos\theta)]^5}\right\}. \tag{6.24}$$

By integration, this corresponds to a mass "scattering absorption coefficient"

$$\frac{\sigma(\alpha)}{\rho} = \frac{\sigma_0}{\rho}\frac{1}{1+2\alpha}. \tag{6.25}$$

Both expressions obviously reduce to Thomson's expressions in the limit of small quantum energies ($\alpha \rightarrow 0$).

Compton displayed equation (6.24) graphically (Fig. 4) and compared it with his Cavendish γ-ray data, concluding; "The beautiful agreement be-

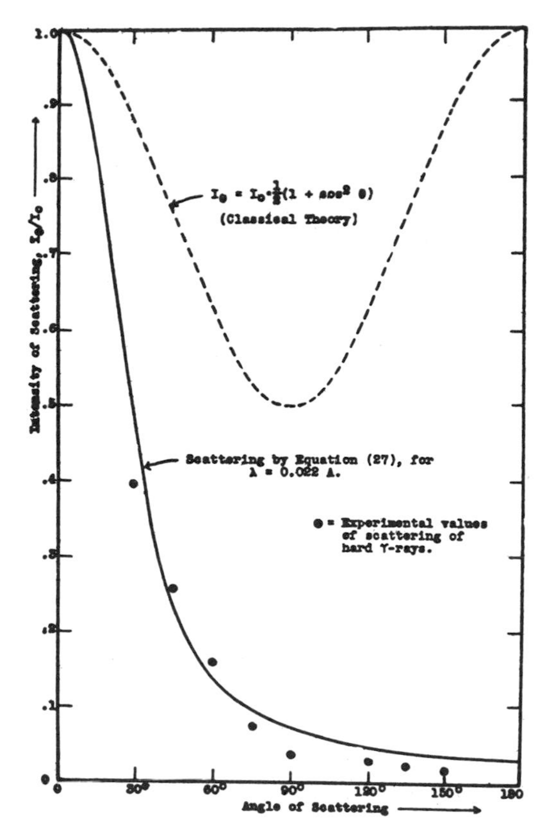

FIG. 4. Comparison of theory and experiment for equation (6.24), which was equation 27 in Compton's paper.

tween the theoretical and the experimental values of the scattering is the more striking when one notices that there is not a single adjustable constant connecting the two sets of values."[34] He also displayed equation (6.25) graphically (Fig. 5) and compared it with C. W. Hewlett's X-ray absorption data[35] and his own γ-ray absorption data. He attributed the deviation at long wavelengths to "excess scattering" effects, concluding that: "For wave-lengths less than 0.5 Å, where the test is most significant, the agreement is perhaps within the experimental error. . . . Thus the experimental values of the absorption due to scattering seem to be in satisfactory accord with the present theory."[36]

Surveying and summarizing all of his past work, Compton came to one of the most fundamental conclusions in the history of recent physics:

This remarkable agreement between our formulas and the experiments can leave but little doubt that the scattering of X-rays is a quantum phenomenon. The

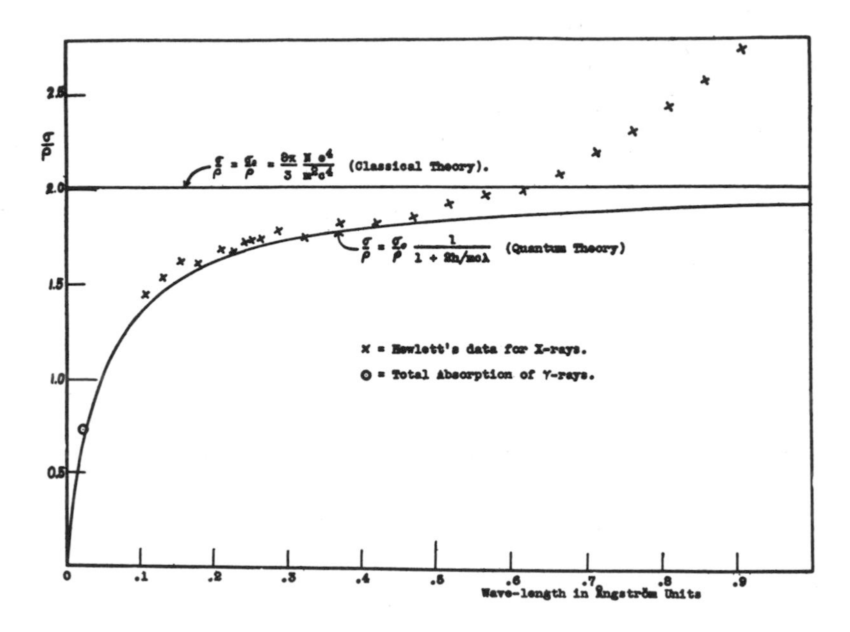

FIG. 5. Comparison of theory and experiment for variation of mass-scattering coefficient with wavelength.

hypothesis of a large electron to explain these effects is accordingly superfluous, for all the experiments on X-ray scattering to which this hypothesis has been applied are now seen to be explicable from the point of view of the quantum theory without introducing any new hypotheses or constants. In addition, the present theory accounts satisfactorily for the change in wave-length due to scattering, which was left unaccounted for on the hypothesis of the large electron. From the standpoint of the scattering of X-rays and γ-rays, therefore, there is no longer any support for the hypothesis of an electron whose diameter is comparable with the wave-length of hard X-rays.

The present theory depends essentially upon the assumption that each electron which is effective in the scattering scatters a complete quantum. It involves also the hypothesis that the quanta of radiation are received from definite directions and are scattered in definite directions. The experimental support of the theory indicates very convincingly that a radiation quantum carries with it directed momentum as well as energy.[37]

4. Compton and the Wave–Particle Dilemma

The decisiveness of Compton's words above strongly suggests that after years of probing, and after the indecision he had experienced as recently as October 1922 in his National Research Council report, he finally and fully embraced the quantum theory of radiation. This conclusion would be very misleading, however, because no one recognized better than Compton the dilemma that he had just introduced into physics. How could quanta and waves be reconciled? For Compton this dilemma was particularly acute, because, as we have seen, shortly before he had discovered his quantum theory of scattering, he had also discovered that X-rays may be totally internally reflected. What could be more particle-like than a billiard ball collision, more wave-like than total internal reflection? Nothing can symbolize Compton's dilemma better than the fact that on December 6, 1922—precisely midway between the date on which Compton presented his quantum scattering paper (December 1 or 2) and the date on which he sent it in for publication (December 13)—Compton sent in a full report on his experiments on the total internal reflection of X-rays to the editor of *The Philosophical Magazine*.[38] Within the space of a few days, therefore, Compton published convincing experimental evidence for *both* the particle *and* the wave nature of X-rays.

That Compton was fully alive to the wave-particle dilemma is evident from his discussion of his total internal reflection experiments:

> The existence of an index of refraction of the magnitude predicted by the classical electron theory would seem irreconcilable with the view that one quantum of X-rays can affect but a single electron. For there would seem to be no possibility of refraction unless the ray can spend a part of its energy in setting in vibration some of the electrons over which it passes so as to give rise to a secondary ray which will combine with the primary train. But this involves the idea that parts of the quantum are capable of affecting the traversed electrons. It is not easy to reconcile this result with such experiments as that of the softening of scattered X-rays, which seem to demand that each electron affected by the primary rays receive both the momentum and the energy of a whole quantum.[39]

In his quantum scattering paper, he speculated on a possible way of resolving the dilemma: "Perhaps if an electron is bound in the atom too firmly to recoil, the incident quantum of radiation may spread itself over a large number of electrons, distributing its energy and momentum among them, thus making interference possible. In any case, the problem of scattering is so closely allied with those of reflection and interference that a study of the problem may very possibly shed some light upon the difficult question of the relation between interference and the quantum theory."[40]

The extent to which Compton dwelled on the wave–particle dilemma at the end of 1922 is indicated in a highly unusual letter which he published in *Science* on December 22.[41] This letter was entitled "Radiation A Form of Matter," and its title suggests that it perhaps dealt with Einstein's mass–energy relationship. Compton in fact did begin in that vein, but then he abruptly changed tack—so much so that a better title for the letter would have been

"Matter A Form of Radiation." Thus, Compton argued that matter cannot be "composed wholly of positive and negative electrons," because the "form of matter known as radiation" contains no charges; nor could matter consist of a combination of positive electrons, negative electrons, and radiation, because that assumption was "more complex than is required." Rather, one should observe "that the fundamental thing in matter is not the electric charge but the electromagnetic field, for the electromagnetic field includes both the electrons and the radiation." The logical conclusion was therefore that:

> According to this point of view, matter is perfectly continuous. It is true that there are certain perhaps limited regions, the electrons, from which electric intensity diverges; but . . . the mass of the matter is associated with the electric intensity and is hence distributed through all space. Similarly, radiation propagated through space . . . is on this view a continuous series of waves of matter. The old argument for the existence of an ether because some medium is necessary to transfer the radiant energy from the sun to the earth has accordingly no weight. For we now see that the radiation may be its own medium, somewhat as the stream of water from a hose acts as the medium for a wave if the nozzle is shaken.[42]

That this is an unusual conclusion—coming from the physicist who had just offered experimental proof for the quantum theory of radiation—is surely an understatement. Compton's letter, above all, illustrates the extent of his groping at the end of 1922. His groping, however, was only a harbinger of what was to occur among physicists generally.

D. Debye Independently Discovers the Quantum Theory of Scattering (1923)

One of the most striking coincidences in the history of physics occurred in early 1923 when Peter Debye, then Professor of Physics at the Federal Institute of Technology in Zürich, independently, and virtually simultaneously, discovered the quantum theory of scattering. Part of the reason that Debye's discovery and Compton's discovery came to the general attention of physicists at about the same time lay in the publishing process. Compton explained this situation, and commented on it, in a letter of July 26, 1923, which he wrote to G. S. Fulcher, Managing Editor of *The Physical Review,* in connection with another paper he had just submitted to Fulcher for publication: "I am very anxious that this paper go to press immediately, as the subject is exceptionally live. My paper in the May number of the Review was preceded by one by Debye on the same subject, written 2 months later, but appearing April 15th in the Physik. Zeitsch . . . . Of course the Physical Review is not designed as a medium for immediate publication, but can we not in some way lessen the interval between pen and printed page?"[43]

To be precise, the complete chronology involved in Compton's and Debye's discoveries was as follows: (*i*) Probably sometime in November 1922, Compton gave a full and detailed exposition of his quantum theory of scattering to G.E.M. Jauncey's large physics class at Washington University;[44] (*ii*) December 1 or 2, 1922, Compton presented his paper on his quantum theory of scattering at the Chicago meeting of the American Physical Society; (*iii*) December 13, 1922, Compton sent the full text of his paper to the editor of *The Physical Review;* (*iv*) March 14, 1923, Debye's paper on his quantum theory of scattering[45] was received by the editor of the *Physikalische Zeitschrift;* (*v*) April 15, 1923, Debye's paper was published in the *Physikalische Zeitschrift;* (*vi*) May 1923, Compton's paper was published in *The Physical Review.* In view of this chronology, it is not difficult to sympathize with Compton's chagrin over the time lags involved in the publishing process.

Prior to 1923, Debye had worked on a variety of research topics, including the photoelectric effect, the specific heats of solids, the kinetic energy of insulators, dispersion, and X-rays. His two best known papers on X-rays had dealt with the phenomenon of "excess scattering" and with the Debye–Scherrer powdered crystal diffraction method. Most recently, however, he had investigated the theory of molecular forces,[46] and the question therefore arises as to why he returned to X-ray studies in early 1923, and thereby was led to develop his quantum theory of scattering.

This question may be answered by examining the introductory section of Debye's paper, where he first discusses Thomson's classical theory of scattering, then his own work on what came to be known as "excess scattering," and finally the evidence conflicting with Thomson's theory in the short-wavelength region. Concerning this region, Debye wrote:

> Four points seem to me to deserve particular attention:
>
> 1. The intensity of the scattered radiation is considerably higher in the direction of the primary radiation ($\theta = 0$) than in the opposite direction ($\theta = \pi$). . . .
>
> 2. It now appears to be certain that the radiation scattered in the direction of the primary beam is harder than that scattered in the opposite direction. Thus the wave length is changed . . . .
>
> 3. The total energy of the scattered radiation sinks below the limiting value corresponding to Thomson's calculation. However this limiting value is in agreement with the experimental value of 0.2, found by Barkla . . . .
>
> 4. Each scattering is accomplished by electron emission. The shorter the wave length, the more the electrons appear to be ejected in the direction of the primary beam.
>
> Not all experimental results are so unequivocal that the assertions 1 to 4 can be considered as absolutely confirmed by experiments. However I recently gained the impression from a survey by A. H. Compton that it is highly probable that they are correct. I will therefore hesitate no longer to present for discussion an

explanation of these effects based on quantum theory which occurred to me as a possibility quite some time ago.[47]

The survey of Compton's that Debye referred to, and explicitly cited, was Compton's October 1922 National Research Council report. Apparently, therefore, it was the appearance of Compton's report that provided the immediate stimulus for Debye to publish his theory, which in its basic assumptions was identical to Compton's. In some cases, however, Debye expressed his results differently, or was led to new ones. Thus he first set up and solved the equations expressing conservation of energy and conservation of momentum, deriving an expression for the ratio ν_θ/ν_0 identical to Compton's. Second, he calculated the *energy* of the secondary electrons, rather than their velocity as Compton had chosen to do. Third—and here was a result not contained in Compton's paper—he calculated the relationship between the scattering angle θ of the secondary quantum and the recoil angle ϕ of the recoil electron. His result was equivalent to

$$\tan\phi = -\frac{1}{(1+\alpha)\tan(\theta/2)}. \tag{6.26}$$

This explicitly proved that the recoil electrons are always scattered in the forward direction, while the secondary quanta are scattered in all directions.[48]

Debye also explained and emphasized that in the ordinary photoelectric effect the parent atom takes up appreciable momentum, and hence the momentum properties of quanta are not apparent here.[49] Finally—and here he diverged entirely from Compton's treatment—Debye used Bohr's correspondence principle[50] to calculate a multiplicative factor which in effect modulated the Thomson scattering coefficient. His result was equivalent to

$$f(\alpha) = \frac{3}{4\alpha^3}[(1 + 2\alpha + 2\alpha^2)\ln(1 + \alpha) - \alpha(1 + 3\alpha/2)], \tag{6.27}$$

where α has its usual meaning. It may be seen that $f(\alpha)$ varies between 0 in the short-wavelength limit ($\alpha\rightarrow\infty$) and 1 in the long-wavelength limit ($\alpha\rightarrow 0$). It therefore produces qualitatively the same variation in the scattering coefficient as Compton's multiplicative factor $1/(1 + 2\alpha)$ in equation (6.25).

Debye's basic assumptions were identical to Compton's, but there were very significant differences between the routes the two men had traveled to arrive at those assumptions. Judging from Debye's statement in point 3 above, as well as from his remarks in an interview[51] that T. S. Kuhn and G. E. Uhlenbeck conducted with him on May 3, 1962, the starting point of Debye's thought was Barkla's observation that the mass-scattering coefficient σ/ρ dipped below Thomson's 0.2 cm^2/g—that is, precisely the same observation that first stimulated Compton's thought. Debye stated in his interview that

Barkla's observation struck him as significant roughly two years before he finally published his paper, which would mean sometime in late 1920 or early 1921—when Compton was just beginning his X-ray experiments at Washington University after returning to the United States. Moreover—and this explains Debye's remark that the theory "occurred to me as a possibility quite some time ago"—Debye apparently carried out the calculation in detail at that time.[52] He stated that he kept telling his colleague P. Scherrer that they should someday do an experiment to see if the change in wavelength actually occurred—which they never did.

To Debye, the scattering problem was fundamentally a theoretical one, and in attacking it he explicitly cited its key concept—Einstein's concept of "needle radiation."[53] The source of Compton's theoretical ideas, as we have seen, was basically his own gradually evolving experimental program—Compton did not once mention Einstein's name in his paper. This striking contrast led to a definite difference in tone in their conclusions. Recall that Compton, on the one hand, concluded: "This remarkable agreement between our formulas and the experiments can leave but little doubt that the scattering of X-rays is a quantum phenomenon."[54] Debye, on the other hand, wrote: "In conclusion I wish to stress that I would like the preceding discussion to be considered as nothing but an attempt to derive as detailed conclusions as possible from the two assumptions: 'energy quanta' and 'radiation quanta,' taking recourse only to general laws: 'law of conservation of energy' and 'principle of conservation of momentum.' By means of experiments which reveal characteristic deviations from this scheme we may hope to secure deeper insight into the laws of quantum theory, particular[ly] as regards their relation to physical optics."[55]

Should it be called the "Compton–Debye effect"? In the interview already referred to, Debye himself answered this question: he vigorously objected to such an idea, noting that a great deal was occurring at the time, and maintaining that the person who did most of the work should get the name.

E. Related Experimental and Theoretical Work in 1923

1. Compton's and Jauncey's Researches

Compton, in fact, never stopped working, even briefly. In some of his subsequent researches he was enthusiastically joined by his colleague, G.E.M. Jauncey (see Fig. 6), the only physicist Compton explicitly acknowledged as contributing to "many of the ideas involved"[56] in his quantum theory of scattering. After Compton's discovery, Jauncey provided a new (but ultimately still unsatisfactory) derivation[57] of the angular distribution $I(\theta, \alpha)$, and in many other ways contributed again and again to the deeper understanding of the

FIG. 6. Washington University physics faculty and staff 1922–1923. From left to right: H. L. May, Carl Eckart, C. F. Hagenow, A. H. Compton, C. A. Rinehart, G. E. M. Jauncey, Lindley Pyle, F. Bubb, C. Wackman, O. K. DeFoe.

Compton effect.[58] Compton, on his part, recognized that the greatest immediate need was not for deeper theoretical insight, but for more X-ray spectroscopic data displaying the angular dependence of the change in wavelength.

To this end Compton constructed a new long and narrow Coolidge tube with a molybdenum target (Fig. 7[59]), which enabled him to achieve

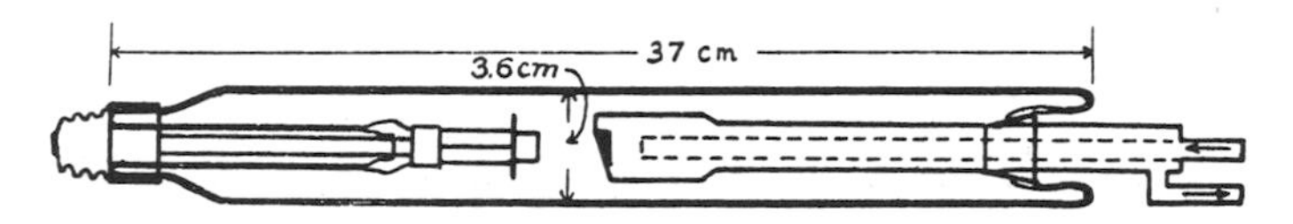

FIG. 7. Compton's improved X-ray tube.

much higher primary intensities and resolution. He also employed, as he later recalled, a new data-taking technique:

> I was afraid of being influenced by finding what I was looking for, so I got to help me in the laboratory an assistant who did not know at all what I had in mind. I made the spectrometer settings while he took the readings. Not knowing what we were looking for, he felt that the readings were very erratic. After the experiment was over, he remarked to me: "It was too bad, wasn't it, Professor, that the apparatus wasn't working so well today?" He was disturbed by the fact that the readings went up and down, and he had no idea that they were just the kind of thing I wanted.[60]

The kind of thing Compton wanted is shown in the beautiful spectra in Fig. 8,[61] which he first reported at the Washington meeting of the American Physical Society on April 21, 1923. Side-by-side are shown two sets of data (corresponding to two different slit settings) for MoK_α X-rays scattered by graphite (carbon) through 0°, 45°, 90°, and 135°. In the future, he definitely would not have to rely on his old γ-ray absorption data to illustrate the angular dependence of the change in wavelength. He explained:

> It is clear from these curves that when a homogeneous x-ray is scattered by graphite it is separated into two distinct parts, one of the same wave-length as the primary beam, and the other of increased wave-length. Let us call these the *modified* and the *unmodified* rays respectively . . . . The wave-length of the modified ray . . . increases with the scattering angle as predicted by the quantum theory, while the wave-length of the unmodified ray is in accord with the classical theory.[62]

The latter ray was probably scattered by tightly bound electrons.[63] There was, therefore, "no indication of any discrepancy whatsoever"[64] with his quantum theory of scattering.

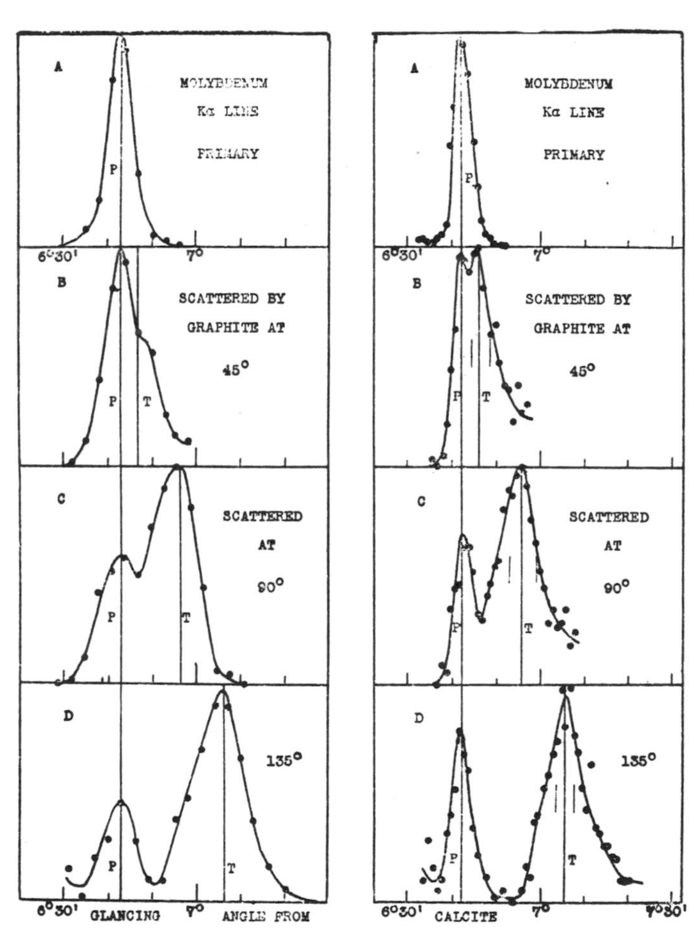

FIG. 8. Spectra showing relative intensity and separation of modified and unmodified lines for MoKα X-rays scattered by graphite through various angles.

2. *Sommerfeld Stimulates Ross's Experiment at Stanford*

Word of Compton's discovery was meanwhile spreading. One of Compton's most effective and enthusiastic propagandizers was Arnold Sommerfeld, who, as it happened, was visiting the University of Wisconsin during the winter semester of 1922–1923 as Carl Schurz Professor of Physics.[65] We have already noted Sommerfeld's willingness to relinquish the wave theory in connection with Einstein's abortive 1921 experiment. Now, on January 21, 1923, while on a lecture tour in California shortly after learning of Compton's discovery, Sommerfeld notified Bohr of it by letter, writing: "The most interesting thing that I have experienced scientifically in America . . . is a work of Arthur Compton in St. Louis. After it the wave theory of Röntgen-rays will become invalid . . . . I am, however, still not completely certain if he is right; I do not yet know if I should already mention his results. I only want to call your attention to the fact that eventually we may expect a completely fundamental and new lesson [*Belehrung*]."[66] Wherever Sommerfeld lectured—Pasadena, Berkeley, the National Bureau of Standards, Harvard, Columbia—he spoke of Compton's discovery.

It was evidently through Sommerfeld's lectures at Berkeley, in fact, that P. A. Ross of Stanford University learned of Compton's discovery, and thereby quite likely became the first experimentalist outside of St. Louis to attempt to confirm it. At first, Ross tried to use interferomeric techniques to photographically detect the change in wavelength for visible light (mercury and helium spectral lines), but by May 25, 1923, he could find no evidence for it.[67] He decided, therefore, to follow Compton more closely and scatter MoK_α X-rays through 90° by paraffin. By early June, using very long exposure times (up to 100 hours), which enabled him to obtain high resolution with very fine slits, he found that the "scattered radiation shows a strong broad line shifted by about .025 Å, which agrees with theory and with Compton's results, and a fainter unshifted line not reported by Compton"[68]—Ross did not know that by this time Compton had also observed the unshifted line.[69] Ross's earlier, negative results could be accounted for by assuming that visible light was not energetic enough to eject electrons from the parent atom, and hence could produce only the unshifted line.

Ross's negative results with visible light actually helped motivate Compton to turn momentarily from spectroscopic to absorption experiments, to "test the theory over a wider range of wave-lengths, and for a greater variety of scattering materials than could be done conveniently by the spectroscopic method."[70] Employing a continuously recording Bragg spectrometer—very similar in design to the one which he had developed in 1916[71]—Compton measured the difference in absorption coefficients (and hence the difference in wavelengths) between the primary and secondary beams at 15° intervals between 30° and 135°. Scattering four different wavelengths (0.024 Å, 0.15 Å, 0.32 Å, 0.71 Å) from a single scatterer, paraffin, he found that theory and experiment agreed well at the shortest wavelength, but that significant divergence occurred at the longer wavelengths. Scattering a single wavelength ($\approx$ 0.16 Å) from five different scatterers (carbon, aluminum, copper, tin, lead), he found that theory and experiment agreed well for the lightest elements, but that significant divergence occurred for the heavier elements.

Now, rather than contradicting his quantum theory of scattering, these results actually confirmed it, because "for short wave-lengths and low atomic numbers nearly all of the energy lies in the modified ray, while for long waves and high atomic numbers the unmodified ray has the greater energy." Hence "for such comparatively great wave-lengths as those used in optics, when the usual materials are used as radiators, the unmodified ray should predominate . . . . This is in accord with the negative result of Ross's experiment . . . ." A closely related observation also could be understood: at the sun's surface, where there is an atmosphere of essentially *free* electrons, even visible light should be energetic enough to be shifted in wavelength when incident upon them. This was the so-called "limb effect," long known to astronomers, but never before understood.[72]

3. *Bothe's and Wilson's Cloud Chamber Photographs of Recoil Electrons*

In Europe other experimentalists were providing direct, visual evidence for the correctness of Compton's quantum theory of scattering—through cloud chamber photographs of the recoil electrons, which Compton's theory had boldly predicted.[73] The first to report such evidence was W. Bothe of the Physikalisch-Technischen Reichsanstalt in Charlottenburg in May 1923,[74] although at this time Bothe actually misinterpreted the origin of the short tracks he had observed. Only in a second paper[75] did he prove that they were produced by recoil electrons and thereby recognize the connection between his experiments and Compton's and Debye's theory.

Almost simultaneously, the inventor of the cloud chamber, C.T.R. Wilson, published a number of impressive cloud chamber photographs, one of which is reproduced in Fig. 9.[76] Wilson was able to distinguish three different types of secondary electron tracks: (1) long tracks having a range of several centimeters and being produced by ordinary photoelectrons; (2) spherically shaped tracks having a range of a few tenths of a millimeter and quite likely being produced by very short range β-rays; and (3) short tracks having a range of one or two millimeters, which Wilson dubbed "fish tracks" because "the tail of the 'fish' is directed toward the source."[77] These "fish tracks," he noted, were "almost certainly due to the direct action of the primary radiation." He discounted an interpretation of them based on Barkla's J-phenomenon, and observed that: "To account for various phenomena relating to the wave-length and distribution of secondary X-rays A. H. Compton has suggested the possibility of just such a forward ejection of electrons as actually occurs in these 'fish' tracks."[78] Wilson's source for Compton's suggestion was, like Debye's, Compton's influential National Research Council report.

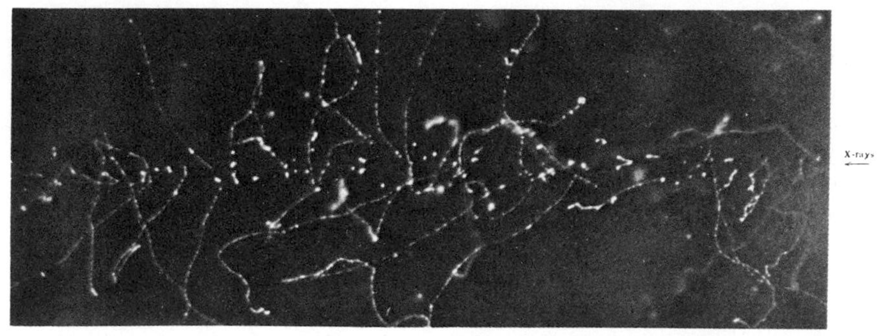

Reproduced by permission of the Royal Society of London

FIG. 9. Cloud chamber photograph taken by C. T. R. Wilson showing secondary electrons produced by X-rays.

Wilson's experiments above, and others which soon followed,[79] immediately attracted Compton's attention. Together, the two physicists sent letters to the editor of *Nature*[80] in August 1923, acknowledging the probable connection between Compton's theory and Wilson's experiments. Compton (who consistently gave Debye full credit for his work) pointed out that his and Debye's theory predicted that the maximum energy of the recoil electrons should be given by

$$E_{\max} = h\nu_0 \, [2\alpha/(1 + 2\alpha)], \tag{6.28}$$

where as usual $\alpha = h\nu_0/mc^2$. Since the range of a charged particle is proportional to the square of its energy, it was then a simple matter for Compton to show that soft X-rays of wavelength 0.5 Å should be able to produce recoil electron tracks 0.11 mm long, while harder X-rays of wavelength 0.242 Å should be able to produce recoil electron tracks 1.7 mm long. Comparing these figures with Wilson's photographs, he concluded that the "quantum idea of X-ray scattering . . . leads to recoil electrons moving in the right direction and processing energy which is of the same order of magnitude as that possessed by the electrons responsible for C.T.R. Wilson's very short tracks."[81]

4. *Wilson Draws Attention to a Critical Point and Compton Reacts to It*

Wilson fully agreed with Compton's conclusion above, but nevertheless drew attention to a very critical point: "The data thus far obtained by this [cloud chamber] method are not sufficient to decide without ambiguity whether a quantum of radiation scattered by an electron is emitted in one direction only or with a continuous wave-front."[82] A few months later, Bohr would push this point to the forefront of physics, but by October 1923 it had already assumed a central importance not only in Wilson's but also in Compton's mind. As the latter wrote: "There are . . . two essentially different methods by which an electron may scatter a quantum. In the postulate as . . . presented it was supposed that an electron receives the radiation quantum from a definite direction and scatters it in a different but equally definite direction. . . . It may be imagined, on the other hand, that while the energy and momentum of the primary quantum are received from a definite direction, the energy thus received is scattered in spherical waves in all directions."[83]

The former interpretation was Compton's and Debye's position; the latter, which involved a Doppler shift explanation of the change in wavelength, was currently favored by Wilson and C. G. Darwin. As we have seen, Darwin had suggested as early as 1919 that conservation of energy did not necessarily hold in individual microscopic interactions. By 1923 he had not changed his opinion, flatly stating that "a critical examination of fundamentals does not by any means justify [our] . . . faith"[84] in it. The key point with respect to the Compton effect was that if the scattered radiation were emitted in spherical waves, each would have to contain the energy stored up over a large number of individual primary interactions, since any small segment of it was capable of once again ejecting an electron. Energy therefore could only be conserved on the average, and not in any given individual interaction.

Compton (with the assistance of J. C. Hubbard) examined these two points of view critically and in detail. He saw that he could decide between them by focussing his attention on the *recoil electrons,* and by comparing their energy to that of ordinary *photoelectrons.* Thus, if the Compton–Debye quantum interpretation were correct, the maximum kinetic energy of the recoil electrons should be given by equation (6.28), while the maximum kinetic energy of ordinary photoelectrons, as usual, is simply on the order of the energy $h\nu_0$ of the incident radiation. Since the range of a charged particle is proportional to the square of its energy, this meant that on the Compton–Debye theory the *ratio* R_{C-D} of the range of the recoil electrons to that of ordinary photoelectrons should be given by

$$R_{C-D} = \left[\frac{2\alpha h\nu_0}{(1+2\alpha)} \cdot \frac{1}{h\nu_0}\right]^2 = \frac{4\alpha^2}{(1+2\alpha)^2}. \qquad (6.29)$$

To calculate the same quantity on the Wilson–Darwin model, Compton, drawing on his own earlier work, argued that to produce the required change in wavelength by the Doppler shift, the secondary electron had to have an "effective velocity" $v = \bar{\beta}c$, where $\bar{\beta} = \alpha/(1+\alpha)$.[85] The maximum kinetic energy E'_{max} of the secondary electrons then followed from the usual relativistic energy expression:

$$E'_{max} = mc^2\left(\frac{1}{\sqrt{1-\beta^2}} - 1\right) = \frac{h\nu_0}{\alpha}\left(\frac{1}{\sqrt{1-\frac{\alpha^2}{(1+\alpha)^2}}} - 1\right)$$

$$= \frac{h\nu_0}{\alpha}\left(\frac{1+\alpha}{\sqrt{1+2\alpha}} - 1\right) = h\nu_0\frac{(\alpha/2)}{1+2\alpha}\,(1 - 1/4\,\alpha^2 + \ldots)$$

$$\doteq \frac{h\nu_0(\alpha/2)}{1+2\alpha}, \qquad (6.30)$$

since $\alpha^2 << 1$. The Wilson–Darwin ratio R_{W-D} of the range of the recoil electrons to that of the photoelectrons should therefore be given by

$$R_{W-D} = \left[\frac{(\alpha/2)\,(h\nu_0)}{(1+2\alpha)} \cdot \frac{1}{(h\nu_0)}\right]^2 = \frac{\alpha^2}{4(1+2\alpha)^2}, \qquad (6.31)$$

which is only 1/16 as large as the Compton–Debye ratio R_{C-D}.

Wilson's own cloud chamber photographs could therefore be used to decide between the two interpretations. Wilson had observed that to produce photoelectron tracks 1.5 cm in length, X-rays of wavelength 0.48 Å, corresponding to $\alpha = 0.05$, were required. Therefore, on the Compton–Debye theory the corresponding recoil electron tracks should be $(1.5\text{ cm})[(4\alpha^2)/(1+2\alpha)^2] = 0.12$ mm in length, while on the Wilson–Darwin theory they should be only (1/16) (0.12 mm) = 0.008 mm in length. Only the former was of the observed order of magnitude.

Other checks, on the relative number of short recoil electron tracks to be expected, and on how this number varies with incident quantum energy,[86] pointed in the same direction. Therefore, the Compton–Debye quantum interpretation agreed with experiment, while, Compton asserted, we "must abandon" the Wilson–Darwin theory: "Both from the standpoint of the experimental evidence and from the internal consistency of the theory *we . . . seem forced to the conclusion that each quantum of scattered x-rays is emitted in a definite direction.*" He added: "It would appear but a short step to the conclusion that all radiation occurs as definitely directed quanta rather than as spherical waves."[87]

On September 18, 1923, even before Compton wrote the above words, P. A. Ross had reported further confirming data at a meeting of the American Physical Society in Pasadena.[88] Ross had taken spectrograms of MoK_α X-rays scattered by paraffin, aluminum, copper, zinc, silver, and lead. For paraffin he had taken readings at four different scattering angles, while for the other scatterers he had taken less extensive measurements. The change in wavelength, he concluded, "was independent of the nature of the scattering substance but the relative intensity of the shifted to the unshifted line decreased with increasing atomic number." In all cases the modified line was "shifting nearly according to theory."[89]

5. *Other Responses in England and Germany to Compton's Discovery*

We have seen that by the fall of 1923 a great deal of evidence had already accumulated which substantiated Compton's discovery. To be sure, some physicists, and especially British physicists it seems, were reluctant to accept Compton's interpretation. We have already noted Wilson's and Darwin's reservations, but others were also hesitant. J. H. Jeans argued, in part quantitatively, that he could avoid Compton's interpretation and also explain the photoelectric effect classically.[90] J. J. Thomson, then in his late sixties, consistently and repeatedly attempted to incorporate Compton's observations into classical electrodynamics.[91] The most adamant British opponent, however, was C. G. Barkla, who at regular intervals published paper after paper[92] well into the 1930's contesting Compton's work—in spite of the fact that in early 1924 Compton had replied directly and in detail to him in a note in *Nature*.[93] Specifically, Compton had characterized Barkla's J-radiation explanation as an "obviously less complete" explanation of the change in wavelength, "since it says nothing regarding the mechanism of the transformation and makes no prediction regarding the magnitude of the change."[94] Nevertheless, Barkla found support—or at least sympathy—for his views from several physicists, for example, from N. M. Bligh, who discussed Barkla's and Compton's theories with equanimity in his 1926 *Evolution and Development of the Quantum Theory*.[95] But most physicists clearly believed Barkla to be in error. As H. S. Allen wrote, "It must have been disheartening for him [Barkla] to find so many other experimenters in disagreement with his own conclusions."[96] There seems to be no clear-cut consensus as to what Barkla was actually observing in his experiments.[97]

At least one British physicist, Ernest Rutherford, almost immediately recognized the importance of Compton's work. On August 14, 1923, Rutherford wrote Compton to that effect,[98] at the same time congratulating him on his recent appointment to the professorship formerly held by R. A. Millikan at the University of Chicago, which Rutherford regarded as a fitting reward for Compton's achievements of the past several years. Continental physicists, too—if we can judge from recent interviews[99] with J. Franck and H. Sponer, W. Heisenberg, and G. E. Uhlenbeck—accepted the validity of the Compton effect very rapidly, if not immediately. It must be remembered that Debye's paper actually appeared in print one month before Compton's, and certainly the persuasiveness of seeing the same theory published independently at virtually the same time, in two different countries, in two different languages, in two major journals, by two productive and highly creative physicists, can scarcely be underestimated.

We also know that Sommerfeld, after returning to Munich, enthusiastically talked to everyone he met about Compton's work. In fact, there exists a very interesting letter Sommerfeld wrote to Compton on October 9, 1923,[100] which is worth quoting in full for the detailed description in it of the situation in Germany.

Munich, October 9, 1923

Dear Colleague:

Your discovery of the change in wavelength of Röntgen rays keeps the scientific world in Germany extremely busy. I was together with Einstein and Kossel in August, and our chief topic of discussion was your effect. Also, at the meeting of physicists in Bonn it was discussed in connection with an α-ray scintillation experiment for visually proving the asymmetry of emission. Joos' report on it appears in the Physikalische Zeitschrift. The result, however, is certainly not yet conclusive. Moreover, in my book [*Atombau und Spektrallinien*], the 4th edition of which I am now preparing, I have inserted a section in the first chapter on the quantum structure of light; I there discuss the "Doppler effect and the Compton effect."

After the beautiful experiments of Ross, there can be no doubt that your observation and theory are completely accurate. I also heard that an English paper confirms your work, but as yet I have not seen it. I interpret the unmodified line in Ross's work in the following way, namely, that the recoil has been taken up by the whole atom, as in optics. The question arises as to how the ratio of intensity between the unmodified and modified line for heavy atoms is influenced, and whether one can possibly find an effect for visible or ultra-violet light with calcium. I hope you will let me know if something important on this subject becomes known. I would also be very interested in the decision on the limb effect of the sun.

I was asked to discuss your results in the Physikalische Zeitschrift. I have not yet had time to do it and was very surprised to see Debye's note on it, which in essence agrees with your theory. Only insofar as the scattering coefficient is concerned, his method seems to be somewhat clearer than yours. I wrote to Debye that you naturally have the priority not only in the experiments, but also in the theory. Debye told me recently that three years ago he discussed the theory with Mr. E. Wagner and asked him to make precision measurements of wavelengths—all of which, however, is naturally a matter of indifference for the question of priority.

How one now has to understand the matter of crystal interferences is very dark. The beginnings of Duane and Breit are perhaps not as absurd as they seem to appear on first glance!

I have published a few notes on the magneton, which I will send you in the near future. I think the integers in "Bohr units" and the space quantization of the magnetic axis will be vindicated by it—even for solid salts and solutions!

München 9. Oktober 23.

Sehr geehrter Herr College!

Ihre Entdeckung der Wellenlängen-Änderung der Röntgenstrahlen beschäftigt auch in Deutschland die wissenschaftliche Welt aufs lebhafteste. Ich war im August mit Einstein und Kossel zusammen und wir diskutierten hauptsächlich über Ihren Effekt. Ebenso wurde auf dem Physikertag in Bonn darüber verhandelt im Anschluss an einen Versuch, mit Lichtblitzen von α-Strahlen, die Einseitigkeit der Emission visuell nachzuweisen. Der Bericht darüber von Joos erscheint in der Physikal. Zeitschr. Das Ergebnis ist aber sicher noch nicht beweisend. Auch in meinem Buch, an dessen 4. Auflage ich arbeite, habe ich im 1. Kapitel einen § eingeschaltet: über die Quantenstruktur des Lichtes: und habe darin über „Dopplereffekt und Comptoneffekt" gesprochen.

Nach den schönen Versuchen von Ross kann ja kein Zweifel sein, dass Ihre Beobachtung und Theorie vollständig richtig ist. Ich höre, dass auch eine englische Arbeit Ihren Effekt bestätigt, habe sie aber noch nicht gesehen. Ich fasse die unveränderte Linie bei Ross so auf, dass hier das ganze Atom den Rückstoss übernommen hat, wie in der Optik. Es entsteht die Frage, wie das Intensitätsverhältnis zwischen unveränderter und verschobener Linie bei schwereren Atomen wird; und ob man nicht auch im Sichtbaren bez. Ultravioletten unter Umständen (bei Cäsium) einen Effekt finden kann. Ich hoffe, Sie werden es mich wissen lassen, wenn etwas Wichtiges hierüber bekannt wird. Auch die Entscheidung über den Randeffekt der Sonne wird mich sehr interessieren.

Sie forderten mich auf, für die Physikalische Zeitschr. über Ihre Resultate zu berichten. Ich bin nicht dazu gekommen und war sehr überrascht, als ich Debye's Note sah, die ja im Wesentlichen mit Ihrer Theorie übereinstimmt. Nur in Bezug auf den Streuungskoefficienten scheint mir seine Methode etwas einleuchtender wie die Ihre. Ich schrieb Debye, dass Sie natürlich nicht nur in den Versuchen sondern auch in der Theorie die Priorität haben. Debye sagte mir übrigens kürzlich, dass er schon vor 3 Jahren die Theorie im Gespräch Hn. E. Wagner mitgeteilt hatte und diesen aufgefordert hatte, Präcisionsversuche über Wellenlängen zu machen — was aber natürlich für die Prioritätsfrage gleichgiltig ist.

Wie man jetzt die Kristallinterferenzen auffassen soll, ist sehr dunkel. Die Ansätze von Duane und Breit sind vielleicht nicht so absurd wie sie auf den ersten Blick scheinen!

Ich habe einige Noten über das Magneton publiciert, die ich Ihnen demnächst schicken werde. Ich denke die Ganzzahligkeit in „Bohr'schen" Einheiten und die räumliche Verankerung der magnetischen Axe wird dadurch — auch für feste Salze und Lösungen! — bewiesen.

Meine Amerikareise hat mir viele interessante und angenehme Eindrücke verschafft. Das Interessanteste aber waren Ihre Mittheilungen. Ich habe überall darüber vorgetragen (Pasadena, Berkeley, Bureau of Standards, Harvard, Columbia N.Y.); einige (etwas verzerrte und missverstandene) Anzeichen davon haben Sie wohl in der Arbeit von Ross bemerkt.

Beste Grüsse von Ihrem A. Sommerfeld

Fig. 10. Sommerfeld's letter to Compton, October 9, 1923.

My trip to America provided me with many interesting and pleasant impressions. The most interesting ones, however, were your reports. I lectured on them everywhere (Pasadena, Berkeley, the Bureau of Standards, Harvard, Columbia N.Y.), a (somewhat slanted and misunderstood) sign of which you will probably have seen in the work of Ross.

Best wishes from your
A. Sommerfeld

Sommerfeld's enthusiasm for Compton's work, his natural preference for Debye's correspondence principle calculation, his awareness of open questions (which will be discussed later), and his sensitiveness toward questions of priority, all come through very clearly in the foregoing letter.

Sommerfeld was convinced that a new vista had just appeared in physics—his specific reference to the "Compton effect" quite likely represents the first time this term was used in the literature. Yet, only a few weeks after Sommerfeld had written to Compton—right in the midst of this atmosphere of enthusiasm and confidence—the most unsettling news began to filter out. Experiments carried out in William Duane's laboratory at Harvard University failed to reveal the anticipated change in wavelength. It is difficult to say how seriously European physicists took this news, although contemporary developments in Europe (which we will discuss in the next chapter) suggest that it had at best marginal influence. But two points are beyond doubt: first, the news created a great stir in the United States, and second, Compton himself took it extremely seriously. He found himself, as he later wrote, "in the most lively scientific controversy that I have ever known."[101] We shall devote the remainder of this chapter to a discussion of that controversy.

F. The Duane–Compton Controversy (1923–1924)

1. Its Origins: Clark and Duane's Experiments and Their Interpretation of Them

William Duane (1872–1935), a direct descendent of Benjamin Franklin (5th generation), educated at the University of Pennsylvania (A.B., 1892, valedictorian of class), Harvard University (A.M., 1895), Göttingen, and Berlin (Ph.D., 1897), formerly Professor of Physics at the University of Colorado (1898–1907), and assistant to Marie Curie (1907–1913), was appointed Assistant Professor of Physics at Harvard University in 1913 and rose rapidly through the academic ranks to become Professor of Bio-Physics in 1917—very likely the first such professorship in the United States.[102] He had made important contributions throughout his career to physics proper,[103] as well as to the therapeutic or medical applications of high-frequency radiation. By 1923 he had achieved an international reputation. It is not difficult,

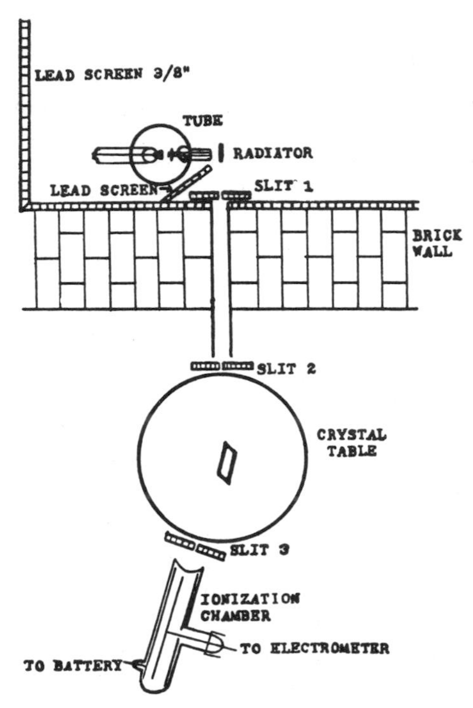

FIG. 11. Clark and Duane's first experimental apparatus.

therefore, to appreciate the fervor created when, at the end of October 1923, G. L. Clark, a National Research Council Fellow working in Duane's laboratory, announced—with Duane's full support—that he had "not been able to detect the presence of any rays in the secondary radiation having wave-lengths a certain fraction of an ångström longer than those of the primary rays, as an interesting theory recently published by A. H. Compton demands."[104] Clark's announcement constituted the beginning of the Duane–Compton controversy.

To all outward appearances, Clark and Duane's apparatus (Fig. 11[105]) was essentially identical with Compton's, although Clark and Duane pointedly emphasized the importance of the lead shielding screen next to their X-ray tube (a standard Coolidge tube with tungsten target), and they placed their tube and scatterer in a completely different room from their spectrometer (calcite crystal). In their first experiments, they scattered the characteristic X-rays of tungsten through 90° using five different compounds. For barium chloride ($BaCl_2$), for example, they obtained the two mirrored secondary spectra shown in Fig. 12[106] (readings were taken on both sides of the zero position of the spectrometer).

Every peak in these spectra (Fig. 12) was identified by Clark and Duane as belonging either to the (unshifted) first- or second-order tungsten

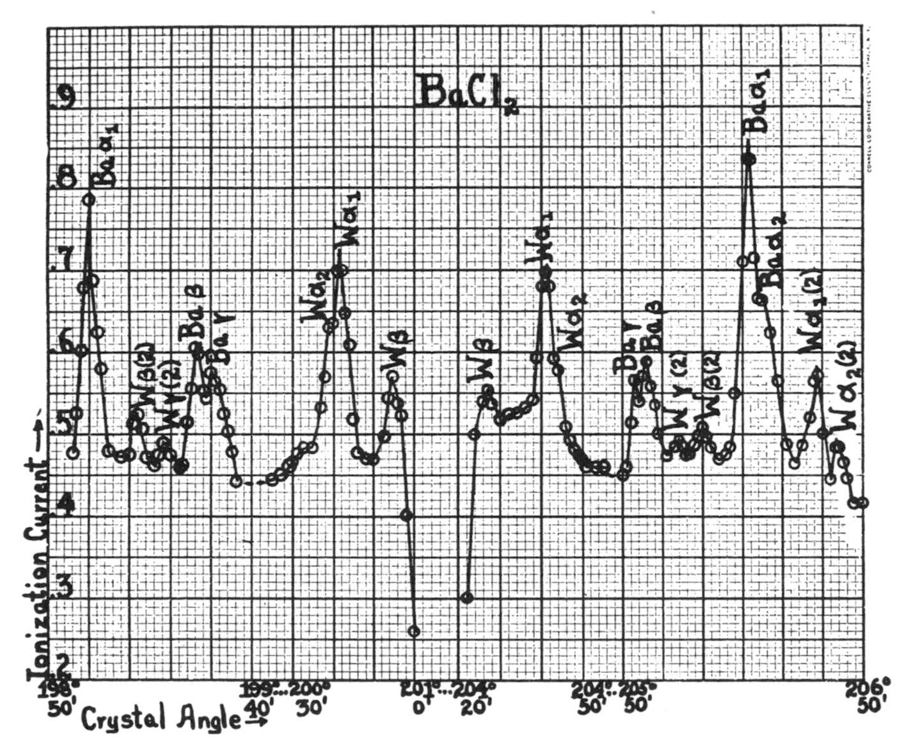

FIG. 12. Clark and Duane's mirrored spectra for tungsten X-rays scattered by $BaCl_2$.

spectrum (W_{α_1}, W_{α_2}, W_β, . . . , W_{α_1} (2), W_{α_2} (2), W_β (2), . . .), or to the usual fluorescent spectrum of barium (Ba_{α_1}, Ba_β, . . .). They associated no peaks whatsoever with radiation shifted by 0.024 Å (or by 13–14 minutes of arc, as calculated from the lattice constant of calcite). "Undoubtedly," they wrote, "there are some rays in the secondary radiation having wave-lengths longer than those of the primary rays, but the amount of this radiation having the definite wave-lengths shift .024 appears from our experiments to be inappreciable as compared with the amount of scattered radiation having wave-lengths precisely equal to those of the primary rays."[107]

Repeating their experiments using as scatterers four elements of low atomic weight, carbon (graphite), aluminum, sulfur, and copper, Clark and Duane found that only with the first did "small irregularities" appear which conceivably might be interpreted as agreeing with Compton's and Ross's data.[108] The other three scatterers yielded only the unshifted tungsten rays, as shown for example in the two mirrored copper spectra in Fig. 13.[109] True, there were "humps" (as Clark and Duane called them) 17 or 18 minutes of arc away from

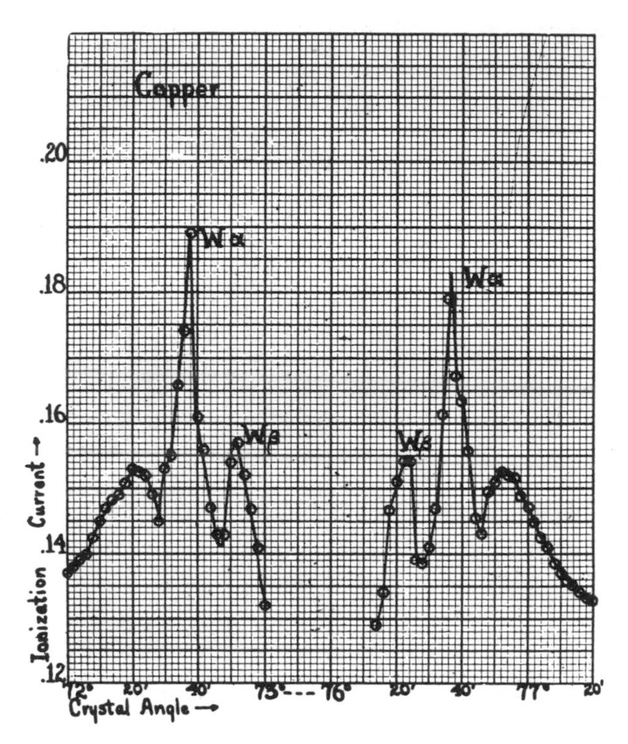

FIG. 13. Clark and Duane's mirrored spectra for tungsten X-rays scattered by copper.

the W_α peak, but this angular displacement was "somewhat greater than the maximum distance demanded by Compton's theory."

These "humps," in any event, could be accounted for by another hypothesis: "there must be tertiary rays having wave-lengths longer than the primary rays by a certain amount. The fundamental idea of this [explanation] is that the photoelectrons produced by the primary rays, when they strike neighboring atoms, must produce radiation just as the electrons in the X-ray tube produce it, when they strike the atoms of the target."[110] Thus, suppose that primary X-rays of energy $h\nu_0$ eject photoelectrons from some level s of energy $h\nu_s$, so that the photoelectrons possess kinetic energy $T = h\nu_0 - h\nu_s$. If these photoelectrons then strike a neighboring atom in the scatterer, they will produce "bremsstrahlung" with a maximum frequency ν' given by the Duane-Hunt law $T = h\nu' = h\nu_0 - h\nu_s$. In terms of wavelengths, $\lambda' = \lambda_0\lambda_s/(\lambda_s - \lambda_0)$, and hence the difference in wavelength between the shortest wavelength "bremsstrahlung" X-rays and the primary X-rays is given by

$$\Delta\lambda = \lambda' - \lambda_0 = \frac{\lambda_0^2}{\lambda_s - \lambda_0}. \tag{6.32}$$

Now, for WK_α X-rays the primary wavelength λ_0 is 0.21 Å. The wavelength λ_s, which is characteristic of a particular energy level in the scatterer, obviously varies from scatterer to scatterer. Clark and Duane assumed that for sulfur, λ_s = 5.0 Å, and for aluminum, a lighter element, λ_s = 7.9 Å. According to equation (6.32), these values correspond to wavelength differences of 0.009 Å and 0.006 Å, respectively, both of which were too small to be measured experimentally. The same would be true for the wavelength difference predicted for carbon, an even lighter element. However, for copper, a relatively heavy element, Clark and Duane assumed that the wavelength λ_s = 1.38 Å, which corresponds to a difference in wavelength of 0.037 Å. And this difference in wavelength, when converted into an angular difference, corresponded closely to the actual angular difference between the WK_α peak and the "humps" Clark and Duane observed, as shown in Fig. 13!

Furthermore, when they applied their ideas to Compton's and Ross's data for MoK_α X-rays (λ_0 = 0.071 Å) scattered from carbon (λ_s = 42 Å), Clark and Duane found the predicted change in wavelength to be 0.012 Å. Hence, they claimed, "there ought to be tertiary radiation beginning at a wave-length something like .012 ångström longer than the $K\alpha$ line . . . and having an average wave-length somewhat longer still," depending upon a "great variety of experimental conditions . . . for instance . . . on the angle of incidence of the primary X-rays." "Possibly," they speculated, "the radiation observed by Compton and Ross may be the peak of this continuous spectrum radiation due to the bombardment of the photoelectrons against neighboring atoms in the secondary radiator."[111]

The general trend of their results indicated to Clark and Duane that to find even more conclusive evidence for the "tertiary radiation," they ought to use even heavier elements than copper as scatterers. Choosing molybdenum and silver, they readily calculated that molybdenum's tertiary radiation peaks should begin at 0.315 Å (the λ'_α peak) and 0.262 Å (the λ'_β peak), with analogous results for silver. They identified every one of these predicted peaks in the secondary radiation they examined. Therefore, they concluded, there was "no doubt but that the humps represent the tertiary radiation."[112]

The most striking consequence of Compton's theory, however, was that the change in wavelength, i.e., the position of the secondary peak, depended upon the angle of scattering. Was this also true for the position of the *tertiary peaks?* To answer this question, Clark and Duane scattered tungsten X-rays from molybdenum, changing the scattering angle from 45° to 90° to 135°. The spectra they obtained[113] are shown in Fig. 14. They noted that the short-

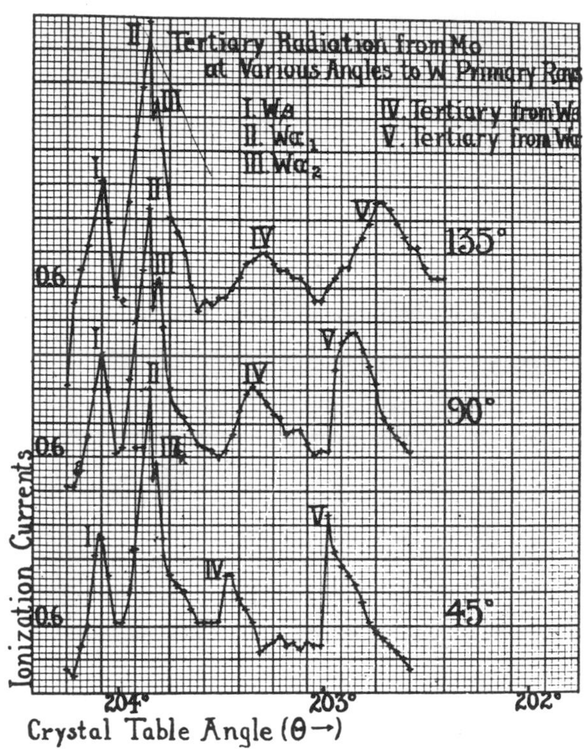

FIG. 14. Clark and Duane's spectra for tungsten X-rays scattered by molybdenum through various angles.

wavelength limits of the tertiary humps (IV and V) all coincide, but that depending upon "such factors as filtration . . . , the shapes and angular positions of the maxima of the humps differ greatly. *The highest points on the humps move towards larger angles (and wave-lengths) with increase in the angle between primary and secondary beams* [italics added]."[114] Once again, analogous results were found for a silver scatterer.

Surveying all of their experiments—they used a total of some *fifteen* different substances as scatterers—Clark and Duane found that they could identify every single peak in every single one of their secondary spectra as either (*i*) an unshifted primary tungsten peak; (*ii*) a fluorescent peak from the scatterer; or (*iii*) a tertiary "hump." Their experimental error, they claimed, was 0.1%. Their general conclusion was unambiguous: "We have not observed any marked characteristics of the radiation that cannot be explained as due to the above-mentioned causes. In particular, we have not observed on any of our curves, peaks which correspond to a shift of exactly .024 ångström . . .

which should occur if the incident X-ray quanta delivered their energy and momentum to single electrons, according to Compton's theory."[115]

2. *The First Compton–Duane Debate (Cincinnati, December 1923) and Subsequent Visits to Each Other's Laboratories*

With the above conviction in mind, Duane traveled to Cincinnati for the Christmas 1923 meeting of the American Physical Society. Compton also attended, as well as a number of other prominent American and European physicists. There was good reason to come, for the most stimulating, and by now the most controversial topic of the day, Compton's quantum theory of scattering, was to be the center of discussion. Since "invited papers" had not yet been invented,[116] a rather formal debate (officially, it was called a "symposium") had been arranged on the subject. The principal participants were Arthur Compton (Chicago), Bergen Davis (Columbia), and William Duane (Harvard). Presiding was W.F.G. Swann, Chairman of Section B.[117]

Compton opened the "symposium" with an address on "The Scattering of X-rays."[118] He confidently reviewed his Washington University experiments and displayed his spectroscopic data, as well as Ross's confirming spectrograms. He argued that Bothe's and Wilson's cloud chamber photographs provided direct visual evidence for the predicted recoil electrons. He repeated his theory and its refinements in detail. Finally, he outlined the implications of his work:

> Unquestionably the most important result . . . is the information which it gives regarding the nature of electromagnetic radiation. We find that the wave-length and the intensity of the scattered rays are what they should be if a quantum of radiation bounced from an electron, just as one billiard ball bounces from another. Not only this, but we actually observe the recoiling billiard ball, or electron, from which the quantum has bounced, and we find that it moves with just the speed it should if a quantum had bumped into it. The obvious conclusion would be that X-rays, and so also light, consist of discrete units, proceeding in definite directions, each unit possessing the energy $h\nu$ and the corresponding momentum h/λ. So in a recent letter to me Sommerfeld has expressed the opinion that this discovery of the change of wave-length of radiation, due to scattering, sounds the death knell of the wave theory of radiation.[119]

The only way to avoid Sommerfeld's grim forecast, to retain the wave theory, was to give up the laws of conservation of energy and conservation of momentum. We must, Compton asserted, "choose between the familiar hypothesis that electromagnetic radiation consists of spreading waves, on the one hand, and the principles of the conservation of energy and momentum on the other. We cannot retain both." Opinions were divided: "The conviction of the truth of the spherical wave hypothesis produced by . . . interference experiments has led Darwin and Bohr in conversation with me to choose . . . the abandonment of the conservation principles." Compton disagreed. He had shown that the application of the conservation laws had led to a theory of scattering which he felt had been fully confirmed by experiment. "For this reason," he said, "I am inclined toward the choice of these principles even at the great cost of losing the spreading wave theory of radiation. I am by this choice confined to the view that radiation consists of directed quanta."[120]

The only glimmer of hope at present for a reconciliation of wave and particle, according to Compton, was "the trail blazed by Duane" in his quantum derivation of Bragg's law. Thus, in a short paper[121] which he had communicated to the *Proceedings of the National Academy of Sciences* on March 2, 1923—when he was on the verge of challenging Compton—Duane had described "an attempt to formulate a theory of the reflection of X-rays by crystals, based on quantum ideas without reference to interference laws." He had been driven to this attempt because certain crystal reflected X-radiation "characteristic of the atoms in the crystal itself, which Dr. G. L. Clark and the writer discovered . . . , does not appear to be explainable in a simple manner by the theory of interference of waves."[122] This unusual radiation resisted other attempts at explanation also,[123] but this did not deter Duane. He pointed out that if a quantum of momentum $h\nu/c$ is incident at a glancing angle ϕ on a crystal, then, as it recoils, it will impart the momentum $(2h\nu/c)\sin\phi$ to the crystal perpendicular to one of its reflecting planes. If, further, the crystal's momentum is quantized—if it can change only by amounts nh/d, where n is an integer and d is the distance between reflecting planes—we then have that $(2h\nu/c)\sin\phi = nh/d$, and Bragg's law, $n\lambda = 2\,d\sin\phi$, follows immediately. A typical interference phenomenon had therefore been explained by invoking quantum considerations.

This result was so encouraging that Compton himself, in November 1923 (one month before the Cincinnati meeting) contributed to its further clarification.[124] He demonstrated that the general quantum postulate, $\oint pdq = nh + \eta$, where η is a constant and where p and q denote a generalized momentum and its conjugate coordinate, leads to Duane's quantum condition for the crystal. At the same time, Compton disagreed with Breit's conclusion that it was "some disturbance traversing the crystal" that was quantized. In general, he felt that "the present quantum conception of diffraction is far from being in conflict with the wave theory," and that "even from the quantum viewpoint electromagnetic radiation is seen to consist of waves"[125]—two remarks which clearly illustrate the note of hope Duane's derivation had injected into the discussion on the wave–particle dilemma.

At the Cincinnati meeting, however, Duane spoke on "The Scattering and Reflection of Short X-Rays,"[126] and vigorously opposed Compton's quantum theory of scattering. Duane's views, however, evidently met with considerable opposition. Bergen Davis cited experiments on "Scattering from other Elements than Carbon,"[127] which confirmed Compton's work. P. A. Ross displayed his spectrograms and wholeheartedly supported Compton. So did Maurice de Broglie, who had traveled all the way from Paris to show the results of his experiments, which were similar to Ross's.[128] The general sentiment was nicely captured in a letter of January 4, 1924, which E. C. Kemble wrote to Bohr:

> You will be interested to know that many of those who went to the recent Cincinnati meeting of the American Physical Society inclined to belittle Compton's corpuscular theory of scattering came away with a very different feeling. In view of the evidence offered it was difficult to avoid the conclusion that the phenomenon of changed wavelength is a real one. It would not be difficult to invent a form of wave theory that would account for Compton's observations and the recoil electrons of C.T.R. Wilson, but when I try to correlate those things with Einstein's discussion of the maintainance [*sic*] of the Maxwell distribution law in an absorbing gas it seems to me hopeless to try to hold on to the wave theory. Perhaps Dr. Slater's recent scheme for a combination of wave theory and corpuscular theory may solve our difficulties.[129]

We shall discuss Einstein's and Slater's work in the next chapter. For the moment we must note that Duane himself remained unconvinced. Nevertheless, the debate was not without profit. Duane graciously extended an invitation to Compton to visit his laboratory at Harvard—"a courtesy," Compton later wrote, "that I should like to think is characteristic of the true spirit of science"[130]—while Compton responded in kind by inviting Duane to visit his newly established laboratory at the University of Chicago. The exchange presumably took place in early 1924. As it happens, while we have no information on Duane's visit to Chicago, except that it actually took place, we have a firsthand account of Compton's visit to Harvard by S. K. Allison, who was then a graduate student in Duane's laboratory.

Allison recalled that when Compton arrived in Duane's laboratory the "situation was rather tense," and Compton himself appeared "completely nontypical," being "disheveled, unshaven, and obviously overtired." The following morning, however, he was "looking like himself—a well groomed, energetic, and clear-thinking physicist." Allison continued:

> [Here] at Harvard in 1924, in the laboratory of a highly respected investigator of x-rays, the crystal spectrometer measurements seemed to give different results. . . .
>
> A peculiar overtone to the situation was Duane's great resistance to accepting a photon theory of scattering. It was Duane and Hunt who, a few years previously, had quantitatively established the relation between the electron kinetic energy and the maximum frequency of the bremsstrahlung . . . . And Duane himself was at the time working on a thought-provoking attempt to explain the crystalline

diffraction of x-rays without recourse to wave theory, using photons only . . . . Nevertheless, Duane had resisted Compton's idea from its first pronouncement and had written Sommerfeld, who was here in the United States at that time, of his doubts and his alternative explanation. Sommerfeld's reply, which Duane duly reported to us, was that after a visit to Compton in his new laboratory at Chicago he remained convinced of the fundamental importance of Compton's discovery.

At the time of Compton's visit I was not working on the scattering problem but was working on some problems Duane had suggested, involving fluorescence radiation. All the excitement, however, was in the next room, and I often wandered in to hear the latest scattering news. Compton's visit did not resolve the difficulty, but his incisive questions and earnestness greatly impressed Duane and his scattering group. The Harvard experiments were continued, with more self-criticism, and Duane, who had been spending most of his time directing the Roentgenology Laboratory at Harvard Medical School, neglected those duties to take readings himself on scattered x-rays.[131]

Since, in fact, Duane's protégé G. L. Clark actually published a paper on the tertiary ray hypothesis under his own name,[132] it seems very likely that it was he, rather than Duane, who was primarily responsible not only for the Harvard experiments, but also for their interpretation.[133] Nevertheless, the experiments were clearly carried out under Duane's auspices, and Duane wholeheartedly supported Clark's interpretation—and not without reason, as Allison suggests: Compton himself, using Clark and Duane's apparatus, could not obtain the Compton effect![134]

The extent of Duane's commitment to the tertiary ray hypothesis may be seen in a letter of March 5, 1924, which he wrote to Bohr.

I am sending you some notes describing experiments on the tertiary radiation produced by the photoelectrons emitted from the K and L levels by primary x-rays. I have no doubt but that the shift in the wave-length that A. H. Compton has been writing about should be ascribed to this tertiary radiation. We find on using Molybdenum rays and secondary radiators containing carbon, oxygen, sodium, aluminum, sulfur and chlorine that the shift in the wavelength increases with the atomic number of the radiator substantially as it should in accordance with the tertiary ray idea. In the curves contained in the notes I am sending you, the humps corresponding to tertiary radiation due to electrons from the K levels are higher than those corresponding to electrons from the L levels. This appears to signify that the probability of x-rays removing electrons from the K level is greater than that of removing electrons from the other levels.[135]

3. Experiments Stimulated at Harvard and Compton's Generalized Quantum Theory of Scattering

Quite likely as a consequence of Compton's visit to Harvard, Duane, Clark, and W. W. Stifler decided, as indicated in Duane's letter to Bohr, to shift from tungsten to molybdenum X-rays, and from heavy to light scatterers. Their experimental conditions were now similar to Compton's. Nevertheless, only with graphite did they find evidence that could "be construed as being favorable to either theory."[136] In particular, they found that the *short-wavelength limit* of the secondary hump appeared at the Clark–Duane tertiary ray position, while its *maximum* was found at about the Compton quantum position. Ice, aluminum, and sulfur all yielded results consistent only with the tertiary ray hypothesis; and the spectra obtained for lithium revealed no shift whatsoever in the position of the secondary peak as the scattering angle increased from 90° to 135°. For rock salt (NaCl), the peak shifted to longer wavelengths, but that was unimportant. The important thing was that the *short-wavelength limit* of the peak did *not* shift, as predicted by the tertiary ray hypothesis. In sum, Duane, Clark, and Stifler could find "no evidence of a line shifted in accordance with the theory of the transfer of quanta to single electrons."[137]

Compton's immediate response to this unwelcome conclusion was to publish[138] a detailed account of his and C. F. Hagenow's old (April, 1921) polarization experiments. He now saw that these experiments were consistent with his quantum theory of scattering, but were inconsistent with Clark and Duane's tertiary ray hypothesis, since *bremsstrahlung* radiation is never completely polarized. But if the tertiary ray hypothesis should therefore be abandoned, why was it apparently consistent with Clark and Duane's experimental results? Did their tertiary ray hypothesis actually offer a clue to a deeper understanding of the scattering process? It was questions such as these that prompted Compton in mid-1924 to re-examine his theory anew, and to attempt to generalize it to include the case of scattering by bound electrons.[139]

Compton saw that if the scattering electron is not essentially free, but bound, the energy and momentum of the incident radiation (wavelength λ) will be taken up not only by the scattered radiation (wavelength λ') and recoil electron (mass m), but also by the parent atom (mass M). In addition, if the electron is originally in the sth level of the atom, some energy hc/λ_s will be expended in tearing it loose from that level. Conservation of energy then shows that

$$\frac{hc}{\lambda} = \frac{hc}{\lambda'} + \frac{hc}{\lambda_s} + mc^2\left(\frac{1}{\sqrt{1-\beta^2}} - 1\right) + \tfrac{1}{2}MV^2, \qquad (6.33)$$

where the parent atom is assumed to move non-relativistically at velocity V. Furthermore, conservation of momentum must be expressed by three equations, since there are three secondary particles whose momenta are not in general coplanar, as follows

$$\frac{h}{\lambda} = \frac{hl_1}{\lambda'} + pl_2 + Pl_3$$

$$0 = \frac{hm_1}{\lambda'} + pm_2 + Pm_3$$

$$0 = 0 + pn_2 + Pn_3, \tag{6.34}$$

where $p = m\beta c/\sqrt{1 - \beta^2}$, $P = MV$, and the symbols (l_i, m_i, n_i) satisfy the equation $l_i^2 + m_i^2 + n_i^2 = 1$ and represent the direction cosines of the momentum of the scattered quantum ($i = 1$), of the recoil electron ($i = 2$), and of the recoil atom ($i = 3$). Under the assumption that the atom's kinetic energy $\frac{1}{2}MV^2$ is small compared to the other energies, Compton proved, finally, that equations (6.33) and (6.34) may be solved simultaneously to obtain a general expression for the kinetic energy of the recoil electron, as well as the following expression for the change in wavelength $\Delta\lambda$:

$$\begin{aligned}\Delta\lambda &= \lambda' - \lambda \\ &= [\lambda/(1 - A)][\alpha(1 - l_1) + s(1 - \tfrac{1}{2}\alpha s) + B(l_1 l_3 + m_1 m_3 - l_3 + B/2\alpha)],\end{aligned} \tag{6.35}$$

where
$\alpha = h\nu/mc^2$, $s = \lambda/\lambda_s$, $A = s(1 + \alpha - \frac{1}{2}\alpha s) - B(l_3 - B/2\alpha)$, and $B = P/mc$.

Note, in the first place, that for the case of a free electron, for which $hc/\lambda_s = P = 0$, or for which $s = A = B = 0$, equation (6.35) reduces to

$$\Delta\lambda = \lambda' - \lambda = \lambda\,[\alpha(1 - l_1)] = (h/mc)(1 - \cos\theta), \tag{6.36}$$

which is simply Compton's original expression. But a second observation is even more significant. If, on the one hand, the recoil electron is bound ($s \neq 0$), and if all of the energy and momentum of the incident quantum is taken up by the secondary quantum and the *atom,* the change in wavelength will be a minimum, and it may be shown either from equations (6.33) and (6.34), or from equation (6.35), that

$$\Delta\lambda_{\mathrm{min}} = \lambda'_{\mathrm{min}} - \lambda = \frac{\lambda^2}{\lambda_s - \lambda}\,. \tag{6.37}$$

On the other hand, if the *recoil electron* takes up all of the energy and momentum of the incident quantum, that is, if $hc/\lambda' = 0$ or $\lambda' = \infty$, the change in wavelength will be a *maximum,* and we obviously have that $\Delta\lambda_{\mathrm{max}} = \lambda'_{\mathrm{max}} - \lambda = \infty$. In other words, the wavelength λ' of the scattered radiation will always lie somewhere between $\lambda + \lambda^2/(\lambda_s - \lambda)$ and ∞. Therefore, Compton concluded, we have "precisely the same wave-length range for the secondary radiation as is predicted by the tertiary radiation theory of Clark and Duane, which also assigns a definite lower limit to the wave-length, but supplies no finite [upper] limit."[140]

Furthermore, consider how the position of the "humps" varies with scattering angle. If in liberating the electron the incident radiation supplies energy hc/λ_s and momentum h/λ_s to the atom, that is, if $B = P/mc = h/mc\lambda_s = \alpha s$, one may set $l_3 = 1$ and $m_3 = 0$ and see that equation (6.35) reduces to:

$$\Delta\lambda = \lambda' - \lambda = [\lambda/(1 - s)][\alpha(1 - s)(1 - l_1) + s]$$
$$= \lambda^2/(\lambda_s - \lambda) + (h/mc)(1 - \cos\theta). \qquad (6.38)$$

Therefore, for a free electron ($\lambda_s = \infty$), one again obtains the ordinary Compton expression, while for a bound electron ($0 < \lambda_s < \infty$) the scattered radiation will have a peak beginning at wavelength $\lambda + \lambda^2/(\lambda_s - \lambda)$, whose maximum will shift with scattering angle according to $(h/mc)(1 - \cos\theta)$—precisely as Clark and Duane had observed. In fact, through a critical analysis of Clark and Duane's very extensive data, Compton demonstrated that with but a single exception, which admitted of another explanation, it could all be accounted for on his present generalized theory.

Clark and Duane's tertiary ray hypothesis was therefore superfluous. To seal its fate entirely, however, Compton noted that very general objections could be raised against it. First, the secondary radiation was observed to be completely polarized, whereas tertiary radiation could not possibly be. Second, a large fraction of the incident energy generally appeared in the modified peak, whereas the efficiency of production of tertiary *bremsstrahlung* radiation was known to be only on the order of 0.1%. Finally, Ross's spectrograms showed that the modified line was very sharp, whereas tertiary *bremsstrahlung* radiation was characterized by a broad maximum. Compton summarized: "In view of the fact that the wave-length of these [shifted] lines can be satisfactorily accounted for by the quantum theory of scattering, and especially in light of the experimental evidence for the existence of the recoil electrons, it is very difficult to avoid the conclusion that the modified rays observed in the spectra of secondary x-rays result from the scattering of whole quanta by individual electrons."[141]

Even as Compton was assembling the above arguments against Clark and Duane's tertiary ray hypothesis, work by other physicists was being reported at an ever increasing rate. Although some of the experiments were inconclusive,[142] and some of the theoretical papers were noncommital,[143] or even critical,[144] most of the work supported Compton.[145] Particularly encouraging were the results coming from Maurice de Broglie's laboratory in Paris,[146] but physicists from New York (Bergen Davis at Columbia) to California (P. A. Ross[147] and his co-workers at Stanford) to Russia (D. Skobeltzyn[148] in Leningrad) were confirming Compton's work.

4. Further Experiments at Harvard, Chicago, and Stanford

Clark and Duane remained unconvinced. In fairness, however, it is clear that unless Compton informed Duane of his general scattering theory by letter, Duane could not have known about it before June 1924, when an abstract of it appeared in *The Physical Review,* and he could not have studied it in detail before August, when the full paper was published. But in any event, Clark and Duane did not intend to abandon their tertiary ray hypothesis as yet. They were currently trying to refute a contention of Ross's colleague, D. L. Webster,[149] who had concluded from intensity considerations that the observed radiation, if it were tertiary radiation, could only be produced by electrons striking and accelerating other electrons in the *same* atom from which it had come, and not by interacting with a *neighboring* atom, as Clark and Duane had assumed. Clark and Duane analyzed Webster's arguments in detail, comparing the wavelength shifts predicted by Webster with their own spectra. They concluded that "there is not the slightest evidence for any radiation that could be attributed to the impact of photoelectrons against other electrons in the atoms from which they come, as assumed by Webster."[150] Once again, they also rejected the results of Compton, Ross, and M. de Broglie. Not much later, S. K. Allison and Duane brought forward additional evidence in favor of the tertiary ray hypothesis.[151]

Webster's response will be discussed in a moment. Meanwhile Compton himself was leaving no experimental stone unturned. He realized that one of the criticisms leveled against his work by Clark and Duane, that he had almost exclusively used the light element carbon as a scatterer, was indisputably valid. He therefore enlisted the aid of his student Y. H. Woo and tested this point[152] by using seven different scatterers at a single scattering angle (roughly 128°). Compton and Woo's results are displayed in Fig. 15,[153] from which Compton concluded:

> The important point in this figure is that the spectra obtained from the various elements are almost identical in character. In every case an unmodified line *P* occurs at the same position as the fluorescent Mo Kα line, and there is a modified line whose peak is within experimental error at the position *M,* calculated from the quantum change of wave-length formula . . . . There is also perhaps some evidence in the cases of sodium and aluminum for a hump at the position *T,* where according to the experiments of Clark, Stifler and Duane the peak of the line due to 'tertiary radiation' should appear. In view of the consistency of the results for the different elements, we feel that these experiments show beyond question the reality of the spectrum shift predicted by the quantum theory of scattering.[154]

Compton's conclusion was immediately corroborated by P. A. Ross, who published spectrograms for X-rays scattered from substances ranging in atomic or molecular weight from paraffin to lead.[155] The entire situation, however,

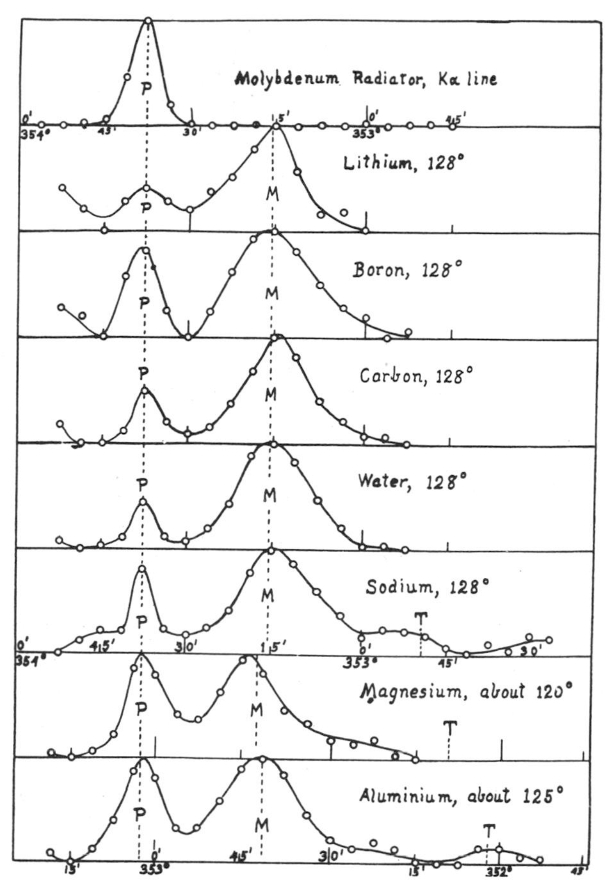

FIG. 15. Compton and Woo's spectra for MoKα X-rays scattered through 128° by various elements.

was at least as perplexing at Stanford as at Chicago, and during the discussions in Ross's laboratory W. R. Smythe suggested that the problem might be basically one of intensity. Accordingly, J. A. Becker, a National Research Council Fellow, designed a series of experiments to test this point.[156] Using a relatively high intensity, and therefore a relatively short exposure time (61 hours!), Becker found "no indication of a Duane shift." However, after reducing the intensity and increasing the exposure time by a factor of three, "a faint new line" appeared, the short wavelength edge of which was "displaced from the α peak by just the amount predicted by Duane's tertiary radiation." This result obviously supported Smythe's conjecture, and Becker concluded: "If in our experiment the intensity could be reduced further by a factor of three it might

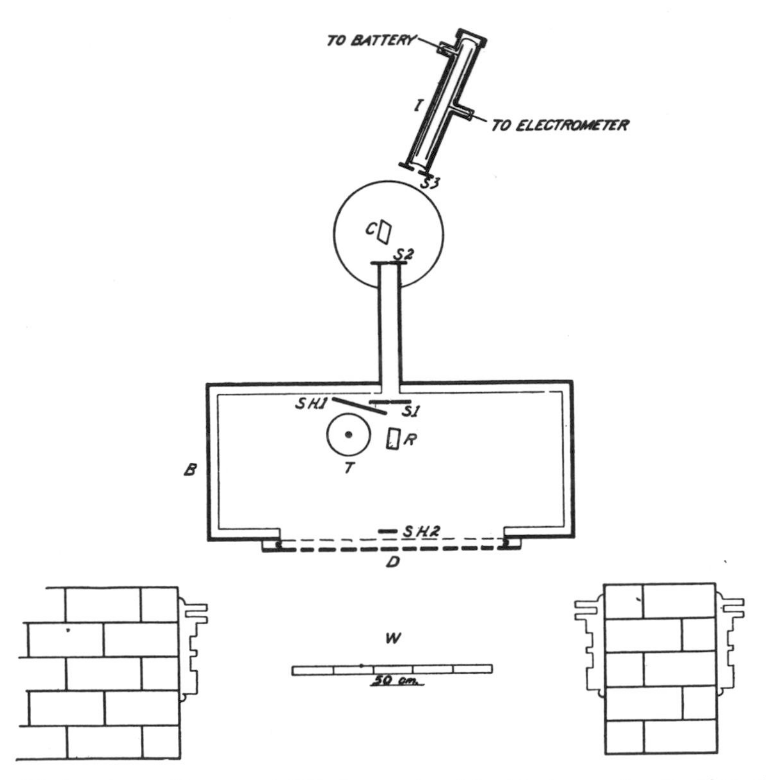

FIG. 16. Armstrong, Duane, and Stifler's experimental setup to examine the "box effect."

very well be that the Duane shift would become much more pronounced compared to the Compton shift. To follow up this clue would require a whole months exposure. The author is unable to continue this work. The suggestion together with its partial experimental support is published in the hope that it will help clear the situation in the controversy."[157]

5. *Duane's "Box Effect"*

Becker's suggestion was never pursued. Instead, Duane himself, aided by Alice Armstrong and W. W. Stifler, injected an entirely new element into the controversy.[158] Duane, too, had asked himself how his work could be reconciled with that of Compton and others.

On thinking over possible differences between the methods of investigating spectra employed in other laboratories and those used in our own laboratory it occurred to us that perhaps the most important difference lay in the fact that we

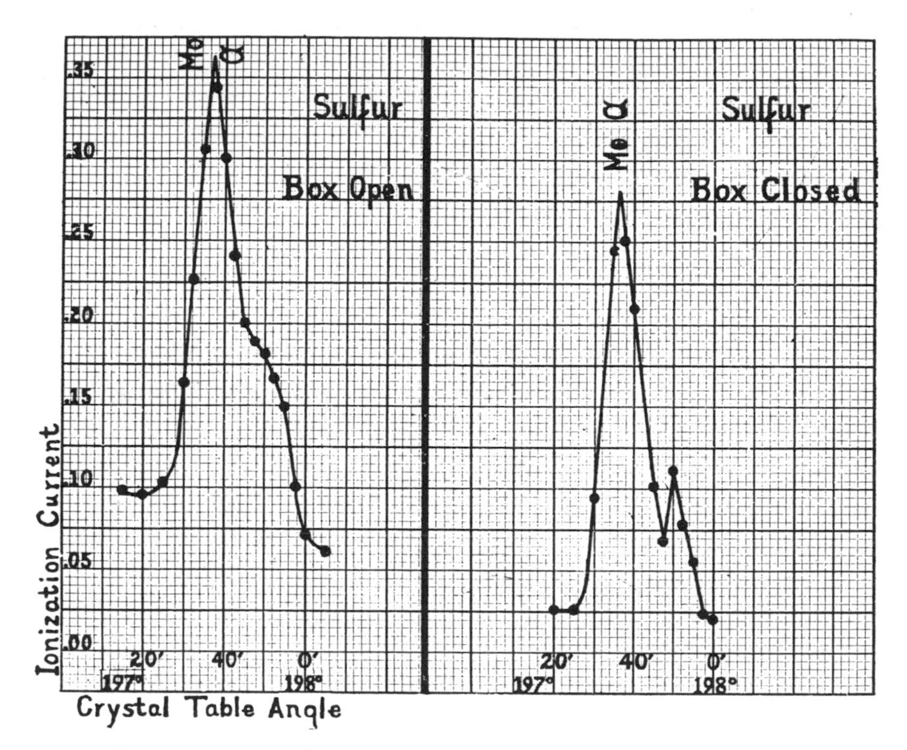

FIG. 17. Armstrong, Duane, and Stifler's spectra illustrating the "box effect."

have never placed our X-ray tubes and radiators in small boxes, whereas others may have used lead boxes with wood or other insulating material composed of carbon and oxygen inside of them. As has been repeatedly stated in the descriptions of our investigations of X-ray spectra, our X-ray tube and radiator lie in one large room and the X-rays to be examined pass through a hole in the wall to the spectrometer in an adjoining room.[159]

Since they had just completed a new X-ray plant at Harvard, Armstrong, Duane, and Stifler decided to use it to examine this "box effect." As shown in Fig. 16,[160] they placed a sulfur radiator *R* and an X-ray tube *T* (molybdenum target) into a wooden box *B* covered with ¼-inch lead plate. The back of the box (facing a window W) had a door D in it which could be opened and closed, the object being to take readings first with with the door open and then with the door closed. This they did—and obtained (Fig. 17[161]) two radically different spectra! They explained: "It appears that the mere closing of the door had some effect on the spectra. With the door closed, there appears very distinct evidence of a small peak on the long wave-length side of the peak representing

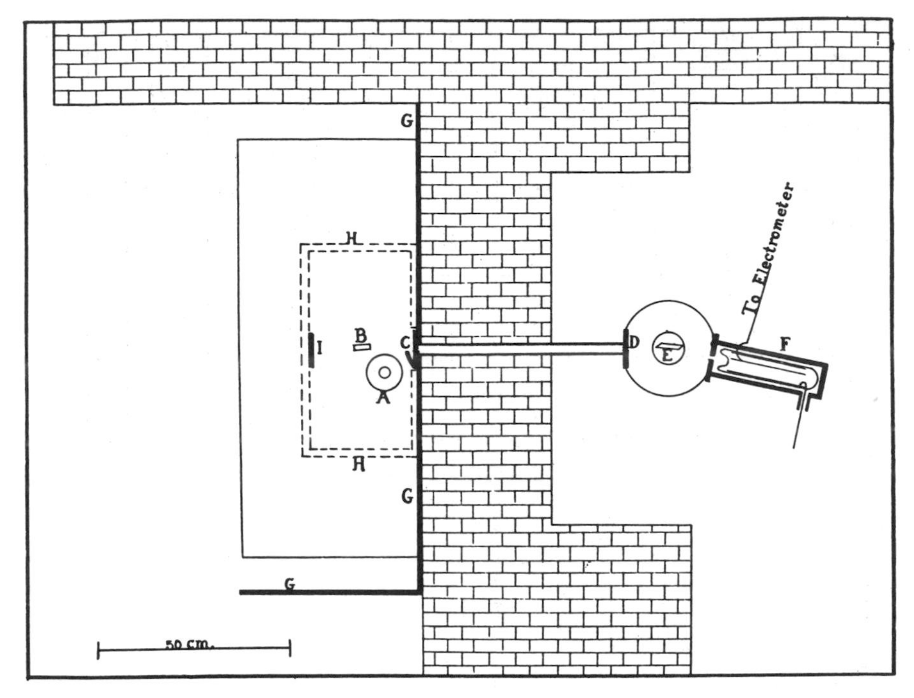

FIG. 18. Allison, Clark, and Duane's experimental setup to further explore the "box effect."

scattered radiation due to the $K\alpha$ doublet of the molybdenum target. The separation between the two peaks is about the same as that reported from other laboratories." Therefore, taken "in connection with the experiments performed with the old X-ray plant . . . in which the second peak did not appear when no box was used, these experiments furnish strong evidence in favor of the view that the radiation with shifted wave-length is due to the box surrounding the tube and radiator and is not characteristic of the secondary-tertiary radiation of the radiator itself." Further experiments revealed that the "intensities of the lines appear to be very complicated functions of the exact structure of the box."[162]

The claim was, therefore, that some of the primary X-rays struck the walls of the box before being scattered from the sulfur radiator in the usual manner, and that this "box radiation" had sufficient intensity to mask the ordinary secondary radiation. Duane remarked that while at first sight this contention might "seem strange," he could point to a helpful analogy from his medical experience which lent it plausibility: "When a roentgenologist takes a photograph of the body of a patient for diagnostic purposes, he uses a device

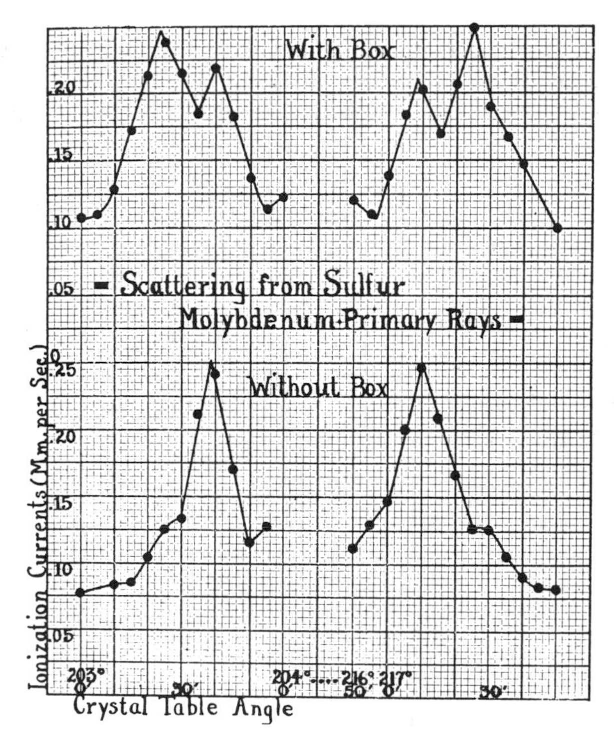

FIG. 19. Allison, Clark, and Duane's spectra illustrating the "box effect."

known as the Potter–Bucky diaphragm. This consists of a number of nearly parallel lead strips, moved across between the patient and the photographic plates during the exposure. The purpose of the lead strips is to cut off the secondary radiation coming from the various tissues in the patients' body. If this radiation were not cut off, it would produce a certain blurring of the photograph."[163]

Further experiments on Duane's "box effect" were immediately carried out by Allison, Clark, and Duane.[164] As shown in Fig. 18,[165] they enclosed an X-ray tube *A* and sulfur radiator *B* within a removable wooden box *H*, everything being isolated from the spectrometer by a wall. Taking readings on either side of the spectrometer's zero position, they obtained the two mirrored spectra shown in Fig. 19[166]—which once again were radically different depending upon whether the box was, or was not, in position. They concluded: "The important thing for our purposes at present is that the peaks representing X-rays with shifted wave-lengths disappear when the wooden box is removed. They must, therefore, be due to a kind of box effect, and since they occupy the position in the spectrum corresponding to tertiary radiation from carbon and oxygen atoms, the presumption is that they are due to the tertiary radiation coming from the carbon and oxygen atoms composing the wooden walls of the box."[167]

6. The Second Compton–Duane Debate (Toronto, Summer of 1924)

With the new foregoing information in mind, Duane traveled to Toronto in the late summer of 1924 to attend a meeting of the British Association for the Advancement of Science. Compton also attended, since one of the highlights of the meeting was to be a continuation of the Cincinnati debate of the previous Christmas. Sir William H. Bragg presided over the physics section, and hence over the second installment of the Duane–Compton debate. According to the reporter for *Nature,* "perhaps no discussion created such widespread interest as that which centred round the papers of Prof. A. H. Compton of Chicago, and Prof. Duane of Harvard, on the scattering of X-rays."[168] Compton himself later recalled: "After discussing it for a solid afternoon, we decided to call the debate a draw. The result was summarized by a comment of C. V. Raman, who was visiting Toronto at the time. As we left, he said to me, 'Compton, you're a very good debater; but the truth isn't in you.' "[169]

The reporter for *Nature* went into more detail on the background and character of the Toronto debate.

> Duane found that, with his apparatus, he was unable to find evidence for the existence of the effects observed by Compton. Compton, on the other hand, could not repeat satisfactorily Duane's experiments. A matter of such importance could not be left in such a position. Each observer investigated the apparatus used by the other and convinced himself of its trustworthiness. Duane observed that the only obvious difference in the experimental arrangements was that Compton's X-ray tube was enclosed in a wooden box covered with lead, while his own tube was not so enclosed, the tube being in one room and the rest of the apparatus in the adjoining room. Improbable as it appeared that such a difference could account for the difference in the experimental results, Duane tried the effect of such a box and found to his surprise that, in addition to the effects he had previously observed, a new peak appeared in approximately the position observed by Compton. The exact position of this effect depended on the orientation of the box. At the time of the Toronto meeting this represented the state of affairs.
>
> In the general discussion various other members took part. Prof. Webster gave a detailed description of the experimental arrangements used by Ross, and contended that the box in which Ross's tube was enclosed could not possibly account for the results he obtained. Prof. Gray described experiments on γ-rays and showed that they were consistent with the Compton theory. Prof. Raman made an eloquent appeal against a too hasty abandonment of the classical theory of scattering.

Compton sketched an extension of his theory in which he considered not only free electrons but also those which were more tightly bound. He showed that the extended theory gave rise in the limiting case to the formula used by Duane and therefore embraced Duane's results. The fundamental difference between the two theories remains; Duane uses only the well-established quantum energy equation, while Compton in addition introduces the idea of conservation of momentum in the interaction between the radiator and matter. There are difficulties in the way of both theories, but at the present stage of the experimental work it is needless to dwell upon them. Before the theoretical side of the question can be satisfactorily discussed, further experimental work must be done. At the time of the meeting each observer appeared to have almost overwhelming evidence in favour of his point of view, and had the audience only had to listen to one side—either side would have done equally well—it would probably have been convinced as to the accuracy and soundness of the views advanced. As it was, however, the average member left the meeting inclined to echo the sentiments of the lover in the 'Beggar's Opera' who sings,

> 'How happy could I be with either
> Were t'other dear charmer away!'[170]

7. Webster and Ross Rebuff Duane

We see that the entire situation still appeared remarkably open to the "average member" of the audience at Toronto. On the one hand, there was clear-cut resistance to Compton's work from Raman[171] and others; on the other hand, there was clear-cut support from Webster, Gray, and others. Webster and Ross, in fact, were particularly piqued by Duane's suggestion of a "box effect," for as soon as they returned to Stanford, they decided to refute it once for all. They described their approach, with undisguised sarcasm, as follows:

> . . . we decided to try experiments under conditions where we . . . could rely on the size of the room containing the tube and scatterer, as Duane did, to get rid of radiation from the walls. Not having a convenient location, nor enough lead, for a large room, we are therefore forced to consider what may distinguish a 'room' from a 'box.' . . . We may . . . define a 'room' as a compartment large enough to make the inverse square law alone suppress the scattered rays from light elements beyond detection, even if the light elements in question return all the rays falling on them from the tube. This definition does not require a very large room. If, for example, the scatterer is 6 cm. from the target and the nearest light atoms struck by the direct rays are 140 cm. away, as in the experiment to be described, then the inverse square law alone will reduce the returned rays to less than $(6/140)^2$, or 0.0018, of the fraction of the primary rays striking such atoms. And whatever may be said about scattering and absorption coefficients, the limitation by the inverse square law cannot easily be denied.[172]

Webster and Ross specifically duplicated Duane's conditions in various important details. They used molybdenum X-rays and a sulfur scatterer, shielding both with lead and isolating both from the spectrometer with a wall. "Special care" was taken to remove from the vicinity all materials containing carbon or oxygen. Finally, out of a large number of spectrograms with exposure times ranging in length from 30 to 222 hours, they selected one for publication. It displayed (*i*) the MoK$_\alpha$ peak at a certain position *A;* (*ii*) the Compton line at a certain position *C,* which agreed within 5% of theory; and (*iii*) an *absolutely unexposed area* around a certain position *D,* the predicted position of "Duane's peak." "As to the position of the theoretical *D* line," Ross and Webster wrote, "there may be some question. The position shown here is taken directly from Duane's own paper, and is doubtful only because of a statement in that paper to the effect that it agrees with his theory." "Applying these results to present theories," they concluded, "obviously no change is required in Compton's. With regard to the tertiary ray theory, however, the case is different."[173]

What most irritated Webster and prompted his sarcasm was that only a short time earlier, as we have seen, he had suggested—in what turned out to be a strictly *ad hoc* manner—a possible modification of the tertiary radiation hypothesis to account for an intensity variation which—it developed—*had been observed only by Clark and Duane.* As Webster wrote, his suggestion owed "its existence entirely to Clark and Duane's tertiary ray spectra. If all such experiments had given results like ours, therefore, it would never have been made and the absence of tertiary radiation would not have been at all surprising." Moreover, the "point that is most surprising is that one set of experiments [Duane's] seems to confirm this hypothesis and the other to deny it. If it were any hypothesis about molar processes, molar differences in apparatus might make this discrepancy. But how can any difference in apparatus, such as we have here, make a difference like this within the atom?"[174]

Ross and Webster made their case airtight by giving three additional reasons for rejecting Clark and Duane's tertiary radiation hypothesis. First, if one box reflection produced the ordinary Compton shifted line, why should not two reflections produce a doubly shifted line? Since the latter had never been observed, this "in itself contradicts their theory when carried to this logical conclusion."[175] Second, a simple calculation proved that the source strength of "the box as a whole" was not greater than 5% of that of the target, which was much too weak to account for the Compton line. But that was not to say that secondary radiations are never important. Applied "to the case of the human body subjected to very penetrating rays,"[176] the same calculation proved that it should emit significant amounts of secondary radiation—which vindicated the practice of roentgenologists in using the Potter–Bucky diaphragm. Finally, Ross and Webster called attention to one experimental detail which Ross had never before mentioned, but which had now "taken on a new importance." In his "numerous experiments," Ross had *never* used an enclosing box containing wood, and hence containing carbon or oxygen. "Altogether," they concluded, "the evidence reviewed here seems to us definitely contrary to Duane's box-effect theory as an explanation of the Compton peaks, either in his experiments or in those of Compton and Ross, and leaves Compton's theory of the scattering of quanta by single electrons as the only theory advanced yet that will explain them."[177] A short time later, Ross and Webster published further experimental results substantiating that conclusion.[178]

8. *Compton and Woo Test Duane's "Box Effect"*

When Compton returned to Chicago from the Toronto meeting in the late summer of 1924, Duane's "box effect" was very much on his mind also. He therefore asked his student Y. H. Woo to test it by enclosing a molybdenum

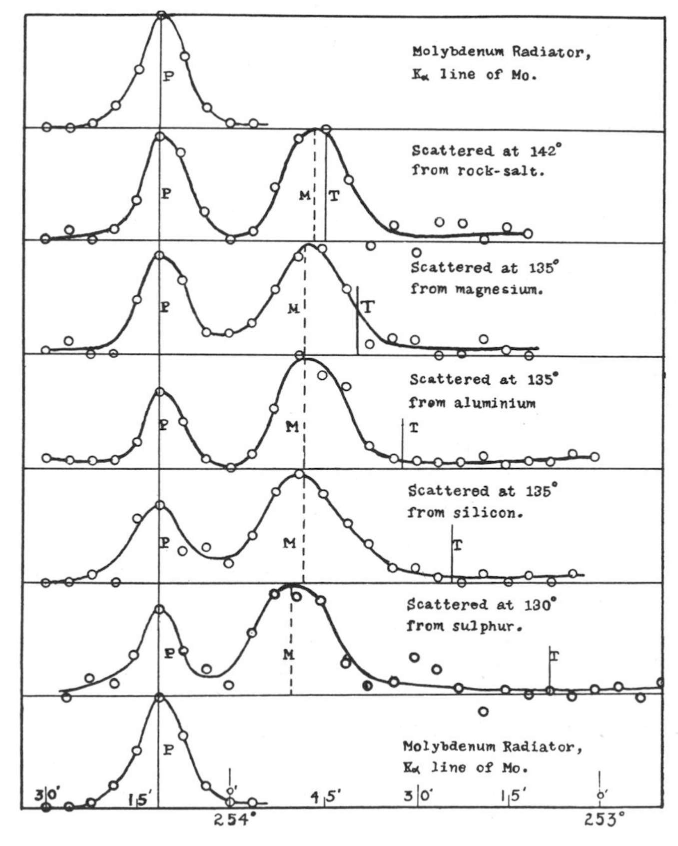

FIG. 20. Woo's spectra for MoKα X-rays scattered through various angles by various scatterers enclosed in a lead box.

target X-ray tube and various scatterers (rock salt, magnesium, aluminum, silicon, and sulfur) inside a lead box. Woo obtained the spectra shown in Fig. 20,[179] which obviously proved that "None of these experiments showed the existence of [Duane's] . . . tertiary peak"[180] at the predicted position *T*. However, in "order to satisfy ourselves completely," Compton wrote, "we set up a water cooled molybdenum target X-ray tube and a sulphur radiator *outside a third floor window, with no surrounding box,* as shown diagrammatically in . . . [Fig. 21]. The face of the target pointed upward, so that nearly all of

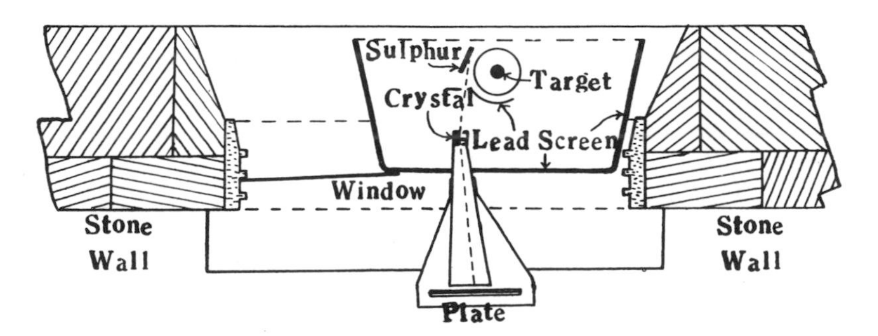

FIG. 21. Compton and Bearden's open-air scattering experiment.

the rays which did not strike the lead screen shielding the window *went to the open sky*."[181] Using a photographic plate as detector, he and his student J. A. Bearden found no "considerable effect on the character of the spectrum obtained due to the presence of a surrounding box."[182]

9. *The Controversy Ends*

As it turned out, the experiments of Ross and Webster, as well as those of Compton and his students, were unnecessary. Duane, after returning to Harvard from Toronto, put S. K. Allison in charge of the experiments, and Allison almost immediately began finding "evidence of radiation agreeing very well"[183] with Compton's theory. Allison used a long, thin X-ray tube after Compton's design, and an improved slit system, which greatly increased both the resolving power of his spectrometer and the intensity of his scattered radiation. Allison and Duane then took a number of spectra at various tube-radiator separations and at various scattering angles. While the shifted peak appeared to be somewhat broader than anticipated, they found "substantial agreement"[184] with the experiments of Compton, Ross and Webster, M. de Broglie, and by now any number of other physicists. A short time later, Allison and Duane made some of the most accurate contemporary measurements on the Compton effect.[185]

Clark and Duane's persistent failure to observe the Compton effect was probably due to insufficient secondary intensity and low resolving power, in some cases combined with a masking effect produced by fluorescent radiations emitted by their scatterers. The entire experiment, as H. Mark noted,[186] was right on the edge of what was technically possible to achieve at the time. Compton himself later concluded that in certain instances "the greater part of the radiation studied by Clark and Duane was not really the secondary rays from the radiator under investigation."[187]

Duane withdrew his objections to Compton's theory, as Allison recalled, "at a memorable meeting of the American Physical Society" during the Christmas holidays of 1924.[188] P. W. Bridgman noted that: "Once convinced of the true state of affairs, Duane was most generous in his admission of error, and the unreservedness of his announcement of his change of position . . . must remain a pleasant memory to all who heard it."[189] In subsequent years, as R. S. Shankland recalls, Duane and Compton remained on the most cordial and friendly terms with each other, and each greatly admired the other.[190]

With the closing of the Duane–Compton controversy the Compton effect was accepted everywhere as a fact of Nature.[191] But well before the controversy ended, well before the experimental situation had been cleared up, the conceptual turmoil generated by Compton's discovery had already begun.

G. References

[1] Any number of examples might be cited here, but a typical account is given in F. K. Richtmyer, E. H. Kennard, and T. Lauritsen, *Introduction to Modern Physics*, 5th ed. (New York: McGraw Hill, 1955), pp. 386–392.

[2] Compton, "Secondary Radiations Produced by X-Rays," *Bull. Natl. Res. Counc.* 4 (1922):25.

[3] A. H. Compton, "Light Waves or Light Bullets?" *Sci. Am.* 133 (1925):246. If Compton did not in fact originate this "photoelectric effect myth," he certainly lent credence to it by repeating it again and again. Thus in "Directed Quanta of Scattered X-Rays" (with A. W. Simon), *Phys. Rev.* 26 (1925):289, he stated: "An increasingly large group of phenomena has recently been investigated which finds its simplest interpretation on the hypothesis of radiation quanta, proposed by Einstein to account for heat radiation and the photo-electric effect." In the *Scientific American* article cited, he also remarked: "Since the idea of light quanta was invented primarily to explain the photoelectric effect, the fact that it does so very well is no great evidence in its favor. The . . . quantum theory could not be given much credence unless it was found to account for some new thing for which it had not been especially designed. This is just what the quantum theory has recently accomplished in connection with the scattering of X rays." In his *X-Rays and Electrons* (New York: Van Nostrand, 1926), pp. 222–223, Compton asserted: "The physical existence of quanta of energy may be said to have been established by studies of the photoelectric effect. In accord with his view that energy must always exhibit itself in quanta, Einstein suggested the possibility that radiation may consist of discrete bundles of

energy of amount $h\nu$. . . ." In "Some Experimental Difficulties with the Electromagnetic Theory of Radiation," *J. Franklin Inst.* 205 (1928):160, Compton claimed that: "It is well known that the photon hypothesis was introduced by Einstein to account for the photo-electric effect." In "What is Light?" *Sigma Xi Quarterly* 17 (1929):24, Compton observed that: "It was considerations of this kind which showed to Einstein the futility of trying to account for the photoelectric effect of the basis of waves." In "The Corpuscular Properties of Light," *Rev. Mod. Phys.* 1 (1929):79, Compton stated: "We have seen that Einstein's hypothesis of corpuscular units of radiant energy gives a satisfactory account of the photoelectric effect. As Jeans has significantly remarked, however, Einstein invented the photon hypothesis just to account for this one effect, and it is not surprising that it should account for it well. In order to carry any great weight the hypothesis should also be found applicable to some phenomena of widely different character. Just such phenomena have recently been found associated with the scattering of x-rays. . . ." And he repeated essentially the same statement in "What Things Are Made Of—II," *Sci. Am.* 140 (1929):235. To see how badly Compton was mistaken, historically speaking, see Martin J. Klein, "Einstein's First Paper on Quanta," *Nat. Phil.* Vol. 2 (New York: Blaisdell, 1963), pp. 57–86, or my discussion in Chapter 1.

[4] At about this same time, C. W. Oseen showed that Maxwell's equations possess solutions which approximate arbitrarily closely the characteristics of Einstein's light quanta. See "Die Einsteinsche Nadelstichstrahlung und die Maxwellschen Gleichungen," *Ann. Physik* 69 (1922):202–204.

[5] A. Einstein, "Über ein den Elementarprozess der Lichtemission betreffendes Experiment," *Sitzber. Preuss. Akad. Wiss.* (1921):882–883. A complete discussion of Einstein's paper in the context of Einstein's thought and contemporary attitudes (especially Bohr's) toward light quanta is given by Martin J. Klein in "The First Phase of the Bohr-Einstein Dialogue" in Russell McCormmach (ed.), *Historical Studies in the Physical Sciences*, Vol. 2 (Philadelphia: University of Pennsylvania Press, 1970), pp. 1–39. My brief treatment is based on Klein's account.

[6] Armin Hermann (ed.), *Albert Einstein/Arnold Sommerfeld Briefwechsel* (Basel: Schwabe, 1968), p. 96.

[7] See Klein, "Bohr-Einstein Dialogue," pp. 10–13, for relevant excerpts from his correspondence and complete references to all published papers.

[8] See also G. Breit, "The Propagation of a Fan-shaped Group of Waves in a Dispersing Medium," *Phil. Mag.* 44 (1922):1149–1152. Einstein acknowledged Ehrenfest's point and gave an analysis of the subtle points involved in the argument in "Zur Theorie der Lichtfortpflanzung in dispergierenden Medien," *Sitzer. Preuss. Akad. Wiss.* (1922):18. Nevertheless, some of his ideas surfaced again in his "Vorschlag zu einem die Natur des elementaren Strahlungs—Emissionsprozesses betreffenden Experiment," *Naturwiss.* 14 (1926):300–301.

[9] Private communication to author, October 11, 1967. R. S. Shankland told the author (private communication, September 9, 1971), he believes that by "east coast circles," Ruark probably meant physicists at Johns Hopkins University and at the Bureau of Standards.

[10] "Dopplerprinzip und Bohrsche Frequenzbedingung," *Phys. Z.* 23 (1922):301–303. An earlier, but less rigorous, derivation had been given by R. Emden in his paper, "Über Lichtquanten," *Phys. Z.* 22 (1921):513–517. Also see G. E. M. Jauncey, "Conservation

of Momentum and the Doppler Principle," *Nature* 117 (1926):343–344. For more recent derivations, see W. C. Michels, "The Doppler Effect as a Photon Phenomenon," *Am. J. Phys.* 15 (1947):449–450; A. van der Ziel, "Note on the Quantum Theory of the Doppler Effect for a Moving Observer," *Am. J. Phys.* 20 (1952):51–52.

[11] See Chapter 2, section D.2. This conclusion was challenged by G. Breit in his paper "Are Quanta Unidirectional?" *Phys. Rev.* 22 (1923):313–319, which also illustrates how surprising Compton's discovery was.

[12] Schrödinger, "Dopplerprinzip," (reference 10) p. 301. Schrödinger gives references to the work, including Bohr's of 1918, which served as the background to his own.

[13] *Ibid.*, p. 303.

[14] See Klein, "Bohr-Einstein Dialogue," pp. 16–23; also see Max Jammer, *The Conceptual Development of Quantum Mechanics* (New York: McGraw-Hill, 1966), especially pp. 69–118.

[15] On July 20, 1919, C. G. Darwin wrote Bohr that the case against conservation was "quite overwhelming." Bohr began composing a sympathetic reply, but never sent it; he responded much later, however, on February 14, 1922. These documents are deposited in the Archive for History of Quantum Physics (Copenhagen, Philadelphia, Berkeley) and are briefly described in Klein, "Bohr-Einstein Dialogue," (reference 5) p. 20.

[16] Lecture reprinted in *Nobel Lectures: Physics,* Vol. II (Amsterdam: Elsevier, 1967), p. 14. Copyright ©, The Nobel Foundation, 1923.

[17] For abstract see *Phys. Rev.* 21 (1923):207; for complete paper see pp. 483–502.

[18] Marjorie Johnston (ed.), *The Cosmos of Arthur Holly Compton* (New York: Knopf, 1967), p. 23.

[19] This is a paraphrase of R. S. Shankland's paraphrase of Einstein's remark; see his "Conversations with Einstein," *Am. J. Phys.* 31 (1963):50.

[20] R. P. Feynman, "The Development of the Space-Time View of Quantum Electrodynamics," *Physics Today* 19 (1966):40.

[21] "The Scattering of X-Rays," *J. Franklin Inst.* 198 (1924):61–62. See also "The Scattering of X Rays as Particles, "*Am. J. Phys.* 29 (1961):817–818.

[22] Compton, "A Quantum Theory of the Scattering of X-rays by Light Elements," *Phys. Rev.* 21 (1923):484.

[23] *Ibid.*, pp. 484–485.

[24] *Ibid.*, p. 485, footnote 1.

[25] *Ibid.*, p. 484–486.

[26] *Ibid.*, p. 486. Compton's theory is found today in virtually every textbook of modern physics.

[27] *Ibid.*, p. 495.

[28] A related result appears in P. Debye, "Zerstreuung von Röntgenstrahlen und Quantentheorie," *Phys. Z.* 24 (1923):165; translated as "X-Ray Scattering and Quantum Theory" in Debye's *Collected Papers* (New York: Interscience, 1954), p. 86. Using the relationship

$\sum_{n=o}^{\infty} x^n = 1/(1-x)$, if $x < 1$, it is easy to verify that, for $\theta = \pi/2$,

$$\begin{aligned}\Delta\lambda &= \{(h/mc)[1-\cos(\pi/2)]\}/(1-h/mc\lambda_o) \\ &= (h/mc)/(1-h/mc\lambda_o) \\ &= (h/mc)/(\lambda_o/\lambda_\theta) \\ &= (h/mc)(\lambda_\theta/\lambda_o), \text{ the expression originally cited.}\end{aligned}$$

[29] Compton, "Quantum Theory of Scattering," p. 494.

[30] *Ibid.*, p. 497.

[31] *Ibid.*, p. 487.

[32] *Ibid.*, p. 497; see also E. Rutherford, *Radioactive Substances and their Radiations* (Cambridge: Cambridge University Press, 1913), p. 273.

[33] *Ibid.*, p. 491. Five years later Compton realized that: "It is a consequence of Ehrenfest's adiabatic principle that photons emitted by a moving radiator will show the same Doppler effect, with regard to both frequency and intensity, as does a beam of waves." See "Some Experimental Difficulties with the Electromagnetic Theory of Radiation," *J. Franklin Inst.* 205 (1928):159.

[34] Compton, "Quantum Theory of Scattering," (reference 22) p. 501.

[35] "The Mass Absorption and Mass Scattering Coefficients for Homogeneous X-Rays of Wave-length between 0.13 and 1.05 Angström units in Water, Lithium, Carbon, Nitrogen, Oxygen, Aluminum, and Iron," *Phys. Rev.* 17 (1921):284.

[36] Compton, "Quantum Theory of Scattering," (reference 22) p. 498.

[37] *Ibid.*, p. 501.

[38] "The Total Reflexion of X-Rays," *Phil. Mag.* 45 (1923):1121–1131.

[39] *Ibid.*, p. 1130.

[40] Compton, "Quantum Theory of Scattering," p. 502.

[41] A. H. Compton, "Radiation A Form of Matter," *Science* 56 (1922):716–717. It is uncertain whether this letter was actually written before or after December 1.

[42] *Ibid.*, pp. 716–717.

[43] Unpublished letter on deposit in the Center for History of Physics, American Institute of Physics, New York City. Quoted by Permission of Mrs. A. H. Compton. An abstract of the paper referred to was published as "Wave-length Measurements of Scattered X Rays," *Phys. Rev.* 21 (1923): 715; the paper itself was published as "The Spectrum of Scattered X Rays," *Phys. Rev.* 22 (1923):409–413. Compton's request was to no avail: the paper was published in the November issue!

[44] I am indebted to Professor R. S. Shankland (private communication, September 9, 1971) for this information.

[45] See reference 28.

[46] See Debye's *Collected Papers* (reference 28) for complete information on all of these researches. Owing to an apparent oversight, Debye's paper on the photoelectric effect (see Chapter 2, section B.3) written with A. Sommerfeld is not included in his *Collected Papers*.

[47] Debye, "X-Ray Scattering," p. 81.

[48] *Ibid.,* pp. 84–85. The same conclusion was also reached shortly thereafter by G. E. M. Jauncey in his paper "A Corpuscular Quantum Theory of the Scattering of X-Rays by Light Elements," *Phys. Rev.* 22 (1923):237.

[49] *Ibid.,* pp. 86–87. See also L. Meitner, "Über eine notwendige Folgerung aus dem Comptoneffekt und ihre Bestätigung," *Z. Phys.* 22 (1923):334–342. For a later analysis, see A. H. Compton, "On the Interaction between Radiation and Electrons," *Phys. Rev.* 31 (1928):59–65.

[50] Bohr's correspondence principle, to Debye, meant that in the long wavelength region, classical theory was valid, or specifically, that the long wavelength probability for scattering in a given direction was also valid in the optical region. Instead of α, Debye worked with a parameter x which in fact was just $1/\alpha$. I have expressed Debye's result in terms of α for the sake of consistency in notation.

[51] The transcript is on deposit in the Archive for the History of Quantum Physics (Copenhagen, Philadelphia, Berkeley).

[52] This is also corroborated in a letter of October 9, 1923, from Sommerfeld to Compton, in which Sommerfeld reported that Debye had discussed the theory with Mr. E. Wagner some three years earlier.

[53] Debye, "X-Ray Scattering," p. 81. This has already been pointed out by Martin J. Klein, "Einstein and the Wave-Particle Duality," *Nat. Phil.* Vol. 3 (New York: Blaisdell, 1964), p. 22.

[54] Compton, "Quantum Theory of Scattering," (reference 22) p. 501.

[55] Debye, "X-Ray Scattering," p. 88.

[56] Compton, "Quantum Theory of Scattering," p. 502. I am very grateful to Professor R. E. Norberg, Chairman of the Physics Department, Washington University, for sending me and granting me permission to reproduce the photograph in Fig. 6.

[57] Jauncey, "The Scattering of X-Rays: A Review," *Wash. U. Stud.* 13 (1926):125. See also A. H. Compton, *X-Rays and Electrons* (reference 3), p. 233. This problem would not be fully understood until the work of O. Klein and Y. Nishina. See "Über die Streuung von Strahlung durch freie Elektronen nach der neuen relativistischen Quantendynamik von Dirac," *Z. Phys.* 52 (1929):853–868; Y. Nishina, "Die Polarisation der Comptonstreuung nach der Diracschen Theorie des Elektrons," *Z. Phys.* 52 (1929):869–877. For a contemporary non-Dirac approach, see F. Sauter, "Intensitätsproblem und Lichtquantentheorie. (Der Comptonsche Streuprozess.)", *Z. Phys.* 52 (1929):225–234.

[58] See Jauncey's papers: "Is There a Change of Wave-length on Reflection of X-rays from Crystals" (with C. Eckart), *Nature* 112 (1923):325–326; "A Corpuscular Quantum Theory of the Scattering of Polarized X-Rays," *Phys. Rev.* 23 (1924):313–317; "Photoelectrons and a Corpuscular Quantum Theory of the Scattering of X-rays," *Nature* 113 (1924):196; "Corpuscular Theory of the Distribution of the Recoil Electrons Produced by Polarized X-Rays," *Phys. Rev.* 23 (1924):580–582; "The Scattering of Polarized X-rays by Paraffin," *Phys. Rev.* 23 (1924):762 (Abstract, with H. E. Strauss); "Theory of the Width of the Modified Lines in the Compton Effect," *Phys. Rev.* 24 (1924): 204–205 (Abstract); "Quantum Theory of the Unmodified Spectrum Line in the Compton Effect," *Phys. Rev.* 25 (1925):314–321; "Theory of the Number of Beta-Rays Associated with Scattered X-Rays" (with O. K. De Foe) *Phys. Rev.* 26 (1925):433–435.

"The Disappearance of the Unmodified Line in the Compton Effect," *Phys. Rev.* 28 (1926):620–624 (with R. A. Boyd); "Theory of the Intensity of Scattered X-Rays," *Phys. Rev.* 29 (1927):757–764.

[59] Compton, "Spectrum," (reference 43) p. 411. The improvement was achieved through the use of the small target-scatterer distances which now become possible. To see how Compton's new X-ray tube fits into the general technical development, see A. E. Lindh, "Bericht über die Entwicklung der Röntgenspektroskopie während der Jahre 1921–1925," *Phys. Z.* 28 (1927):24–28.

[60] Johnston (ed.), *Cosmos,* (reference 18) pp. 35–36.

[61] Compton, "Spectrum," (reference 43) p. 411.

[62] *Ibid.*, p. 412.

[63] Compton, "Quantum Theory of Scattering," (reference 22) p. 502.

[64] Compton, "Spectrum," p. 413.

[65] Max Born, "Arnold Johannes Wilhelm Sommerfeld," *Biog. Mem. Fell. Roy. Soc.* 8 (1952):287. On his return trip, Sommerfeld wrote up his extremely interesting "American Impressions," which are on deposit in the Archive for History of Quantum Physics.

[66] Letter on deposit in the Archive for the History of Quantum Physics; quoted by permission of A. Bohr and H. Sommerfeld.

[67] P. A. Ross, "Change of Frenquency on Scattering," *Science* 57 (1923):614.

[68] "Change in Wave-length by Scattering," *Proc. Natl. Acad. Sci.* 9 (1923):248.

[69] This fact has led to some discussion concerning priority for the discovery of the unmodified line. See for example the transcript (p. 44) of a taped interview with D. L. Webster on deposit at the Center for History of Physics, American Institute of Physics, New York.

[70] Compton, "Absorption Measurements of the Change of Wave-Length accompanying the Scattering of X-Rays," *Phil. Mag.* 46 (1923):897.

[71] Compton, "A Recording X-ray Spectrometer, and the High-frequency Spectrum of Tungsten," *Phys. Rev.* 7 (1916):646–659; 8 (1916):753.

[72] Compton, "Absorption Measurements," pp. 908–909.

[73] In "The Scattering of X-Rays," *J. Franklin Inst.* 198 (1924):68, Compton noted: "In view of the fact that these recoil electrons were unknown at the time this theory was presented, their existence and the close agreement with the predictions as to their number, direction and velocity supplies strong evidence in favor of the fundamental hypotheses of the quantum theory of scattering."

[74] W. Bothe, "Über eine neue Sekundärstrahlung der Röntgenstrahlen I. Mitteilung," *Z. Phys.* 16 (1923):319–320. It is not certain that Bothe saw the connection here between his work and Compton's, since he does not cite Compton's work in this paper.

[75] W. Bothe, "Über eine neue Sekundärstrahlung der Röntgenstrahlen II. Mitteilung," *Z. Phys.* 20 (1924):237–255, especially pp. 250ff.

[76] Wilson announced his work in "Investigations on X-rays and β-rays by the Cloud Method," *Nature* 112 (1923):26–27, and published a full report in "Investigations on X-Rays and β-rays by the Cloud Method. Part I.—X-Rays," *Proc. Roy. Soc. (London)* 104 (1923):1–24. The photograph is Plate 3 accompanying the latter paper.

[77] Wilson, "Part I.—X-Rays," p. 10.

[78] *Ibid.*, p. 15.

[79] "Investigations on X-Rays and β-Rays by the Cloud Method. Part II.—β-Rays," *Proc. Roy Soc. (London)* 104 (1923):192–212.

[80] These two letters are both published under the heading "Recoil of Electrons from Scattered X-Rays," *Nature* 112 (1923):435.

[81] *Ibid.* Later Compton realized that the "most convincing reason for associating these short tracks with the scattered x-rays comes from a study of their number." See "The Corpuscular Properties of Light," *Rev. Mod. Phys.* 1 (1929):82. See also D. H. Loughbridge, "The Direction of Ejection of Photo-electrons Produced by X-Rays," *Phys. Rev.* 26 (1925):697–700.

[82] "Recoil of Electrons," (reference 80) p. 435. See also Wilson, "Part I.—X-Rays," (reference 76) p. 15.

[83] Compton, "The Recoil of Electrons from Scattered X-Rays," *Phys. Rev.* 23 (1924):440–441.

[84] Darwin, "A Quantum Theory of Optical Dispersion," *Proc. Natl. Acad. Sci.* 9 (1923):25. See also A. Smekal, "Zur Quantentheorie der Dispersion," *Naturwiss.* 43 (1923):873–875.

[85] Compton, "Recoil of Electrons," p. 442.

[86] *Ibid.*, p. 448.

[87] *Ibid.*, p. 449.

[88] P. A. Ross, "The Wave-length and Intensity of Scattered X-rays," *Phys. Rev.* 22 (1923):524–525.

[89] *Ibid.*, p. 525.

[90] "Electric Forces and Quanta," *Suppl. Nature* (1925):361–368; *Atomicity and Quanta* (Cambridge: Cambridge University Press, 1926).

[91] See, for example, J. J. Thomson, "The Structure of Light," *Phil. Mag.* 50 (1925): 1181–1196, especially pp. 1190–1196. For a full discussion see R. McCormmach, "J. J. Thomson and the Structure of Light," *Brit. J. Hist. Sci.* 3 (1967):362–387.

[92] See, for example, "The 'J' Phenomena and X-ray Scattering," *Nature* 112 (1923):723–724. For a complete list of Barkla's papers, see the bibliography of Barkla's work at the end of H. S. Allen's obituary notice, "Charles Glover Barkla," *Roy. Soc. Obit. Not. (London)* 5 (1945–1948):366.

[93] A. H. Compton, "Scattering of X-Ray Quanta and the J Phenomena," *Nature* 113 (1924):160–161.

[94] *Ibid.*, p. 160.

[95] (London: Arnold, 1926), pp. 106–107.

[96] H. S. Allen, "Barkla," p. 360.

[97] *Ibid.*, pp. 356–360.

[98] Letter on deposit at Washington University Archives, St. Louis.

[99] All of these interviews, mostly conducted by Professors T. S. Kuhn and J. L. Heilbron, are on deposit in the Archive for the History of Quantum Physics (Copenhagen, Philadelphia, Berkeley).

[100] Letter on deposit at Washington University Archives, St. Louis; quoted by permission of Mrs. A. H. Compton and Dr. Ing. H. Sommerfeld. For Joos' report see "Ein Versuch zum Nachweis einer etweigen einseitigen Intensitätsverteilung beim Emissionsprozess," *Phys. Z.* 24 (1923):469–473.

[101] "The Scattering of X Rays as Particles," *Am. J. Phys.* 29 (1961):818. One must view this statement in the light of Compton's well-known later controversy with Millikan over the nature of cosmic rays.

[102] For a more complete account of Duane's life, see P. W. Bridgman, "William Duane 1872–1935," *Natl. Acad. Sci. Biog. Mem.* 18 (1936):23–41; P. Forman's biography in *Dict. Sci. Biog.* Vol. 4 (New York: Scribners' 1971), pp. 194–197; and G. W. Pierce, P. W. Bridgman, and F. H. Crawford, "Minute on the Life and Service of William Duane, Professor of Bio-Physics, Emeritus," *Harvard Gaz.* (April 30, 1935).

[103] The most well-known is probably the discovery of the "Duane-Hunt law," although D. L. Webster (see interview in AHQP) has disputed the magnitude of Duane's contribution to this discovery.

[104] "The Wave-Lengths of Secondary X-Rays," *Proc. Natl. Acad. Sci.* 9 (1923):418.

[105] *Ibid.*, p. 414.

[106] *Ibid.*, p. 415.

[107] *Ibid.*, p. 418.

[108] "The Wave-Lengths of Secondary X-Rays (Second Note)," *Proc. Natl. Acad. Sci.* 9 (1923):421.

[109] *Ibid.*, p. 420.

[110] *Ibid.*, pp. 421–422.

[111] *Ibid.*, pp. 423–424.

[112] "On Tertiary X-Radiation, Etc.," *Proc. Natl. Acad. Sci.* 10 (1924):44.

[113] These spectra were soon published in "On Secondary and Tertiary X-Rays from Germanium, Etc.," *Proc. Natl. Acad. Sci.* 10 (1924):93.

[114] "Tertiary X-Radiation," p. 45.

[115] *Ibid.*, p. 47.

[116] See Samuel K. Allison, "Arthur Holly Compton," *Natl. Acad. Sci. Biog. Mem.* 37 (1965):85.

[117] For minutes of the meeting see *Phys. Rev.* 23 (1924):287.

[118] Compton's address was later published in *J. Franklin Inst.* 198 (1924):57–72. Also see Johnston (ed.), *Cosmos,* (reference 18) p. 36.

[119] Compton, "Scattering of X-Rays," (reference 21) pp. 69–70.

[120] *Ibid.*, pp. 70–72.

[121] "The Transfer in Quanta of Radiation Momentum to Matter," *Proc. Natl. Acad. Sci.* 9 (1923):158–164.

[122] *Ibid.*, p. 158. See also Clark and Duane, "The Reflection by a Crystal of X-Rays Characteristic of Chemical Elements in it," *Proc. Natl. Acad. Sci.* 9 (1923):126–130; "On The Abnormal Reflection of X-rays by Crystals," *Proc. Natl. Acad. Sci.* 9 (1923):131–135; "The Abnormal Reflection of X-Rays by Crystals," *Science* 58 (1923):400–402.

[123] As is evident from letters deposited in the AHQP, Bohr himself attempted to account for Clark and Duane's "discovery," but finally withdrew his paper to the *Physical Review* on it after convincing himself of his error. See also A. P. Weber, "Über die Nichtexistenz der Clark-Duaneschen Sekundärspektren bei Verwendung fehlerloser Kristalle," *Z. Phys.* 33 (1925):767–769.

[124] "The Quantum Integral and Diffraction by a Crystal," *Proc. Natl. Acad. Sci.* 9 (1923): 359–362; see also "A Quantum Theory of Uniform Rectilinear Motion," *Phys. Rev.* 23 (1924):118 (Abstract).

[125] Compton, "Quantum Integral," p. 362. For more recent comments on Duane's quantum condition, see Alfred Landé, "Quantum Fact and Fiction," *Am. J. Phys.* 33 (1965):123–127.

[126] See the minutes of the symposium published in *Phys. Rev.* 23 (1924):287.

[127] *Ibid.* In a letter of March 24, 1924 (AHQP), Davis informed Bohr of his results, which confirmed Compton's theory. For more information on Davis see his obituary notice by Harold W. Webb in *Natl. Acad. Sci. Biog. Mem.* 34 (1960):65–82. Also see A. H. Compton, "A General Theory of the Wave-length of Scattered X-Rays," *Phys. Rev.* 24 (1924):168; 23 (1924):763 (Abstract).

[128] Johnston (ed.), *Cosmos,* p. 36.

[129] Letter on deposit in AHQP.

[130] Compton, "X Rays as Particles," (reference 101) p. 818.

[131] Arthur Holly Compton, Research Physicist," *Science* 138 (1962):794–795. Copyright © 1962 by the American Association for the Advancement of Science.

[132] "The Excitation, Reflection, and Utilization in Crystal-Structure Analyses of Characteristic Secondary X-Rays," *J. Am. Chem. Soc.* 46 (1924):372–384, especially p. 383.

[133] This is corroborated in recent interviews with D. L. Webster (A. I. P. Center for History of Physics transcript p. 46) and L. Pauling (AHQP, transcript p. 19). In 1927 Clark apparently still accepted the "Duane effect," for he described it in the following terms: "This effect, while for a time confused with the Compton shift, is entirely independent from it and is characterized by very much smaller intensities." See his book *Applied X-Rays* (New York: McGraw-Hill, 1927), p. 43.

[134] Johnston (ed.), *Cosmos,* p. 37.

[135] Letter on deposit in AHQP.

[136] "The Secondary and Tertiary Rays from Chemical Elements of Small Atomic Number due to Primary X-Rays from a Molybdenum Target," *Proc. Natl. Acad. Sci.* 10 (1924):150.

[137] *Ibid.*, p. 151.

[138] "A Measurement of the Polarization of Secondary X-Rays," *J. Opt. Soc. Am.* 8 (1924):487–491.

[139] Compton, "General Theory," (reference 127) pp. 168–176; one symbol has been changed to maintain a consistent notation.

[140] *Ibid.*, p. 171.

[141] *Ibid.*, p. 176.

[142] See, for example, H. Kulenkampff, "Die Wellenlänge gestreuter Röntgenstrahlen," *Z. Phys.* 19 (1923):17–19; E. Bauer, "Sur le changement de longueur d'onde accompagnant la diffusion des rayons X," *Compt. Rend.* 177 (1923):1031–1033; G. E. M. Jauncey and H. L. May, "The Intensity of X-Rays Scattered from Rocksalt," *Phys. Rev.* 23 (1924):128–136; F. W. Bubb, "Direction of Ejection of Photo-Electrons by Polarized X-Rays," *Phys. Rev.* 23 (1924):137–143; W. Seitz, "Über die Asymmetrie der Entladung von Röntgenelektronen," *Phys. Z.* 25 (1924):546–550; H. Fricke and O. Glasser, "Über die durch Röntgenstrahlen in Elementen niederer Atomgewichte ausgelösten sekundären Elektronen," *Z. Phys.* 29 (1924):374–382; "The Secondary Electrons Produced by Hard X-Rays in Light Elements," *Proc. Natl. Acad. Sci.* 10 (1924):441–447; H. Fricke, "Compton's Theory of X-ray Scattering," *Nature* 116 (1925):430–431.

[143] See, for example, G. A. Schott, "On the Scattering of X-Rays by Hydrogen," *Phys. Rev.* 23 (1924):119–127; L. Meitner, "Über eine notwendige Folgerung," (reference 49) pp. 334–342.

[144] See, for example, E. Friedel and F. Wolfers, "Les variations de longueur d'onde des rayons X par diffusion et la loi de Bragg," *Compt. Rend.* 178 (1924):199–200.

[145] See for example sections of the papers in reference 142, as well as F. Dessauer and R. Herz, "Zur Härteverteilung der gestreuten Röntgenstrahlen," *Z. Phys.* 27 (1924):56–64; E. A. Owen, N. Fleming, and W. E. Fage, "Absorption and Scattering of Gamma-Rays," *Proc. Phys. Soc. (London)* 36 (1924):355–366.

[146] See for example "Sur le changement de longueur d'onde par diffusion dans le cas des rayons K du tungstène," *Compt. Rend.* 178 (1924):908; "The Phenomena of High-frequency Radiation," *Proc. Phys. Soc. (London)* 36 (1924):423–428; "Recherches complementaires sur l'effet Compton," *Compt. Rend.* 179 (1924):11–14 (with A. Dauvillier); earlier, Dauvillier had found conflicting results—see "Recherches spectrographiques sur l'effet A. H. Compton," *Compt. Rend.* 178 (1924):2076–2078.

[147] "Photographic Investigations of Scattered X-Radiation," *J. Opt. Soc. Am.* 11 (1924):217–220.

[148] See "Über eine Art der Sekundärstrahlung der γ-Strahlen," *Z. Phys.* 24 (1924):393–399; "Über den Rückstosseffekt der zerstreuten γ-Strahlen," *Z. Phys.* 28 (1924):278–286; a letter of November 8, 1924, from Skobeltzyn to Compton, informing Compton of his favorable results, is deposited in the Washington University Archives, St. Louis.

[149] "Tertiary Line Spectra in X-Rays," *Proc. Natl. Acad. Sci.* 10 (1924):186–191.

[150] "On the Theory of the Tertiary Radiation Produced by Impacts of Photo-electrons," *Proc. Natl. Acad. Sci.* 10 (1924):193.

[151] "Absorption Measurements of Certain Changes in the Average Wave-length of Tertiary X-rays," *Proc. Natl. Acad. Sci.* 10 (1924):196–199.

[152] "The Wave-length of Molybdenum K_α Rays when Scattered by Light Elements," *Proc. Natl. Acad. Sci.* 10 (1924):271–273 (with Y. H. Woo). Also see *Phys. Rev.* 23 (1924):763.

[153] *Ibid.*, p. 272.

[154] *Ibid.*, p. 273.

[155] "Scattered X-Rays," *Proc. Natl. Acad. Sci.* 10 (1924):304–306.

[156] "The Compton and Duane Effects," *Proc. Natl. Acad. Sci.* 10 (1924):343; see also "The reality of the Compton effect," *Phys. Rev.* 23 (1924):761 (with E. C. Watson, W. R. Smythe, R. B. Brode, and L. M. Mott-Smith).

[157] *Ibid.*, pp. 344–345.

[158] "The Influence on Secondary X-Ray Spectra of Placing the Tube and Radiator in a Box," *Proc. Natl. Acad. Sci.* 10 (1924):374–379.

[159] *Ibid.*, pp. 375–376.

[160] *Ibid.*, p. 376. For photographs of the impressive equipment Duane had at his disposal see S. K. Allison and L. Clark, "An Improved Apparatus for Precision Researches with X-Rays," *J. Opt. Soc. Am.* 8 (1924):681–691; Alice H. Armstrong and W. W. Stifler, "A New Laboratory for Precision X-Ray Research," *J. Opt. Soc. Am.* 11 (1925):509–517.

[161] "Influence of Box," (reference 158) p. 377.

[162] *Ibid.*, p. 376–377.

[163] *Ibid.*, p. 378.

[164] "The Influence on Secondary X-Ray Spectra of Placing the Tube and Radiator in a Box," *Proc. Natl. Acad. Sci.* 10 (1924):379–384.

[165] *Ibid.*, p. 381.

[166] *Ibid.*

[167] *Ibid.*, p. 382.

[168] "The Scattering of X-ray," *Nature* 114 (1924):627.

[169] Johnston (ed.), *Cosmos,* p. 37.

[170] "Scattering of X-rays," (reference 168) pp. 627–628.

[171] Much later Raman rather surprisingly remarked regarding his own well-known discovery: "the possibility that the corpuscular nature of light might come into evidence in scattering was not overlooked and was in fact elaborately discussed in the essay of February 1922 which was published at least a year before the well-known discoveries of Compton on X-ray scattering." See *Nobel Lectures: Physics,* Vol. 2 (Amsterdam: Elsevier, 1964), p. 269. He therefore implied that he accepted the light quantum hypothesis before Compton's discovery. Compton himself noted [Johnston (ed.), *Cosmos,* p. 37] that it was probably the Toronto debate that "led him [Raman] to discover the Raman effect two years later. . . ."

[172] "The Compton Effect with no Box around the Tube," *Proc. Natl. Acad. Sci.* 11 (1925):56.

[173] *Ibid.*, pp. 59–60.

[174] *Ibid.*, p. 60. See also "The Compton Effect," *Nature* 115 (1925):51.

[175] "Compton Effect: Evidence on its Relation to Duane's Box Effect," *Proc. Natl. Acad. Sci.* 11 (1925):62.

[176] *Ibid.*, pp. 62–63.

[177] *Ibid.*, p. 64.

[178] "The Compton Effect with Hard X-Rays," *Proc. Natl. Acad. Sci.* 11 (1925): 224–227; also see "The Compton Effect," *Nature* 115 (1925):51.

[179] "The Compton Effect and Tertiary X-Radiation," *Proc. Natl. Acad. Sci.* 11 (1925):124.

[180] *Ibid.*, p. 125.

[181] "The Effect of a Surrounding Box on the Spectrum of Scattered X-Rays," *Proc. Natl. Acad. Sci.* 11 (1925):117–118; italics added. See also "Tests of the effects of an enclosing box on the spectrum of scattered x-rays," *Phys. Rev.* 25 (1925):236 (Abstract, with J. A. Bearden and Y. H. Woo).

[182] *Ibid.*, p. 118.

[183] "On Scattered Radiation due to X-Rays from Molybdenum and Tungsten Targets," *Proc. Natl. Acad. Sci.* 11 (1925):25. Clark and Duane published one further paper together—on fluorescent radiation. No mention was made of Compton. See "The Relative Intensities of Fluorescent and Scattered X-Rays," *Proc. Natl. Acad. Sci.* 11 (1925):173–175.

[184] *Ibid.*, p. 26.

[185] "Experiments on the Wave-lengths of Scattered X-Rays," Phys. Rev. 26 (1925): 300–309.

[186] H. Mark, "Der Comptoneffekt. Seine Entdeckung und seine Deutung durch die Quantentheorie," *Naturwiss.* 23 (1925):494. Mark asserts, however, that the specific reasons for Clark and Duane's failure were not uncovered (p. 499). G. Wentzel, "Die Theodien des Compton-Effektes. I.," *Phys. Z.* 26 (1925):437, claimed that in Clark and Duane's experiments the primary radiation was incident on the scatterer at too large a solid angle, and hence the shifted line was washed out.

[187] A. H. Compton, "The Spectrum and State of Polarization of Fluorescent X-Rays," *Proc. Natl. Acad. Sci.* 14 (1928):550.

[188] Allison, "Arthur Holly Compton, Research Physicist" (reference 131), p. 795.

[189] Bridgman, "Duane," (reference 102) p. 32.

[190] Private communication to author, September 10, 1971.

[191] Any number of other confirmations appeared after the close of the Compton-Duane controversy. See for example P. A. Ross, "X-Rays Scattered by Molybdenum," *Proc. Natl. Acad. Sci.* 11 (1925):567–569; "Ratio of Intensities of Unmodified and Modified Lines in Scattered X-Rays," *Proc. Natl. Acad. Sci.* 11 (1925):569–572; G. E. M. Jauncey, "The Compton and Duane Effects," *Nature* 115 (1925):456–457; H. Kallmann and H. Mark, "Zur Grösse und Winkelabhängigkeit des Comptoneffektes," *Naturwiss.* 13 (1925):297–298; "Über einige Eigenschaften der Compton-Strahlung," *Ibid.*, pp. 1012–1015; H. M. Sharp, "A Precision Measurement of the Change of Wave-length of Scattered X-Rays," *Phys. Rev.* 26 (1925):691–696; M. de Broglie and J. Thibaud, "Réflexion totale et variation de l'indice de refraction des radiations X au voisinage d'une discontinuité d'absorption du miroir," *Compt. Rend.* 181 (1925):1034–1035; M. de Broglie and A. Dauvillier, "Recherches Spectrographiques sur l'effet Compton," *J. Phys. Radium* 6 (1925):369–375. G. Hoffmann, "Über den Comptoneffekt bei γ-Strahlen," *Z.*

Phys. 36 (1926):251–258. At least one experimenter still had difficulty: see G. Hagen, "Versuche über den Comptoneffekt," *Ann. Physik.* 78 (1925):407–420; this was Hagen's dissertation and his work was promptly reexamined and reinterpreted by his thesis professor, F. Kirchner, in "Bemerkungen zur vorstehenden Arbeit des Hrn. G. Hagen," *Ann. Physik.* 78 (1925):421–422. A comprehensive summary of work on the Compton effect was provided by G. Wentzel. See "Die Theorien des Compton-Effektes. I," *Phys. Z.* 26 (1925):436–454.

CHAPTER 7

Turning Point in Physics

A. Introduction

Werner Heisenberg, in his June 1929 review article "The Development of the Quantum Theory 1918–1928," outlined the successes and failures of the old quantum theory (1918–1923) as a prelude to an examination of the "really interesting stage," "The Crisis of the Quantum Theory 1923–1927." He recalled:

> At this time [1923] experiment came to the aid of theory with a discovery which would later become of great significance for the development of the [quantum] theory. Compton found that with the scattering of X-rays from free electrons the wavelength of the scattered rays was measurably longer than that of the incident light. This effect, according to Compton and Debye, could easily be explained by Einstein's light quantum hypothesis; the wave theory of light, on the contrary, failed to explain this experiment. With that [result] the problems of radiation theory, which had hardly been advanced since Einstein's works of 1906 [*sic*], 1909, and 1917, were opened up. Since physics had meanwhile also been confronted with fundamental difficulties in atomic theory, one turned with renewed interest to the unsolved questions in radiation theory, in order to perhaps learn something by comparing the difficulties in these two different fields.[1]

Heisenberg therefore saw the discovery of the Compton effect as representing not only a turning point in radiation theory, but a turning point in physics as well.[2]

It was in radiation theory, however, that its significance was most immediately recognized. Compton asserted at the end of 1923 that his discovery presented "to us a revolutionary change in our ideas regarding the process of scattering of electromagnetic waves";[3] and Einstein noted in the *Berliner Tageblatt* (April 20, 1924) that although theoretical considerations (mostly his own, he modestly neglected to say) had indicated for a long time that a light quantum, like a projectile, carries momentum as well as energy, Compton's experiment proved the validity of that supposition. Therefore, one now had two theories of light, "both indispensable and—as one has to admit today in spite of twenty years of immense efforts by theoretical physicists—without any logical connection."[4]

Other physicists preferred a more drastic interpretation. Arnold Sommerfeld, as we have seen, felt that Compton's discovery sounded the "death knell" of the wave theory; and he emphasized its importance by immediately incorporating a full discussion of the "Compton effect" into the fourth edition of his *Atombau und Spektrallinien.*[5] Sommerfeld and many of his contemporaries were struck by the apparent quirk of history that enabled first Newton's corpuscular theory, and then the Young-Fresnel wave theory, to dominate thought for roughly one century.[6]

This huge swing of the pendulum taught physicists important lessons. R. A. Millikan, for example, in a rather extravagant statement, wrote that Compton's discovery was "a discovery of the very first magnitude, one of whose chief values may be to keep the physicist modest and undogmatic, still willing, unlike some scientists and many philosophers, not to take himself too seriously and to recognize that he does not yet know much about ultimate realities."[7] H. A. Lorentz, in mid-1923, made the same point unoffensively:

> One of the lessons which the history of science teaches us is surely this, that we must not too soon be satisfied with what we have achieved. The way of scientific progress is not a straight one which we can steadfastly pursue. We are continually seeking our course, now trying one path and then another, many times groping in the dark, and sometimes even retracing our steps. So it may happen that ideas which we thought could be abandoned once for all, have again to be taken up and come to new life.[8]

In this final chapter, I shall attempt to sketch some of the avenues of research, especially in theoretical physics, which that rebirth in radiation theory precipitated.

B. Theoretical Responses to Compton's Discovery (1923–1925)

1. By Pauli and de Broglie

One immediate consequence of Compton's discovery was that it enabled Wolfgang Pauli to point the way to the understanding of a long unresolved difficulty, the fact that classical electromagnetic theory is not capable of providing an interaction mechanism leading to the establishment of thermal equilibrium between radiation and free electrons.[9] This result followed from the researches of H. A. Lorentz (1911) and of Lorentz's student, A. D. Fokker (1913). Now, a decade later, Pauli saw that the Compton effect could provide the appropriate interaction mechanism, and, guided by Einstein's 1917 paper, he proved that equilibrium would obtain between black-body radiation (Planck spectral distribution) and free electrons (Maxwell–Boltzmann energy distribution) provided that the (appropriately defined) probability per unit time associated with the light quantum–electron interactions was of the form $A\rho + B\rho\rho'$, where ρ and ρ' are the black-body spectral distributions corresponding to the incident and scattered frequencies ν and ν'.

This was an apparently paradoxical result, because it meant that the probability of the Compton scattering process depended upon the spectral distribution of the *scattered* radiation, or, as Compton himself explained, it meant that "the energy absorbed by an electronic oscillator in a field of radiation depends not only upon the radiation reaching the electron, but also upon the radiation which the electron is about to emit. In other words, the action of the electron is conditioned not only by present and past events, but also by events which have not yet happened."[10] This paradox was illuminated in 1923 by Einstein and Ehrenfest, who pointed out that since the scattering process involves both the disappearance (absorption) and appearance (emission) of a light quantum, both of whose directions are fixed by the conservation laws, the interaction probability should be of the form $(b\rho)(a' + b'\rho')$, where the first factor involves an induced term only, while the second factor involves both a spontaneous and an induced term.[11] Pauli's conclusion therefore represented a natural extension of Einstein's 1917 analysis.[12]

In addition to facilitating the solution of an old theoretical enigma, Compton's discovery encouraged the formation of a profoundly new one—de Broglie's conception of matter waves. Of course, since de Broglie had already used the concept of "light molecules" in November 1922,[13] it is certain that he took Einstein's light quantum hypothesis seriously before—and perhaps years before—Compton made his discovery.[14] Therefore, Compton's discovery by no means initially stimulated de Broglie's thinking; rather, it confirmed his already formulated belief in "the actual reality of light quanta." As he wrote in October 1923, in a paper which we will later discuss in more detail:

> The experimental evidence accumulated in recent years seems to be quite conclusive in favour of the actual reality of light quanta. The photoelectric effect, which is the chief mechanism of energy exchange between radiation and matter, seems with increasing probability to be always governed by Einstein's photoelectric law. Experiments on the photographic actions, the recent results of A. H. Compton on the change in wave-length of scattered X-rays, would be very difficult to explain without using the notion of the light quantum. On the theoretical side Bohr's theory, which is supported by so many experimental proofs, is grounded on the postulate that atoms can only emit or absorb radiant energy of frequency ν by finite amounts equal to $h\nu$, and Einstein's theory of energy fluctuations in the black radiation leads us necessarily to the same ideas.[15]

Now, since de Broglie's conviction that the Einsteinian dualism was "absolutely general" grew out of his conviction that one had to accept the "actual reality of light quanta," it is seen that Compton's discovery came at a rather critical time in the development of de Broglie's thought. The question of how de Broglie subsequently extended and developed his revolutionary new ideas on matter waves for his Sorbonne thesis (submitted November 25, 1924); of how de Broglie's ideas were seized upon and profoundly exploited by Einstein, who immediately saw their connection with Bose's work on the statistics of light quanta; and of how Einstein drew Schrödinger's attention to de Broglie's work, with the most profound consequences—all of these questions have already been discussed, especially by Martin J. Klein[16] and Max Jammer.[17] The question of why Schrödinger was particularly receptive to de Broglie's revolutionary new ideas has already been discussed by V. V. Raman and Paul Forman.[18]

These questions, in any event, go beyond the scope of this book, and I will not treat them here. It is, however, worthwhile to point out—looking at the other side of the coin—that shortly after Schrödinger discovered wave mechanics, the Compton effect was interpreted on the basis of that theory. The first to do so was W. Gordon in Berlin; in late September 1926, he calculated the frequency and attempted to calculate the intensity of the scattered radiation.[19] A few months later, in January 1927, Schrödinger himself calculated the direction and frequency of the scattered radiation.[20] The general principles underlying Schrödinger's treatment have been succintly summarized by Compton, who explained that the "incident photon is represented by a train of plane electromagnetic waves. The recoiling electron is likewise represented by a train of plane de Broglie waves propagated in the direction of recoil. These electron waves form a kind of grating by which the incident electromagnetic waves are diffracted. They are increased in wave-length by the diffraction because the grating is receding, resulting in a Doppler effect." Another aspect of the problem, the calculation of the intensity distributions of the secondary quanta and recoil electrons, although attacked by Compton, Jauncey, Breit, Dirac, Wentzel, and Schrödinger,[21] was not fully solved until Klein and Nishina applied Dirac's relativistic theory of the electron to it in 1929.[22]

2. *Semiclassical Theories of the Compton Effect*

The immediate reactions of some physicists to Compton's discovery were more conservative than those of Pauli or de Broglie. C. T. R. Wilson, C. R. Bauer, K. Försterling, and O. Halpern all attempted during 1923 and 1924 to incorporate the Compton effect into classical electrodynamics by means of semiclassical theories.[23] These theories all started from the fact, first pointed out by Compton himself, that the wavelength of the scattered radiation corresponds to that of Doppler shifted radiation emitted in spherical waves by an electron moving at high velocity in the direction of the primary radiation. As Wentzel insightfully remarked in 1925, "The relationships here lie completely similarly as in the theory of Planck's cavity [radiation] spectrum: the validity of Planck's formula is generally recognized, but it may be derived in different ways, based on different models and quantum hypotheses."[24] A major drawback of these theories, as Kallmann and Mark pointed out in 1926,[25] was that they could not explain the sharpness of the modified line without introducing some supplementary assumption to bring in a quantumlike element. Furthermore, measurements on the ranges of the recoil electrons, and on their directions of emission, contradicted the predictions of these theories.[26] In general, they seem to have had only limited influence on contemporary physics.

3. *Slater's Virtual Oscillator Concept and the Bohr-Kramers-Slater Paper*

A very different reception was accorded to an "essentially new" idea conceived in 1923 by John C. Slater; he had received his Ph.D. degree from Harvard University in June, and in the fall was on a Sheldon traveling fellowship in Cambridge, England, en route to Copenhagen.[27] Slater's idea, the "virtual oscillator" concept, when assimilated by Niels Bohr and H. A. Kramers, led directly to the highly original and programmatic Bohr-Kramers-Slater paper, which was published in *The Philosophical Magazine* in May 1924.[28] In Heisenberg's judgment: "This investigation represented the real high point in the crisis of quantum theory, and, although it could not overcome the difficulties, it contributed, more than any other work of that time, to the clarification of the situation in quantum theory."[29] Because of its great importance, we will examine the origin, nature, and impact of this investigation in some detail before describing other theoretical responses to Compton's discovery.

The fundamental question forced on physicists by Compton's discovery was how to unite the particle and wave theories, the essentially discontinuous aspects of radiation and the essentially continuous ones. One approach, which C. G. Darwin especially advocated well before Compton's discovery, was to explore the consequences of relinquishing the conservation laws,[30] but this approach did not bear fruit until after it had been coupled with Slater's virtual oscillator concept. Slater was convinced, as he explained in *Nature* in January 1924, that the "two apparently contradictory aspects of the mechanism" governing the interaction of radiation and matter "must be really consistent, but [that] the discontinuous side is apparently the more fundamental, and for this reason any attempt at a consistent interpretation in the present state of science must inevitably appear rather formal. Nevertheless, on the basis of Bohr's correspondence principle, it seems possible to build up a more adequate picture of optical phenomena than has previously existed, by associating the essentially continuous radiation field with the continuity of existence in stationary states, and the discontinuous changes of energy and momentum with the discontinuous transitions from one state to another."[31] To accomplish this, any "atom may, in fact, be supposed to communicate with other atoms all the time it is in a stationary state, by means of a virtual field of radiation, originating from oscillators having the frequencies of possible quantum transitions. . . ." This "idea of the activity of the stationary states" suggested itself, Slater noted, while he was considering a way in which to combine classical electrodynamics and light quanta; for example, by considering the classical Poynting vector as a guide for quanta.[32]

This was evidently about as far as Slater's thoughts had progressed in Cambridge, where, as he wrote Kramers on December 8, 1923, he began "rather plaguing Mr. [R. H.] Fowler with my speculations on quantum theory."[33] Further developments occurred a few weeks later in Copenhagen, where his entry into Bohr's institute had been prepared as early as March through recommendations of T. Lyman and E. C. Kemble.[34] There, Kramers pointed out to Slater that the quoted interpretation "scarcely suggested the definite coupling between emission and absorption processes which light quanta provide, but rather indicated a much greater independence between transition processes in distant atoms than I had perceived."[35] Slater's virtual radiation field, as Kramers and Bohr saw it, was consistent with the statistical conservation of energy and momentum in that it could determine an atom's transition probabilities—that part of the virtual radiation field originating from a given atom being associated with its own spontaneous transitions, while those parts of the virtual radiation field originating from all other atoms were associated with the given atom's induced transitions, "much as Einstein had suggested." A discontinuous transition would then have "no other external significance than simply to mark the transfer to a new stationary state. . . ."[36]

The "idea of statistical conservation of energy and momentum," as Slater wrote B. L. van der Waerden much later, "was put into the theory by Bohr and Kramers, quite against my better judgment."[37] Slater had arrived in Copenhagen believing light quanta to be "real entities" whose probability for being at a given place at a given time was determined by the virtual radiation fields, but whose interactions with material particles such as atoms obeyed the conservation laws. "Bohr and Kramers," however, "opposed this view so vigorously that I saw that the only way to keep peace and get the main part of the suggestion published was to go along with them with the statistical idea."[38] Bohr and Kramers, of course, could point out "that no phenomena at that time known demanded the existence of corpuscles. Under their suggestion, I [Slater] became persuaded that the simplicity of mechanism obtained by rejecting a corpuscular theory more than made up for the loss involved in discarding conservation of energy and rational causation. . . ."[39]

Bohr, on his part, explained to Slater in January 1925 that the "essential point" he (Bohr) had made earlier in a criticism of Darwin's theory of dispersion dealt precisely with his

> . . . discussion of the limitation of the principle of conservation of energy and momentum. It was just the completion which your suggestion of radiative activity of higher quantum states apparently lent to the general views on the quantum theory with which I had been struggling for years which made me welcome your suggestion so heartily. Especially I felt it was far more harmonious from the point of view of the correspondence principle to connect the spontaneous radiation with the stationary states themselves and not with the transitions.[40]

On the one hand, Bohr had been toying with the idea of statistical conservation at least as early as the summer of 1919 when C. G. Darwin forcefully maintained in correspondence with him that "the case against conservation [is] quite overwhelming"—an opinion Darwin continued to hold for years.[41] On the other hand, while Bohr ignored the radiation problem in print before 1922, he fully appreciated the importance of Einstein's ideas on induced and spontaneous transitions, and he was able to incorporate these concepts into his own evolving ideas on atomic structure—so that he came to view the quantum theory of spectra in "a certain sense" as a "natural generalization of our ordinary ideas of radiation."[42] All in all, Bohr was so delighted to learn of Slater's idea that he could not resist reporting it almost immediately (February 7, 1924) to A. A. Michelson in a letter that he wrote primarily to thank Michelson for the hospitality he had received at the recent meeting of the American Physical Society in Chicago.[43]

Bohr, with the help of Kramers, then (still in January 1924) assimilated Slater's virtual oscillator concept, transformed it, and wrote up the resulting theory—actually an entirely qualitative program for research[44]—for publication. As applied to the Compton effect, only one of a variety of physical phenomena which Bohr discussed, the theory yielded the following picture instead of the usual one:

> . . . the scattering of the radiation by the electrons is . . . considered as a continuous phenomenon to which each of the illuminated electrons contributes through the

emission of coherent secondary wavelets. Thereby the incident virtual radiation gives rise to a reaction from each electron, similar to that to be expected on the classical theory from an electron moving with a velocity coinciding with that of the above-mentioned imaginary source and performing forced oscillations under the influence of the radiation field. That in this case the virtual oscillator moves with a velocity different from that of the iluminated electrons themselves is certainly a feature strikingly unfamiliar to the classical conceptions. In view of the fundamental departures from the classical space–time description, involved in the very idea of virtual oscillators, it seems at the present state of science hardly justifiable to reject a formal interpretation as that under consideration as inadequate. On the contrary, such an interpretation seems unavoidable in order to account for the effects observed, the description of which involves the wave-concept of radiation in an essential way. At the same time, however, we shall assume, just as in Compton's theory, that the illuminated electron possesses a certain probability of taking up in unit time a finite amount of momentum in any given direction. By this effect, which in the quantum theory takes the place of the continuous transfer of momentum to the electrons which on the classical theory would accompany a scattering of radiation of the type described, a statistical conservation of momentum is secured in a way quite analogous to the statistical conservation of energy in the phenomena of absorption of light. . . . In fact, the laws of probability for the exchange of momentum by interaction of free electrons and radiation derived by Pauli are essentially analogous to the laws governing transition processes between well-defined states of an atomic system. Especially the considerations of Einstein and Ehrenfest . . . are suited to bring out this analogy.[45]

The virtual oscillators, therefore, as Compton expressed it, "scatter as if moving in the direction of the primary beam, accounting for the change of wavelength as a Doppler effect."[46]

4. *Negative and Positive Reactions to the Bohr-Kramers-Slater Theory*

Even before the Bohr–Kramers–Slater theory was fully formulated, word of Bohr's statistical conservation idea had spread, provoking reactions. As we saw in the last chapter, Arthur Compton was one of the first to respond, in an address before the AAAS on December 28, 1923. "It seems to me," he stated, "that the very fact that the energy and momentum principles may be applied to the problem of the scattering of radiation with results in accord with experiment, constitutes a test of their validity for phenomena of this type. For this reason I am inclined toward the choice of these principles even at the great cost of losing the spreading wave theory of radiation."[47]

Compton's attitude was shared by any number of his contemporaries, including, not surprisingly, Arnold Sommerfeld, who after learning of the Bohr–Kramers–Slater theory could not suppress his skepticism of this "compromise" theory in a February 1924 article in *Die Naturwissenschaften*.[48]

One of the earliest and most incisive skeptics, however, was Wolfgang Pauli, who wrote Bohr on February 21, 1924, that while he knew what both of the words "communicate" and "virtual" meant, try as he might he had not as yet succeeded in guessing the content of Bohr's theory.[49] While an Easter visit to Copenhagen resolved that difficulty for Pauli, and while Bohr even succeeded in stilling his "scientific conscience" then, in succeeding months his qualms again became more and more pronounced, and by October, as he wrote in a long letter to Bohr,[50] he stood "as a physicist, completely opposed" to the Bohr–Kramers–Slater theory. This view, moreover, was shared by "many other physicists, perhaps even the majority"—the foremost being Einstein, whom Pauli had finally met recently in Innsbruck. Pauli remarked with tongue in cheek that Einstein's opposition to Bohr's theory proved that even if it were *psychologically* possible for him to form his scientific opinions on the basis of some kind of belief in authority, it would obviously be *logically* impossible for him to do so in this case.

According to Pauli, Einstein's basic arguments against Bohr's theory were:

(1) That if one assumes the statistical independence of elementary processes in spatially separated atoms, one can show from fluctuation considerations that, for example, the total kinetic energy of radiation in a cavity with perfectly reflecting walls can take on arbitrarily large values as time goes on—something that Einstein found "disgusting."

(2) That the more one learns of elementary quantum processes, including electron collision processes, the more they reveal the "strict validity" of the conservation laws.

(3) That Einstein was convinced that future developments would lead to the ascription "of a higher reality" to a light quantum "as a carrier of energy and momentum than to the wave field (even if it is tasteless to calculate the dimensions of light quanta)." As soon as the idea was relinquished that the frequency of light emitted by an atom corresponds to a frequency of motion of the emitting electron, "the undulatory character of light had for him [Einstein] something intuitively shadowy attached to it."

(4) That Einstein objected to the way in which Bohr's theory accounted for spectral natural line widths, because, on the one hand, they must be related both to the decay times and to the sharpness of the stationary states of an atom, but, on the other hand, since there was no logical connection between the two, Bohr's theory would have to assume the existence of a "pre-established harmony"—a "most distasteful" feature of *any* theory to Einstein.

These objections of Einstein partially overlap the objections he conveyed to Ehrenfest in a letter of May 31, 1924, which was discussed by Martin J. Klein.[51] In that letter, points (1) and (2) are essentially identical, but then Einstein added:

> (3) A final abandonment of strict causality is very hard for me to tolerate.
>
> (4) One would also almost have to require the existence of a *virtual* acoustic (elastic) radiation field for solids. For it is not easy to believe that quantum *mechanics* necessarily requires an electrical theory of matter as its foundation.
>
> (5) The occurrence of ordinary scattering (not at the proper frequency of the molecules), which is above all standard for the optical behavior of bodies, fits badly into the scheme. . . .

To Einstein's objections, Pauli added his own. In the first place, he was not really disturbed over Einstein's point about the statistical independence of quantum processes in spatially separated atoms—it was "all right" with Pauli however Geiger's experiment turned out. Rather, the "great stumbling block" to Pauli was that he was convinced that the Einstein coefficient $B\rho\nu$ corresponded to the total energy abstracted from waves incident on atoms, and hence he was also convinced that Bohr's distinction between fluorescent (or resonance) radiation emitted by two different types of atoms, excited and unexcited atoms, was "not in the least real"; and that there was in principle no experiment—not even a certain thought experiment that Bohr had proposed at Easter—which enabled one to separate the two types of atoms by means of scattered radiation. Nor was it "meaningfully possible" to distinguish, for example, Compton transitions from ordinary atomic transitions—the total average frequency of occurrence for the two processes had to be given by the Einstein coefficient $B\rho\nu$. In general, one was led to contradictions if one attempted to use the classical field concept to describe average values associated both with individual atoms and with many atoms.[52]

Negative reactions came from other quarters as well. In Amsterdam, J. D. van der Waals, Jr., pointed out an inconsistency in Bohr's theory.[53] In Cambridge, England, E. C. Stoner, although lamenting the "vagueness" of the theory (a view not shared for example by C. D. Ellis[54]), nevertheless pointed out a "number of [its] unsatisfactory features" in a critical lecture very early in 1925.[55] He insightfully remarked that if the emitting and absorbing atoms were uncoupled, "There seems [to be] no reason, in fact, why an absorption switch should not actually occur before the 'corresponding' emission switch"—and he then discussed certain cloud chamber photographs of C. T. R. Wilson which indicated a definite temporal sequence of events, and hence contradicted that supposition. "As to conservation," he added, "it may at least be said that it seems unwarranted to assume that it does not hold in individual processes when there is no definite evidence of its breakdown, unless the supposition leads to a much more complete and satisfying explanation of observed phenomena than has hitherto been put forward."[56] He concluded that radiation consists of directed quanta, and then went on to develop a new theory of radiation, which we will discuss later. Wolfgang Pauli, as Heisenberg wrote to Bohr on January 8, 1925,[57] was still at that date grumbling "over the 'virtualization' of physics," and it is certain that Einstein still shared Pauli's attitude.

Negative reactions to Bohr's theory therefore abounded. Positive reactions, however, were by no means absent. R. H. Fowler, Stoner's mentor, attempted to develop Bohr's theory quantitatively in early 1925 in a Cambridge lecture.[58] Richard Becker in Berlin attempted to use the Bohr–Kramers–Slater approach in July 1924 to derive the Ladenburg and Kramers dispersion formula in a certain limit,[59] although his theory was soon criticized by Slater.[60] P. Jordan, influenced by Bohr's theory, attempted to demonstrate in his November 1924 Göttingen dissertation that through "a simple generalization of Einstein's theory" it was possible to avoid the assumption that atomic recoils of magnitude $h\nu/c$ are associated with an emitting or absorbing atom in thermal equilibrium with radiation.[61] Einstein himself immediately replied to Jordan,[62] acknowledging "the logical correctness" of Jordan's considerations, but very simply proving that certain assumptions in Jordan's "generalization" were incompatible with the existence of an absorption coefficient for material media—they were incompatible, therefore, with "the most elementary facts of absorption." Einstein traced Jordan's error to the latter's assumption that bundles of radiation incident on an atom from different directions are not independent of each other. If the contrary assumption were made, Einstein stated, "one necessarily comes to the conclusion that with each elementary absorption or radiation process momentum of the absolute magnitude $h\nu/c$ is transferred to the molecule."

By far the most enthusiastic—and in retrospect the most prominent—advocate of Bohr's theory was Erwin Schrödinger. Immediately after reading the Bohr–Kramers–Slater paper in May 1924, Schrödinger wrote a long congratulatory letter to Bohr,[63] explaining that ever since he had been a student of Franz Exner at Vienna, he had been friendly to the idea that our concept of statistical probability was based on "absolute chance," and that perhaps even the conservation laws were only statistically valid—themes, in fact, which Schrödinger had chosen for his Zürich inaugural lecture (1921). That O. W. Richardson was probably the first to advocate these ideas, as Bohr had pointed out in his paper, was extremely interesting for Schrödinger to learn.

While it was obvious to Schrödinger that Bohr's theory signified a "far reaching return to classical theory," he could not, he continued, go along with it completely if Bohr insisted on a strict interpretation of the term "virtual." What, after all, was "real" radiation if it was not the radiation which "causes" the transitions, that is, the radiation which creates the transition probabilities? Indeed, from a philosophical point of view, one could even question which of the two electron systems—the "real" one describing the stationary states, or the "virtual" one emitting and scattering the "virtual" radiation—possesses the greater "reality."

There was one specific difficulty with Bohr's theory which troubled Schrödinger, and he illustrated it by considering the linear harmonic oscillator with its different energy levels. In the first place, from correspondence principle considerations one probably had to assume that such an oscillator emits more virtual radiation in a higher energy state than in a lower one—its energy content was no doubt proportional to the quantum number n. In the second place, one knew from classical theory that for radiation emitted from such an oscillator, the ability to interfere was independent of the oscillator's amplitude—or, quantum theoretically, that the probability for the oscillator to undergo a spontaneous transition from one energy level to the next lower level was the same for all levels. Finally, one would like to assume that an isolated oscillator at an energy level $n\epsilon$ would, in returning to the lowest energy level, radiate on the average its total energy $n\epsilon$.

These three assumptions, Schrödinger proved, were incompatible. Thus, if $\bar{t}$ is the average decay time for each level, and if $m\sigma$ designates the average energy radiated by the mth level, it must then be true that

$$n\sigma\bar{t} + (n-1)\sigma\bar{t} + \ldots \sigma\bar{t} = n\epsilon, \qquad (7.1)$$

or that

$$\frac{n(n+1)}{2}\sigma\bar{t} = n\epsilon. \qquad (7.2)$$

That is, contrary to the second assumption cited, it must be true that t is a function of n. A similar difficulty arose when one considered the presumed close relationship between the relative widths of spectral lines and the slight, relative fuzziness in the atomic energy levels; on the one hand, observation revealed that the latter was roughly the same for all levels, while, on the other hand, the former theoretically should increase with increasing n. The only way out of these difficulties, Schrödinger felt, was to recognize that the usual picture of the radiating electron as a quasi-elastically bound harmonic oscillator was "much too schematic"—that in actual cases, the situation was much more complicated.

Schrödinger did not confine his positive reaction to correspondence; he published a paper expressing it in September 1924.[64] Asserting that the "most exciting" aspect of Bohr's theory, which "claims equally great interest from the physical and philosophical points of view," was its "fundamental violation of the laws of conservation of energy and momentum in each radiation process," Schrödinger gave a thorough explication of the theory and its applicability to ionization processes, the Compton effect, and other phenomena (thereby proving the definiteness of the program in Schrödinger's mind). However, in keeping with his earlier doubts, he completely avoided the term "virtual." He discussed the statistical aspects of atomic transitions—for example, the associated energy fluctuations—in detail, drawing on an analogy from the rampant inflation of the day. ("Consider a man who possesses a fortune of 10,000 gold marks, or 10^{16} paper marks. . . .") He found, as Einstein had realized, that the energy fluctuations of an isolated system increase indefinitely with the time of observation, but unlike Einstein, Schrödinger was not unduly disturbed by this result. He closed his paper on a note which mirrored both the concept of communicating atoms and a fundamental tenet in his general philosophy (the "Vedantic vision"):

> Thus one can also say: a certain stability in the world order sub specie aeternitatis can only exist through the interrelationship of each individual system with the rest of the whole world. The disconnected individual system would be, from the point of view of unity, chaos. The interrelationship is necessary as a continuous regulative factor, without which, with respect to energy considerations, the system would aimlessly wander about.—Is it idle speculation if one is reminded by this of a similarity in social, ethical, and cultural phenomena?[65]

5. *The Bothe-Geiger and Compton-Simon Experiments*

The private and public reactions by theorists to the Bohr–Kramers–Slater paper were widespread, varied—and conflicting. This situation reflected in no small degree the fact that it presented both a complex and an entirely qualitative program (the only equation in the 18-page paper was $h\nu = E_1 - E_2$). Nevertheless, one of its key assumptions, that the conservation laws are only valid statistically on the micro-level, was clearly and universally understood—and for very good reason. Everyone was forced to admit, on reflection, that no experiments had ever been carried out which definitely excluded this possibility. This point rapidly became the central issue in the debate, and as C.D. Ellis observed: ". . . it must be held greatly to the credit of this theory that it was sufficiently precise in its statements to be disproved definitely by experiment."[66] The decisive reactions to the Bohr–Kramers–Slater paper came not from theorists, but from experimentalists.

The first announcement of a planned experimental test came only about one month after the Bohr–Kramers–Slater paper had been published in May 1924. W. Bothe and H. Geiger, working in the Physikalisch-Technischen Reichsanstalt in Charlottenburg, had designed an experiment employing two oppositely positioned point Geiger counters to enable them to detect coincidences between the recoil electron and (indirectly) the scattered quantum of radiation in a Compton scattering process.[67] Such coincidences would not be expected to occur if energy and momentum were only statistically conserved in elementary processes. To achieve a definite experimental decision on this point, however, was long and difficult in practice, and it was not until April 18, 1925, that Bothe and Geiger felt sufficiently confident in their results to report them to the editor of the *Zeitschrift für Physik*.[68] (They provided complete details of their work one week later in a second report.[69])

The care Bothe and Geiger took is nicely illustrated in a letter Max Born wrote to Bohr on January 1, 1925,[70] three and one-half months earlier, in which Born reported that "everyone" in Berlin was already taking Bothe and Geiger's results to mean that "Einstein triumphs"—"even if," Born added in a footnote, "Geiger and Bothe themselves do not yet regard their results as final." Interestingly enough, Born in this same letter proposed an alternate interpretation for the observed coincidences, suggesting that if the primary X-rays consisted of pulses followed by relatively long pauses, the release of a recoil electron, and hence the chance of a coincidence, existed only during the period of excitation. "The counting, however, has nothing whatsoever to do with light quanta." Bohr responded in an open-minded way to Born's idea,[71] but added that it would be difficult to come to definite conclusions until after it was known whether or not the coincidences occurred only if the two counters were placed at the Compton secondary quantum and recoil electron conjugate positions.

Bothe and Geiger had calculated, taking account of unavoidable experimental errors, that on the Compton-Debye theory one recoil electron should be detected in coincidence with a secondary quantum for about every ten secondary quanta that were produced, while on the Bohr–Kramers–Slater theory no coincidences at all were to be anticipated. They actually found about one coincidence for every eleven secondary quanta that were produced, and they calculated that the probability for this result to be pure chance was about 10^{-5}. They concluded, therefore, that their results were "not in accord with Bohr's interpretation of the Compton effect. . . . It is recommended therefore to retain until further notice the original picture of Compton and Debye. . . . One must therefore probably assume that the light quantum concept possesses a high degree of reality, as is assumed in this theory."[72]

Geiger had graciously (perhaps also diplomatically) given Bohr advance notice of his and Bothe's conclusions by letter on April 17.[73] Four days later Bohr responded:

Thank you very much for the great friendliness with which you have informed me of your important results. I was completely prepared [for the news] that our proposed point of view on the independence of the quantum process in separated atoms should turn out to be incorrect. The whole thing was more an expression of an attempt to achieve as great as possible application of classical concepts, rather than a completed theory. Not only were the objections of Einstein very unsettling; but recently I have also felt that an explanation of collision phenomena, especially Ramsauer's results on the penetration of slow electrons through atoms, presents difficulties to our ordinary space-time description of nature similar to those presented by a simultaneous understanding of interference phenomena and a coupling of the changes in state in separated atoms through radiation. In general I believe that these difficulties so far exclude the maintaining of the ordinary space-time description of phenomena, so that in spite of the existence of coupling, conclusions concerning an eventual corpuscular nature of radiation lack a satisfactory basis.[74]

On precisely the same day, April 21, Bohr added the following postscript to a letter which he had just written to R. H. Fowler: "Just in this moment I have received a letter from Geiger, in which he tells, that his experiment has given strong evidence for the existence of a coupling in the case of the Compton-Effekt. It seems therefore that there is nothing else to do than to give our revolutionary efforts as honourable a funeral as possible."[75]

The finality of Bohr's decision, as Bohr himself emphasized in his previously mentioned reply to Born, did not detract from the importance of completely independent experiments to test Bohr's statistical conservation idea which were simultaneously underway at the University of Chicago in Arthur Compton's laboratory. The "possibility of such a test," Compton recalled, had been suggested by W. F. G. Swann in conversation with Bohr and himself in November 1923[76]—that is, even before Slater arrived in Copenhagen with his "essentially new idea" and provided Bohr with the key concept necessary to unite his thoughts on atomic structure with those on statistical conservation. To aid him with his experiments, Compton enlisted the help of A. W. Simon. The first fruit of Compton's and Simon's collaboration, which they described on November 28, 1924, at the American Physical Society meeting in Ann Arbor—a full year after Swann had made his suggestion—involved stereoscopically photographing cloud chamber tracks of β-rays ejected from moist air by strongly filtered X-rays of "effective wave-length" 0.1–0.9 Å.[77] In agreement with the early work of Wilson and Bothe, discussed in the last chapter, Compton and Simon observed two different types of tracks which could be quantitatively interpreted as photoelectron and recoil electron tracks; and measurements on the range and angular distributions of the latter tracks agreed "in every detail" with Compton and Hubbard's earlier predictions based on the quantum theory of scattering. Compton and Simon emphasized in their full report (published March, 1925) that the evidence was therefore "very strong" that one recoil electron was associated with each quantum of scattered radiation, and that "this electron possesses, both in direction and magnitude, the vector difference of momentum between the incident and the scattered x-ray quantum."[78]

On June 23, 1924, just two months after Bothe and Geiger had completed their experiments, Compton and Simon reported much more definite evidence for the cited conclusion.[79] As such, they took it to represent a "crucial test" between the Bohr–Kramers–Slater and Compton interpretations of the scattering process. The purpose of their experiments was to settle precisely the point Bohr raised in his letter to Born—to determine if a definite relationship existed between the quantum scattering angle θ and the conjugate recoil electron scattering angle ϕ, in other words, to see if Debye's equation (6.26) held. They therefore sent a collimated beam of X-rays into a relatively large cloud chamber (18 cm in diameter, 4 cm high) and took roughly 1,300 stereographic photographs. They found that, out of the last 850, 38 revealed the presence of *both* recoil electrons and scattered X-rays; and in 18 of these 38 cases, the experimentally determined angle θ was within 20° of the calculated value. This number was four times as large as would have been expected if the distribution were a matter of pure chance (Bohr–Kramers–Slater theory). Moreover, by a simple statistical argument, Compton and Simon proved that the probability that this agreement with theory was accidental was 1/250. However, while they definitely ruled out the possibility that "the observed coincidences are the result of an unconscious tendency to estimate the angles falsely," they nevertheless were not able to explain the origin of about half of the random scatterings they had observed. Hence, they concluded conservatively that "at least a large part of the *scattered x-rays proceed in directed quanta of radiant energy*." These results were in any event "in direct support" of the conservation laws, but did not "appear to be reconcilable with the view of the statistical production of recoil and photo-electrons proposed by Bohr, Kramers and Slater."[80]

6. *Postmortem*

The Bothe-Geiger and Compton-Simon experiments laid to rest Bohr's statistical conservation explanation of the Compton effect. Nevertheless, in the next few years a number of other, directly relevant experiments were suggested and carried out, especially by Bothe.[81] However, by the time the Bothe-Geiger and Compton-Simon experiments were reported in April and June, 1925, respectively, an irreversible trend had been established in theoretical physics by the private and public discussions engendered by the Bohr–Kramers–Slater research program.[82] In March 1924 in Copenhagen,

Kramers, using the virtual oscillator concept, made a first step en route to the discovery of matrix mechanics by generalizing Ladenburg's 1921 dispersion formula,[83] and within a few months, Kramers' work was extended, first by Born[84] and then by Kramers himself.[85] Later in the year, J. H. Van Vleck at the University of Minnesota showed how to relate the absorption of radiation and Kramer's dispersion formula to the correspondence principle.[86] At about the same time, J. C. Slater, the originator of the virtual oscillator concept, before leaving Copenhagen in mid-1924, drafted a paper on optical theory which both employed virtual oscillators *and* embodied his conviction "that all actions take place in time and space as we ordinarily think about them, and that the behavior outside the atom is described by Maxwell's equations, while the atom is described according to the theory of energy levels."[87] These requirements made his opposition to Bohr's and Kramers' views explicit. Nevertheless, when Bohr, on January 10, 1925, acknowledged receipt of a preprint of Slater's finished paper,[88] he could not refrain from stating that he and Kramers were "still very sceptical as to the essential reality of your suggestions."[89] In the same letter, Bohr called Slater's attention to the most recent development in dispersion theory, the joint paper which Kramers and Heisenberg had just sent off to the *Zeitschrift für Physik*.[90]

It was roughly at this point that Bothe and Geiger's results became known, first in a preliminary way (recall Born's letter to Bohr of January 1, 1925), and then definitely (April 18 and 25, 1925). These experiments directly displayed, as Heisenberg put it, "the reality of light quanta."[91] Of course, as Einstein wrote Ehrenfest: "We both had no doubts about it"[92]—and even before the results were in, Einstein had begun searching for new ways in which to incorporate light quanta into a causal space-time description of atomic processes.[93] Slater immediately concluded that the "simplest solution of the radiation problem then seems to be to return to the view of a virtual field to guide corpuscular quanta"[94]—a view which was similar to one independently suggested by W. F. G. Swann.[95] Pauli was jubilant over the new state of affairs, as clearly indicated in a letter which he wrote to Kramers on July 27, 1925:

In general, I think it was a magnificent stroke of luck that the theory of Bohr, Kramers and Slater was so rapidly refuted by the beautiful experiments of Geiger and Bothe, as well as the recently published ones of Compton. It is certainly true that Bohr himself, even if these experiments had not been carried out, would not have maintained this theory. But many excellent physicists (e.g. Ladenburg, Mie, Born) would have maintained this theory, and this ill-fated work of Bohr, Kramers and Slater would perhaps for a long time have become an obstacle to progress in theoretical physics! Because it moves in a completely false direction: *the energy con-*

cept should not be modified, but rather the motion-and-force concept. Certainly for cases in which interference phenomena are present, no definite trajectories for light quanta can be defined (one can also not define such trajectories for electrons in the atom), and just as little as it would therefore be justified to doubt the existence of electrons, would it be justified to doubt the existence of light quanta because of interference phenomena. It can now be taken for granted by every unprejudiced physicist that light quanta are as much (and as little) physically real as electrons. Classical concepts in general, however, should not be applied to either.[96]

Pauli's remark on the "many excellent physicists" was not entirely unjustified. Mie, for example, published papers in June and September, 1925, in which he questioned the finality of the Bothe-Geiger and Compton-Simon experiments;[97] Born wrote a long letter to Bohr on April 24, 1925, expressing the same point of view and suggesting a way in which to retain what was valuable in the Bohr–Kramers–Slater theory, "namely, the emission of waves *during* the stationary states";[98] and Schrödinger wrote Sommerfeld on July 21, 1925, that although Sommerfeld would call him "incorrigible," he could not come so rapidly to the conclusion that these experiments had "buried everything that smells of classical waves"—he "had the feeling" that "something completely, completely different" was now indicated, and he suggested that the way to pick up its scent was to critically examine several classical optical phenomena to isolate their typical and compensatory behavior, such as the way in which individual wavelets mutually cancel each other out to produce a well defined shadow when passing through a wide single slit.[99] Since these speculations of Schrödinger occurred only about six months before he submitted his first paper on wave mechanics to the editor of the *Annalen der Physik,*[100] it is quite likely that his route to that discovery was more complex than is generally assumed.[101]

At the same time, Schrödinger's intuitive speculations ran parallel with the research program on dispersion theory which culminated in Heisenberg's creation of matrix mechanics. Heisenberg's classic paper[102] was received by the editor of the *Zeitschrift für Physik* on July 29, 1925; Born and Jordan's joint paper[103] was received on September 27, 1925; and the "drei-Männer Arbeit"[104] of Born, Heisenberg, and Jordan was received on November 16, 1925. This research program became uncoupled, so to speak, from one of the central hypotheses of the Bohr–Kramers–Slater theory, the statistical conservation hypothesis, which had been definitely disproved by the Bothe-Geiger and Compton-Simon experiments several months before Heisenberg's discovery occurred. This fact naturally does not diminish the importance or necessity of those experiments, for the statistical conservation hypothesis itself had *not* yet been subjected to experimental test, and any other, further developments in theoretical physics would therefore always have to take account of its denial. This point was made explicitly by Slater in a letter to Bohr on May 27, 1926,[105] after Bohr had admitted to Slater that he still had "a bad conscience in persuading you to our view."[106]

It is worth repeating that these developments (as Heisenberg for example explicitly recognized[107]) followed in the wake of Compton's discovery—truly a turning point in physics. Conversely, just as Compton's discovery had been interpreted on the basis of wave mechanics after Schrödinger's discovery, so Compton's discovery was interpreted on the basis of matrix mechanics after Heisenberg's discovery. A first attack on this problem was made by G. Beck in March 1926, but a much more satisfying derivation was provided a few months later by Dirac, who showed how to understand the Compton effect by considering the incident radiation and the atomic electron as forming a coupled and perturbed dynamical system.[108] Soon thereafter, in 1927, Dirac quantized the radiation field, thereby inaugurating the study of quantum electrodynamics.[109]

C. Developments in Radiation Theory 1922–1928

1. *Introduction: Background to Complementarity*

We have already noted the decisiveness with which Bohr personally reacted to the Bothe-Geiger experiment[110]—how, even before Heisenberg discovered matrix mechanics, he had already begun exploring those physical phenomena and those fundamentally philosophical questions which ultimately convinced him of the generality and pervasiveness of the wave–particle duality in physics and culminated in the enunciation of his principle of complementarity in the fall of 1927. However, it is not my purpose here to trace in detail the evolution of Bohr's thought or, for example, Dirac's. Rather, in the remainder of this chapter, I would like to sketch some less well known developments in radiation theory that were initiated by other physicists, some famous, some not so famous, after Compton's discovery. Indeed, unless the existence of these other developments is also recognized, the difficulty and complexity of Bohr's route to complementarity, and the controversiality of that principle, cannot be fully grasped. These other endeavors took the form of attempts, first, to theoretically and experimentally determine the physical properties of light quanta, and second, to wed particle and wave; i.e., to formulate a constructive theory of radiation—a problem, of course, with which Einstein had wrestled since 1905, and one with which he would wrestle for the rest of his life.[111] I would like to survey the nature of these endeavors in a preliminary way here, leaving for future research the problems of isolating their origins, influences, and interrelationships.

2. *Attempts to Determine the Dimensions of Light Quanta*

The first physicists to attempt theoretically to determine the actual physical dimensions of light quanta—perhaps the most natural question of all to ask if one accepted their physical reality—were apparently L. S. Ornstein and H. C. Burger of the University of Utrecht in mid-November 1923.[112] Their point of departure, as they explicitly stated, was "the hypothesis of point radiation, which has found such brilliant confirmation through Compton's experiment."[113] Basically, what they tried to do was determine the collision cross section of a light quantum by examining in detail the conditions for statistical equilibrium between free electrons and black-body radiation. They concluded that within the range of validity of Wien's law, the quantum's cross section was proportional to the square of its wavelength, with the constant of proportionality being on the order of 1/10. Shortly thereafter, at the end of their second paper on the subject, they recognized that they had attacked essentially the same problem that Pauli had explored earlier, but they concluded that some of Pauli's assumptions were faulty.[114] Pauli immediately replied, criticizing Ornstein and Burger's analysis in turn, not only because it was restricted to the range of validity of Wien's law, but, in general because Pauli felt that their attempt to make such a definite model of a light quantum was misguided.[115] Ornstein and Burger, however, had meanwhile already extended their work and—in spite of Pauli's criticism—continued to do so until late 1924.[116] Concurrently, E. Marx in Leipzig was also trying to determine theoretically the physical dimensions of quanta;[117] and K. Schaposchnikow in Iwanowo-Wosenessensk and V. S. Vrkljan in Zagreb exchanged arguments on this same subject in the *Zeitschrift für Physik* until early 1927.[118] By this time, of course, quantum mechanics had been created, and in the light of that knowledge a number of other theoretical papers appeared.[119]

The earliest *experiments* on the physical dimensions of quanta which attracted considerable attention were carried out in 1923 by G. P. Thomson.[120] Thomson sent hydrogen "positive rays" transversely past two aligned slits placed in front of a spectroscope in such a way that the transit time allowed the transmission through the slits of a wave train of maximum length 3.5 cm for the H_β line. Now, since Thomson had calculated that any wave train shorter than 6 cm would produce definite changes in the (Doppler shifted) spectroscopic pattern, and since none were actually observed, he concluded that the length of a quantum must be less than about 3 cm.

Thomson's conclusion was corroborrated in much more sophisticated experiments in late 1926 by E. O. Lawrence and J. W. Beams at Yale University.[121] Using a Kerr cell shutter, Lawrence and Beams succeeded in producing pulses of light on the order of 10^{-10} seconds in duration (i.e., on the order of a few centimeters in length), whose photoelectric effects they then observed. They found that per unit incident energy these effects were independent of the length of the pulses, and therefore, *"if light quanta are of the commonly understood wave nature, they are less than 3 cm. in length and an electron absorbs a light quantum photo-electrically in less than 10^{-10} sec."*[122] Thus,

> . . . quanta as bundles of energy of considerable dimensions do not exist and [these results] suggest that the effects of radiation on atoms are independent of the length of the pulses to the extent that the effects would be independent on the classical theory. This conclusion harmonizes well with a theory advanced by Professor W. F. G. Swann [who probably suggested Lawrence and Beams' experiment in the first place] in which he developed the consequences of the postulate that quanta of negligible dimensions follow the Poynting flux of the classical theory. A similar idea is incorporated in the more recent theory of the structure of light put forward by Sir J. J. Thomson. . . . Both theories are supported by the present experiment.[123]

Shortly after Lawrence and Beams had published their paper, G. Breit observed in *Nature*[124] that experiments "were made by me in the spring and summer of 1926 on the same subject with a similar result, though by a different method." Breit, however, did not publish his results because "a theoretical consideration showed that a positive result would be very improbable. . . ." Other, related experiments were also carried out at the University of Chicago in 1926 and 1927 under the direction of A. J. Dempster.[125]

3. *Hopes for Reconciliation of Wave and Particle Theories; Constructive Theories of Radiation*

The cited theoretical and experimental attempts to determine the physical dimensions of light quanta, coming on the heels of Compton's discovery, and taken in conjunction with the Bothe-Geiger and Compton-Simon experiments, led to the widespread conviction, as C. D. Ellis put it in 1926, that the "wave-theory itself cannot be correct, but except for its greater age it has no greater claims than the light-quantum view."[126] Of course, the "great problem" of the period, in Planck's characterization, was to discover in "what relation . . . the corpuscular laws stand to the laws of wave-motion in the general case. . . ."[127] H. A. Lorentz's belief, which he had held as early as mid-1923, and which was shared by many, was that

> . . . it must after all be possible to reconcile the different ideas. Here is an important problem for the physics of the immediate future. We cannot help thinking that the solution will be found in some happy combination of extended waves and concen-

trated quanta, the waves being made responsible for interference and the quanta for photo-electricity.[128]

One hint for a possible reconciliation of waves and quanta was offered at the end of 1923 by Compton, who suggested that one should search for a quantitative criterion for the transition from Compton to Thomson ("modified" to "unmodified") scattering, "for it is this criterion which determines whether or not interference is possible."[129] A second and even earlier hint, and one which appeared very promising to many physicists, arose from Duane's 1923 proof that the Bragg equation for crystal reflection of X-rays follows directly from the assumption that the crystal's momentum changes by $h\nu/c$ normal to one of its reflecting planes—that is, Duane proved that a wave equation could be derived from a quantum condition. As we saw in the last chapter,[130] Compton subsequently proved that Duane's quantum condition could be justified on the basis of the general Sommerfeld quantum conditions. Both Duane's and Compton's discussions, however, were restricted to the Fraunhofer diffraction case (collimated incident and reflected beams), and to the case of an infinite grating (which yields absolutely sharp Laue spots).

That was the situation when, in February 1924, P. S. Epstein and P. Ehrenfest (who was then visiting the California Institute of Technology) generalized the analysis one step further.[131] Employing Fourier analysis techniques and Bohr's correspondence principle, Epstein and Ehrenfest proved that the Duane-Compton quantum condition was also applicable to finite gratings, and hence that it was possible to bring about *"the complete translation of the theory of Fraunhofer diffraction into the language of the quantum theory."*[132] R. A. Millikan, Epstein's colleague and Ehrenfest's host, noted that this result represented "a little progress towards removing the apparent contradiction with interference," but added that "the way is yet very dark."[133] Nevertheless, the approach was so promising that Epstein and Ehrenfest immediately began to study the case of Fresnel diffraction (uncollimated incident and/or reflected beams). This study, however, turned out to be ultimately disappointing. It became "clear that the phenomena of the Fresnel diffraction cannot be explained by purely corpuscular considerations. It is necessary to attribute to the light quanta properties of phase and coherence similar to those of the waves of the classical theory."[134] This negative conclusion, and the fact that they became "busy with other work," prompted Epstein and Ehrenfest to leave their calculations unpublished until mid-1927—until the "recent discovery made by Davisson and Germer" gave "to the problem of corpuscular diffraction a new interest and importance."[135] Precisely the same considerations motivated Duane in 1927 to recall and discuss his old work;[136] and even before the Davisson-Germer experiment became known, in April 1926, P. Jordan discussed theoretically the connection between Duane's work and the interference of de Broglie waves, concluding for example with W. Elsasser that material particles like electrons ought to be diffracted by crystals.[137]

These hints for a reconciliation of the wave and particle theories, however, only represent specific instances of a very pervasive trend in theoretical physics after Compton's discovery. By the summer of 1925, J. H. Van Vleck for example could remark that "Modern physics certainly is passing through contortions in its attempt to explain the simultaneous appearance of quantum and classical phenomena; but it is not surprising that paradoxical theories are required to explain paradoxical phenomena."[138] In a similar vein, E. B. Wilson noted one year later: "To cover all the recent speculations on the nature of light would be fairly well to cover the interesting and disputed parts of modern physics."[139]

These "speculations" arose out of the very widespread feeling among physicists, as C. G. Darwin had emphasized in mid-1923, that: "It must be taken as absolutely certain that both the electromagnetic theory and the quantum theory are valid in their respective fields, and equally certain that the two descriptions are incompatible. We can only conclude that they are parts of an overriding system, which would give rise to mathematical formulae identical with those of the present theories."[140] I would now like to briefly sketch some of the contemporary attempts to formulate such overriding constructive theories of radiation. In this way, I hope to convey some sense of the great flux in radiation theory following Compton's discovery, and to indicate the extent to which Bohr's principle of complementarity, when Bohr announced it in the fall of 1927, ran counter to the trend of thought of many of his contemporaries.

4. *De Broglie Postulates Light Quanta of Finite Mass*

Even before Compton's and Debye's papers were published, Louis de Broglie, as we have seen, had taken Einstein's light quantum hypothesis seriously—his conviction may even have germinated as early as 1911, after his elder brother Maurice told him of the papers and discussions at the first Solvay conference in Brussels.[141] By November 1922, his ideas had clarified sufficiently to enable him to give a derivation of Planck's law on the assumption that black-body radiation consists of "light molecules" of magnitude $nh\nu$ (where $n = 0, 1, 2 \ldots \infty$); and less than a year later (September–October, 1923), after his convictions were confirmed by Compton's discovery, he published much more definite and extended views in three short notes in the *Comptes rendus*. Today, most physicists are familiar with these notes (and their English summary published in *The Philosophical Magazine* the following February) because they contain the first hints of de Broglie's radically new conception of "matter waves." This conception, however, as de Broglie's papers make abundantly clear, grew out of his ideas on the nature of radiation, specifically on the nature of light quanta, the "real existence" of which he took for granted. Taken together, these ideas constitute the first constructive theory of radiation published after the discovery of the Compton effect.

De Broglie postulated that all light quanta are identical in nature; that all possess an "internal binary symmetry" to account for polarization phenomena; and—most significantly—that all have an "extremely small," but nevertheless finite, rest mass m—which implied that their velocities, while "very nearly equal to the Einstein's limiting velocity c," are *not equal* to c.[142] However, as de Broglie argued, since their velocities cannot be experimentally distinguished from c, one could invert the procedure and thereby estimate that their rest masses m "should be at most of the order of 10^{-50} gr." It will be understood that this model, which of course differs radically in principle from present-day conceptions, enabled de Broglie to extend his ideas on light quanta to all other material particles.

De Broglie now further assumed that a light quantum (or any other material particle) of rest mass m possesses an "internal energy" equal to mc^2, to which may be ascribed a "periodical phenomenon" whose frequency is $\nu_0 = (1/h)mc^2$. If the body moves at velocity v, this internal frequency, according to special relativity, will appear to a fixed observer to be lowered to $\nu_1 = \nu_0 \sqrt{1 - \beta^2}$, where $\beta = v/c$. In other words, the fixed observer will see the body's internal periodic phenomenon varying as $\sin 2\pi\nu_1 t$. At the same time, he will conclude that the body's total energy equals $mc^2/\sqrt{1 - \beta^2}$, corresponding to a frequency $\nu = mc^2/h\sqrt{1 - \beta^2} = \nu_1/(1 - \beta^2)$, that is, corresponding to a frequency which in general is very different from the "internal" frequency ν_1.

De Broglie now proved that if at time $t = 0$ the moving body coincides with an energyless "phase wave" of frequency ν which spreads with velocity $c/\beta = c^2/v > 1$, and which is locked in phase with the body's internal periodic motion of frequency ν_1, then *"this harmony of phase will always persist"*—a result which is implicitly contained in the Lorentz time transformation. Thus, "any moving body may be accompanied by a wave," and "it is impossible to disjoin motion of body and propagation of wave." From the known relationship between phase and group velocities, de Broglie easily calculated that the group velocity of the energyless phase waves is equal to βc, i.e., equal to the velocity $v = \beta c$ of the body itself. Hence, the *energy* of the moving body must be transported at the *group* velocity of the energyless phase waves. The actual trajectory of the body, de Broglie proved, is determined by the principle that the *"rays of the phase wave are identical with the paths [of the moving body] which are dynamically possible."* "I think," he added, "that these ideas may be considered as a kind of synthesis of optics and dynamics."[143]

While de Broglie found that these ideas led to a very simple interpretation of Bohr's quantum postulate, it was not primarily the behavior of electrons, but rather the behavior of light quanta which concerned him at this time. He therefore immediately asked himself how he might account for the coherence of quanta, how two or more quanta (moving with velocity $v < c$) and their associated phase waves (spreading with velocity $v > c$) could be "parts of the *same* wave"? De Broglie based his answer to this question on the postulate that: "When a phase wave crosses an excited atom, this atom has a certain probability of emitting a light quantum determined at each instant by the intensity of the wave."[144] Hence, after passing over many atoms, the "non-material phase wave will carry many little drops of energy which slide slowly upon it and whose internal phenomena are coherent." In passing obstacles, for example in diffraction phenomena, these "little drops of energy" (quanta of finite mass) do not interact gravitationally with the obstacle as Newton had assumed two centuries earlier, because their phase rays (i.e., the normals to the phase waves) and not the inertia principle govern their directions of motion. Nor could a given phase ray be identified with a Poynting-like vector from classical theory, since experiments could be cited (such as Wiener's experiment) where the phase ray permitted the passage of quanta through points where the square of the Poynting vector was zero.

Interference and diffraction phenomena could, however, be accounted for, first, by noting that all means of detecting light "can, in fact, be reduced to photoelectric actions and scattering"; and, second, by supposing "that, for a material atom, the probability of absorbing or scattering a light quantum is determined by the geometrical sum of one of the defining vectors of the phase waves crossing upon it." For example, in a Young's two-hole experiment, the

> . . . capacity of photoelectric action will vary from point to point according to the interference state of the two phase waves which have crossed the two holes. We shall then see interference fringes, however small may be the number of diffracted quanta, however feeble may be the incident light intensity. The light quanta do cross all the dark and the bright fringes; only their ability to act on matter is constantly changing. This kind of explanation, which seems to remove at the same time the objections against light quanta and against the energy propagation through dark fringes, may be generalized for all interference and diffraction phenomena.[145]

De Broglie was fully aware that his ideas, which he related to other optical phenomena as well, "if they are received, will necessitate a wide modification of the electromagnetic theory." In his view, of course, such a modification was inevitable, and several times during the next three years—first in response to points raised by Wilhelm Anderson and G. E. M. Jauncey;[146] then in his extraordinary Ph.D. thesis;[147] and finally in a brief note in which he proved that the density of light quanta is proportional to the classical intensity[148]—he attempted to further this modification. In 1927, however, W. Bothe reported experimental results which he took to be detrimental to de Broglie's theory, but the interpretation of Bothe's results is not entirely clear, since he considered them to be detrimental to Schrödinger's wave packets also.[149] In any event, by 1927, de Broglie himself was much less concerned with light quanta than with matter waves.

5. *Bateman Postulates Neutral Doublets as Light Quanta*

It is not surprising that a second constructive theory of radiation was formulated by Harry Bateman—the Cambridge-educated mathematical physicist at the California Institute of Technology—who, in E. B. Wilson's judgment, was "probably the most powerful and consistent mathematician now working in electromagnetic theory."[150]

While some of the elements in Bateman's theory actually appeared in his publications prior to Compton's discovery, and remnants of it survived in his publications for years thereafter,[151] it was evidently Compton's discovery that occasioned Bateman's immediate concern with the radiation problem when he attacked it in November 1923.[152] (Recall that Sommerfeld had personally carried the news of Compton's discovery to the West Coast within a few weeks after it had been made.) In many respects, Bateman's theory constituted a mathematization of W. H. Bragg's old "neutral pair" hypothesis on the nature of X-rays and γ-rays. As Bateman wrote:

> The original idea of the "neutral pair" of electric charges was that it consists of an electron neutralized by a positively-charged particle, such as a proton, but on account of the fact that, according to the usual ideas, electrons and protons cannot travel with the velocity of light, it seems better to regard any electric charges that travel with waves of light as consisting of an entirely different form of electricity which can travel with the velocity of light and still be associated with finite amounts of energy and momentum. There is nothing unreasonable in this supposition, for the laws which determine the structures of the electron and proton may quite likely determine also a third form which can travel with the velocity of light and have either a positive or negative charge.[153]

It was on this "third form" of electricity, which "is not essentially different from the 'aether' particles which are the carriers of electromagnetic fields of the ordinary type," that Bateman focused his attention. He specified the motion of this electricity mathematically through the equations:

$$\begin{aligned} E_x &= \quad H_x = 0 \\ E_y &= \quad H_z = \frac{\partial\Omega}{\partial y} f\left(t - \frac{x}{c}\right) \\ E_z &= -H_y = \frac{\partial\Omega}{\partial z} f\left(t - \frac{x}{c}\right) \end{aligned} \tag{7.3}$$

where Ω, a funcion of y and z, may be zero in the yz-plane, and where the derivatives $\partial\Omega/\partial y$ and $\partial\Omega/\partial z$ are large at some point or points in the yz-plane and small elsewhere. Bateman proved that these equations and requirements describe a dipole of electricity in the yz-plane which travels at velocity c in the x-direction. Moreover, since the E and H vectors are normal to each other and to the direction of propagation, and since the equations satisfy Maxwell's equations for free space, he concluded that "many of the familiar properties" of light were reproduced. Bateman also proved that this type of dipole radiation may "be obtained theoretically in a perfectly natural manner by the reflexion of 'ordinary' plane waves of light at the surface of a mirror in the form of a paraboloid of revolution."[154] Its concentration at a given point in space and time could be described mathematically by either superimposing "a number of fields of [the above] type . . . so as to obtain something like a continuous volume distribution of electricity, or . . . [by adopting] some method of averaging similar to that adopted by Lorentz in his theory of a dielectric." In both cases, if the factor f were assumed to be for example $f(t - x/c) = [\sin 2\pi\nu(t - x/c)]/(t - x/c)$, the total energy of the dipole could be shown to be proportional to the frequency ν; and hence if the total energy were "regarded as a universal constant . . . we obtain a simple model of a light-quantum."[155]

Bateman's light quantum, as described by equations (7.3), as well as by other "reasonable" ones which he specified for the second case mentioned above, also underwent a Doppler shift when reflected normally from a moving mirror. For oblique reflection, however, the situation was not so clear—and understandably so: "This question of the interaction between fields is a difficult one; some light may be shed upon it, however, by the recent researches of A. H. Compton and W. Duane."[156] With respect to interference and diffraction phenomena, Bateman concluded that in his theory the problem "is in some respects analogous to the hypothesis of inflected rays adopted by Young in his theory of Diffraction," analogous, that is, to the interference of the direct and reflected rays, so that it could be solved by Fresnel-Kirchhoff diffraction theory. All in all, Bateman's theory could account for a number of common optical phenomena; and he pursued and modified it in succeeding years.

6. *Some Miscellaneous Programmatic Suggestions*

In addition to (and at times concurrently with), the detailed and quantitative development of such constructive theories of radiation as de Broglie's and Bateman's, a number of other qualitative and largely programmatic suggestions for wedding particle and wave were advanced by a number of other physicists. H. S. Allen, for example, in June 1924—building on ideas he had already held in 1921—criticized E. T. Whittaker's "calamoids" (four-dimensional electromagnetic tubes of force) with the remark that further consideration "has inclined me to believe that, even though there may be advantages in regarding calamoids as Quanta, there is still something to be learnt from the more familiar magnetic tubes in three dimensions"—an idea he then pursued in some detail.[157] Allen's basic point of view, which he had expressed some months earlier, was that "In the present stage of the development of physics, when we seem forced to believe in two mutually contradictory theories of light . . . at the same time, the wildest guess at a solution may be permitted."[158]

In a similar spirit, and motivated by the Bohr–Kramers–Slater theory, H. A. Senftleben at the end of 1924 proposed a model for an absorbing and radiating "molecule" or singularity in the electromagnetic field.[159] He argued that the singularity must be surrounded by a concentric field-free region, the two of which together constituted a field molecule which is embedded in, but separated from, the field itself. The absorption of radiation, then, according to Senftleben, consisted in the passage of *quanta* from the field through the field-free region into the singularity; while the emission of radiation consisted in the passage of *waves* from the singularity through the field-free region into the field. Significantly enough, while Senftleben drew detailed pictures of these processes, he did not specify quantitatively how they could occur, and hence how the emitted waves and absorbed quanta could be reconciled with each other.

Other qualitative suggestions were made in 1925 by F. Russell Bichowsky of Johns Hopkins University,[160] and in 1926 by L. V. King of McGill University.[161] Even as late as 1928, G. Wataghin of Turin developed a quantitative dynamics for quanta;[162] and Johannes Stark, who had by then resigned from academic life and was directing his own laboratory near Munich, illustrated his alienation from the physics community by discussing a connection he himself claimed to have established between a light vortex and a quantum vortex (the electromagnetic field of an electron).[163] Significantly enough, Stark made no reference whatsoever to any of Einstein's work.

One largely qualitative program of the period that attracted considerable attention was proposed by J. H. Jeans, first in the second edition (1924) of his well-known *Report on Radiation and the Quantum-Theory,*[164] and then in two prestigious lectures, his Kelvin Lecture of February 5, 1925,[165] and his Rouse Ball Lecture of May 11, 1925.[166] Jeans' discussion proved, in general, that while he had been converted to the quantum theory after the first Solvay Conference (1911), he had remained a conservative in matters concerning radiation theory. Thus, after developing a rudimentary probabilistic interpretation of the photoelectric effect in the first two of the cited publications, Jeans treated the Compton effect in the third by starting from the suggestion that if "we discard the bullet-theory of radiation, the phenomenon admits of almost as simple an explanation." To wit: we should imagine that the atom first absorbs "a quantum $h\nu$ in the ordinary way," but "that in some way the process is checked and that before it is complete the atom emits a new quantum of energy $h\nu'$." If the "nucleus and remainder of the atom remain unmoved by the double process," we would then, Jeans claimed, "obtain equations which would be identical with those leading to Compton's formula except in one respect," that is, the coefficient multiplying the geometric factor would not be precisely Compton's h/mc.[167] Jeans explained that this discrepancy reflected the fact that on his theory the wavelength of the freely spreading secondary quantum would not be "strictly monochromatic." He admitted that his explanation "may seem somewhat artificial," but asserted that it was "far less at variance with the other phenomena of radiation than the usual explanation which postulates a bullet-like structure of radiation." Moreover, he felt that it could "hardly be claimed that experiments are as yet precise enough to decide between the two formulae." He also argued that since the photoelectric effect involves a flow of energy between two atoms it could be explained by analogy to a "hydrodynamical flow in a field in which there is a source and a sink of equal strength," all of which proved to Jeans' satisfaction that the undulatory theory "has nothing to fear from exact mathematical analysis, from which indeed it receives nothing but encouragement."[168]

Jeans' ideas were not warmly received. E. C. Stoner commented on Jeans' initial probabilistic interpretation of the photoelectric effect with the remark: "It is here suggested that such a treatment virtually begs the whole question."[169] The point was that according to Jeans the atom had to be able to somehow absorb *waves,* which "seems no more comprehensible for probability waves than electromagnetic ones." Moreover, if the wave system did not carry energy, "there seems little value in postulating its existence—for our interest in waves depends fundamentally on their energy-carrying properties." (This point also led Stoner to reject the Bohr–Kramers–Slater theory.) In sum, "Jeans' treatment . . . , while of value in bringing out difficulties, is purely formal." Stoner's judgment was also apparently shared by R. C. Tolman,[170] but an anonymous reviewer was much more harsh. The later claimed that Jeans, throughout his theory, "resorts to arguments which seem unnecessarily crude."[171]

7. *Stoner Postulates Localized and Coherent Light Quanta*

The give-and-take and flux of ideas characterizing the post-Compton effect period is nicely illustrated by the fact that E. C. Stoner, who in January 1925 criticized Jeans' theory (as well as the Bohr–Kramers–Slater theory), simultaneously offered an alternative himself.[172] Stoner was fully convinced of the physical reality and linear directedness of quanta. Where he differed from others was in his belief that quanta are localized not only laterally, but *longitudinally* as well; in other words, that they have a point rather than a line structure. He had several grounds for his belief. First, he pointed out that in the usual derivation of Planck's law black-body radiation is assumed to be concentrated in standing waves, i.e., that a "whole quantum must be supposed to be capable of localisation within a volume λ^3"—or longitudinally within a distance λ. Second, if one recalls that atoms typically emit wave trains on the order of one meter or 10^6 wavelengths, it must then be the case, Stoner argued, that this distance is "indefinitely" greater than the length of the emitted quantum, "for otherwise there would not be that sharp distinction between stationary and intermediate states which forms so essential a feature of the quantum picture of the atom." Finally, G. P. Thomson's recent experiments, and Ornstein and Burger's recent theoretical papers, indicated that a quantum had an "effective sphere" on the order of λ^3—or a linear dimension on the order of λ. In sum, "there can be little doubt that the evidence so far reviewed, if interpreted in any straightforward manner, leads definitely to the conclusion that radiation is propagated by linearly directed, spatially localised quanta."[173] This model constituted the fundamental point of departure for Stoner's attempt to reconcile particle and wave in his constructive theory of radiation.

Stoner recognized that the basic challenge he faced was to account for interference and diffraction phenomena; and he knew that his model had to overcome the "main objection" pointed out years earlier by H. A. Lorentz in his cogent analysis of telescopic resolving power.[174] The fact that Bohr had not accepted light quanta because they excluded "in principle the possibility of a rational definition of the conception of a frequency" did not "carry great weight" with Stoner, since, he argued, if one assumes that a single quantum has periodic properties "a conception of frequency and also of 'wave length' can at once be arrived at," although only by analogy, because the very "adoption of the quant involves abandonment of the wave picture." He therefore concluded that one could introduce without contradiction the concepts of frequency, wavelength, and phase of a "quant," the later being fixed "by the time which has elapsed since the quant was in some arbitrarily defined state." Tentatively, he suggested that the interpretation of these states be sought in the variation in the magnitude and direction of the E and H vectors which, he assumed, characterize the "quant," and "presumably" maintain its total energy at a constant value. Indeed, the "nomenclature and general wave theory methods can largely be taken over and applied to the 'phase field' surrounding point sources."[175]

Given such quanta, Stoner had to postulate that they interact mutually with each other to account for interference and diffraction. However, he argued that their relative phases need not be associated with a single emitting atom, but only with a common geometrical point of origin such as a distant star, for it "seems not unnatural or far-fetched to assume that their phase at the point of origin is the same—that the 'phase at emission' for similarly polarised quanta, at least, is constant, and hence that different quanta travelling by two paths arrive at a point of observation in the same phase if the path difference is a whole number of 'wave lengths,' as above defined." The relative intensities at different points in an interference pattern could be explained by assuming the existence of a "coupling" between the incident quanta and the absorbing atoms determined by the relative phases of the two.

Thus the greater the number of quanta in the same phase which pass an absorbing system, the greater the probability of absorption of one of them; the passage of quanta of opposite phase, producing opposite impulses, will diminish the probability of absorption. It must of course be supposed that the effect of the impulse persists, unless the state is modified by external disturbances, for Taylor's classical experiment shows that interference effects still occur even for very weak light.[176]

Stoner suggested that perhaps the impulses "are given in the process of scattering"; and that they add vectorially to determine, not the number of quanta actually passing over the atoms, but the probability of throwing the atoms' coupling switches. The intensities at various points, then, "if defined in terms of the number of absorption switches induced, will . . . depend not only on the number of quanta which pass it, but also on their relative phases."

Stoner concluded his paper by discussing other implications of his theory for diffraction and dispersion, and for the changes in angular momentum that must accompany radiative transition processes. His whole theory—actually it represented more of a program for research—was criticized shortly after its publication in an unpublished lecture delivered at Cambridge by his mentor, R. H. Fowler.[177] One reason at least seems fairly obvious: Fowler at this time, in contrast to Stoner, accepted the Bohr–Kramers–Slater theory.

8. *Thomson Postulates Ring Light Quanta*

For those familiar with the life and work of Sir J. J. Thomson, it will hardly come as a surprise to learn that this Grand Old Man of English Physics also offered his views on the current situation in radiation theory—and that he characteristically proposed (in October 1924) a "mental picture, based on the idea of tubes of electric force" which he "found useful in reconciling the optics with the electricity."[178]

Imagine, Thomson argued, that a positive and a negative charge (presumably the nucleus of an atom and one of its electrons) is joined together by a tube of force; and further imagine that the negative charge (the electron) jumps from a higher to a lower energy level. The connecting tube of force according to Thomson (who drew detailed pictures of the process) will then collapse into a loop which closes upon itself and which will be thrown off as a ring—Thomson's light quantum model—traveling "with the velocity of light in a direction at right angles to its plane, like a circular vortex ring." The energy of Thomson's ring light quantum, which was roughly equal to the drop in energy of the electron (a small part was presumably converted into ordinary electromagnetic radiation during the ring's formation) "remains constant as long as the ring is unbroken." Moreover, as the "ring is emitted from the atom, and indeed for some time before and after, the electric field in the atom is changing rapidly. This change will produce vibrations in the ring and give rise to electrical waves, so that before and after the ejection of the ring, electric waves are coming out of the atom; . . . these, like the ring, travel with the velocity of light. . . ."[179] The ring therefore was the "centre of a system of electrical waves of the normal type" but of very small energy (the difference between the energy lost by the atomic electron and that contained in the ring itself). In the inverse (absorption) process, depending upon how much energy was transferred from the incident ring to the electron in the atom, the electron would either be emitted as a photoelectron or simply be excited to a higher energy level.

With respect to interference and diffraction phenomena, Thomson argued that if the ring–wave system meets for example a single narrow slit, the accompanying "Poynting vector, *i.e.* the direction of the flow of energy, will change in direction from place to place in the neighbourhood of the slit.

Thus the flow of energy gets diverted when the wave passes through the slit; it is no longer always in one direction, but spreads out fanwise after leaving the slit. The result of this spreading is shown by the ordinary theory of diffraction. . . ." "Thus . . . the diffraction patterns will be the same when the energy of the light is concentrated in a ring as when it is, as in the ordinary theory, spread throughout the whole of the wave."[180] To relate his theory to the quantum theory more directly, Thomson proved that the ring's electrical energy, which he assumed to be vibrating at the same frequency as that of the low-energy waves surrounding the ring, was inversely proportional to the ring's radius, and hence to its circumference. Thus, under the further assumption that the ring's circumference was equal to the wavelength of its internal vibrations, the energy of the ring turned out to be proportional to the frequency of these internal vibrations, as well as to the frequency of the low-energy guiding waves. The constant of proportionality, under proper additional assumptions of course, turned out to be equal to Planck's constant h. Everything therefore worked out very nicely, including for example the fact that visible light and X-rays (rings of different sizes) would interact differently with atoms.

When E. B. Wilson summarized Thomson's theory in May 1926, he did so with skepticism, even with a certain amount of derision. He remarked that in contrast to Thomson himself, "I can not see that Thomson's [theory] of 1903 has any similarity to Thomson's of 1924, and I must confess to having had that difficulty before with some of Thomson's papers separated by an interval much shorter than twenty-one years."[181] Instead of appearing as a bright speck against a dark background, "What the Faraday tube now does is to stay in the atom, but occasionally to get snarled up and throw off the snarls as light corpuscles." The Faraday tube therefore appeared to be a "very obliging if not changeable fellow." Wilson proved from reasonable assumptions that although the ring somehow emerged from the atom, once outside, the "doughnut would contain 50,000,000 atoms." Moreover, the Faraday tube evidently moved slowly enough *inside* the atom to get looped, while *outside* the atom the loop moved with the velocity of light. "The leopard [therefore] has changed his spots."[182]

9. *Whittaker Mathematizes Thomson's "Smoke Rings"*

Wilson's transparently negative reaction to Thomson's theory was in sharp contrast to E. T. Whittaker's reaction. Whittaker, who compared Thomson's system to "a procession of rings of cigarette-smoke, moving forward through the air in front of the mouth of the smoker," published a paper in 1925 whose object was (Whittaker's italics) *"to bring Sir J. J. Thom-*

son's theory within the compass of the Maxwell-Lorentz theory of the electromagnetic field, or to be more precise, *within the compass of an extended form of the Maxwell-Lorentz theory.*"[183]

To carry out this program, Whittaker first set up expressions for the three components of the electric field describing a ring of electricity confined to a circular region $r < a$ and traveling in the x-direction with the velocity of light. He noted that these equations collectively satisfied Maxwell's first equation, div E = 0, and used this fact to determine the three corresponding components of the magnetic intensity from Maxwell's second equation, curl H = $(1/c)(\partial E/\partial t)$. At this point he introduced his *"essentially new feature . . . into the Maxwell-Lorentz equations in order to make them capable of representing Thomson's ideas regarding light," viz.,* instead of using div H = 0 for Maxwell's third equation, the existence of the closed rings of electricity demanded that one use div H = μ, where μ is the volume density of the associated "magnetism"; and by integrating over the circular region $r = 0$ to $r = a$, using the value of μ required for consistency with the known components of H, Whittaker easily proved that the *"total quantity of magnetism in the pulse is zero."* This modification of Maxwell's third equation, first pointed out as long ago as 1885 by Heaviside, was also consistent with Maxwell's fourth equation, provided it too was modified to read curl E = $-(1/c)(\partial H/\partial t) + \mu\mathbf{v}$, where $\mathbf{v}$ denotes the velocity of the magnetism μ. Finally citing the mathematical methods employed by Wolfke in 1913, Whittaker concluded, with Thomson, that the energy of the ring or light quantum was equal to Planck's constant times its frequency, and that its propagation in a dielectric also satisfied known relationships. In general, and mindful of Thomson's basic motivation, Whittaker remarked that: "It seems as if magnetic currents have the special function of making it possible to reconcile the quantum theory with the classical theory."[184]

Whittaker's full mathematization of Thomson's theory, like Thomson's theory itself, did not ring true to E. B. Wilson. Wilson recalled that an earlier paper of Whittaker's on a magneton interpretation of photoelectric phenomena had been "howled down" when Wilson had presented it at a Harvard physics seminar. "I rather liked the model," said Wilson—"it was so definite, but I shall not revert to it." With respect to Whittaker's present theory, Wilson made two main criticisms. First, he demonstrated that Whittaker's equations did not necessarily describe a ring of electricity, but only a circularly symmetrical configuration about its line of motion. Second, he noted that its velocity was evidently measured with respect to a highly unrelativistic, fixed ether. "The magneton," concluded Wilson, "may help in reconciling heat and light, quanta and classical theory, free propagation of radiant energy and interchange of energy between radiation and matter. It may be doubted whether it will be sufficient."[185]

Thomson himself, however, entertained few doubts over the basic validity of his model, for even as Whittaker was mathematizing it, Thomson was also extending it. Eventually, in December 1925, he published a lengthy paper on his ring model in which he applied it, for example, to the problem of the interaction of a light quantum and an electron, that is, the Compton effect.[186] He found that an increase in the ring's "wavelength" was indeed predicted. But his expressions, both for the actual increase in wavelength and for the angle of deflection of the rings, contained constants which had to be specified through auxiliary assumptions—not a very satisfying situation, although Thomson concluded his paper on his customery positive note. One year later, N. P. Kasterin, too, was still discussing Thomson's theory.[187]

10. Swann Postulates Light Quanta Guided by a Virtual Poynting Vector Field

Attempts to reconcile particle and wave, i.e., to formulate constructive theories of radiation, were not confined to English and European physicists. In the United States, W. F. G. Swann, who had gone to Yale after being first at Minnesota and then at Chicago, speculated on a possible extension of the Bohr–Kramers–Slater theory at the end of 1924,[188] before it had been disproved by the Bothe-Geiger and Compton-Simon experiments. (We have already seen that Swann helped conceive the latter experiment.) His point of departure was the suggestion, therefore, that the electrons in an atom should be "replaced by virtual linear oscillators in the sense prescribed by Bohr, Kramers and Slater," although the "use of these virtual oscillators must, of course, be regarded as a temporary expedient designed to enable us to proceed in our speculations without awaiting a complete solution of the story of how the atoms influence each other." He assumed that during a transition, "a quantum, in the form of a ball if we like, is emitted from the electron, and let us suppose that its probability of emission in a fixed element of time, at any instant, and from any element of surface, is proportional to the magnitude, at that time and place, of the vector, which in classical theory would be the Poynting vector associated with the *irreversible radiation field* of the virtual oscillator, the magnitude of the quantum being, of course $h\nu$. . . . Let us suppose that the quantum then follows the path of the Poynting vector emanating from its point of emission. . . ." If the quantum subsequently "comes into the vicinity of one of the electrons of a virtual oscillator" of another atom, it may be "caught" by the electron and induce a transition, whose effects "at least in the limiting case of ionization, are the things physically observable as photographic, photoelectric or visual effects."[189]

These ideas, Swann recognized, were "related in some degree" to de Broglie's and to Slater's, both of which he learned about only after arriving at his own.[190] Regarding their "practical aspects," all interference phenomena "follow as in the classical theory, except that there is no interference of quanta. At the places where the classical theory indicates darkness there is darkness on the present view because no quanta succeed in getting there. Polarization phenomena do not invoke any polarization in the quanta themselves, but are provided for by the polarization of the vectors E and H as calculated by the classical process." The Compton effect could be explained "by supposing that the energy of the original quantum, having been received by the system, becomes available for total or partial reemission in varied amounts. . . ." The probability of emission of the secondary quantum from its virtual oscillator was to be determined by the corresponding Poynting vector, and the quantum's magnitude was to be "determined by the frequency of the wave emission" from a given point on the surface of the oscillating electron (virtual oscillator). The quantum, of course, had to be endowed "with the characteristics of momentum in such a way that the energy and direction of emission of the electron associated with the scattering follows as in Compton's calculations." The whole picture, in Swann's view, provided "a physical interpretation" of a transition probability, and was consistent with the instantaneous emission of quanta and electrons from atoms.

We have already seen that E. O. Lawrence and J. W. Beams, working under Swann's direction at Yale, took their experiments to confirm Swann's (as well as Thomson's) theory.[191] These experiments may have been one of the reasons, in fact, that Swann espoused essentially the same reconciliation of quanta and waves at least as late as October 1929, when he delivered an address before the Optical Society of America at the Thomas Young Memorial Meeting at Cornell University.[192] He argued for his old ideas in spite of the fact that by this time he recognized that they involved a certain "rather alarming and some will say, even fantastic" conception of the conservation of Poynting flux in the vicinity of an absorbing or radiating electron. He had no doubt, however, that everything could be consistently formulated, even if some of the requirements would "rather grate upon that particular region of sensitivity in our intellectual makeup, the region where resides what we call physical intuition."[193] It is significant, with respect to Swann's assessment of the state of affairs in optics at that time (he entitled his address "Contemporary Theories of Light") that while he discussed rather thoroughly both matrix mechanics and wave mechanics, he mentioned neither Heisenberg's uncertainty principle nor Bohr's principle of complementarity. Evidently Swann still believed in 1929 that it was possible to formulate a constructive theory of radiation.

11. Lewis's Direct Interparticle Interaction Theory

Quite likely the most original proposal of all to resolve the paradoxes in radiation theory was published in 1925 by the American physical chemist, G. N. Lewis, who spent most of his professional life at Berkeley. Lewis's interest in radiation theory grew out of his conviction, an unusual one in 1925, that time is symmetric:

> If we should consider any one of the elementary processes which are occurring in a system at equilibrium, and could, let us say, obtain a moving-picture film for such a process, then this film reeled backward would present an equally accurate picture of a reverse process which is also occurring in the system and with equal frequency. Therefore in any system at equilibrium, time must lose the unidirectional character which plays so important a part in the development of the time concept. In a state of equilibrium there is no essential difference between backward and forward direction in time, or, in other words, there is complete symmetry with respect to past and future.[194]

The first conclusion Lewis drew from his belief in the symmetry of time was that there must exist in physics a "principle of detailed balancing," as it came to be called, or a "principle of entire equilibrium," as Lewis preferred to call it. He preferred the latter designation largely because it encompassed both radiative and non-radiative microphysical processes. Its validity in 1925, however, was still controversial, and hence Lewis felt constrained to point out that: "I believe that the law of entire equilibrium will ultimately be recognized as one of the most fundamental of natural laws, and that whatever consequences may be drawn from it must be accepted, even though they contradict well established beliefs."[195]

To Lewis one of the most important of those "well established beliefs" was Einstein's concept of "induced emission," because he believed that while "induced absorption" had an exact inverse process, namely "spontaneous emission," "induced emission" had none. Lewis's conclusion, however, was immediately contested privately and in print by R. C. Tolman[196] and R. H. Fowler,[197] the latter pointing out the relevance of Van Vleck's recent work to the question. Einstein himself, after Lewis had initiated a correspondence with him,[198] pointed out Lewis's error[199] by showing that all three terms are necessary to obtain complete reversibility of the energy exchanges. Hence, Einstein in effect proved that nothing in his derivation of Planck's law contradicted either the principle of detailed balancing or the symmetry of time. Lewis, however, remained skeptical on this point, and even five years later he attempted to avoid Einstein's "induced emission" concept by seeking alternate derivations of Planck's law.

Although some months earlier he had actually calculated the cross section of a quantum, it was not until the end of 1925 that Lewis brought his belief in the symmetry of time and in the principle of entire equilibrium to bear directly on the nature of radiation. The occasion Lewis chose to explain his ideas was in his Silliman Lectures at Yale University,[200] and in these lectures, as J. H. Hildebrand wrote, he did "his best to shock scientific prejudices in several fields."[201] In radiation theory, Lewis first dismissed Swann's approach (with Swann probably in the audience), and then presented his own—a highly original interpretation based on the special theory of relativity, a subject with which Lewis had been thoroughly familiar since 1909, when he became the first American scientist to speak on this topic before the American Physical Society.[202] In essence, Lewis suggested that an emitting and an absorbing atom should be viewed as coupled together through their relativistic light cones in a completely symmetric way—". . . that it is as absurd to think of light emitted by one atom regardless of the existence of a receiving atom as it would be to think of an atom absorbing light without the existence of light to be absorbed."[203]

Lewis emphasized two very striking features of his picture. First, the complete symmetry between emitter and absorber involves an apparent causal paradox: in precisely the same way that one ordinarily thinks of an event at a given time as altering the course of future events, so one now has to think of an event at a given time as altering the course of past events. Second, the entire picture was readily understandable in terms of a Minkowskian space–time diagram, where the space–time locus of the absorbing atom is simply a point on the light cone of the emitting atom, for which the four-dimensional distance $ds = 0$. "This is an idea," said Lewis, "of which much use has been made in the mathematics but none in the physics of relativity. . . . The proposals which I am making are tantamount to assuming that such a distance is also zero in a physical sense, and that [the] two atoms . . . may be said to be in *virtual contact* at any two points . . . which are connected by singular lines [$ds = 0$]."[204] In other words, as he remarked elsewhere, "I may say that my eye touches a star, not in the same sense as when I say that my hand touches a pen, but in an equally physical sense."[205] To test his theory, and to distinguish it from Poynting vector theories such as Swann's or Slater's, Lewis proposed a certain "crucial experiment" which he described in detail.[206]

Lewis's theory, while discussed approvingly by E. B. Wilson,[207] was incisively criticized both by R. C. Tolman, who repeatedly questioned the validity of Lewis's proposed experiment, and by Albert Einstein.[208] What was "sad" about Lewis's theory, Einstein wrote Lewis in August 1926, was that if one considered the case of light from an emitting atom *A* being reflected from a mirror *B* before striking an absorbing atom *C*, one could easily show that the two four-dimensional distances S_{AB} and S_{BC} could both be zero, while the direct four-dimensional distance S_{AC} was *not* in general zero. This meant that the "degree of coupling" so to speak between emitter and absorber depended upon whether or not the light was reflected from a mirror, a highly unsatisfactory feature of Lewis's theory.

Lewis took Einstein's criticism very seriously,[209] for in spite of a published promise to the contrary, he never pursued the consequences of his theory further. Instead, his researches took another turn. He began developing an "extended thermodynamics," one main object of which was to derive Planck's law by postulating the conservation of "a new kind of atom," "an identifiable entity, uncreatable and indestructible, which acts as the carrier of radiant energy and, after absorption, persists as an essential constituent of the absorbing atom until it is later sent out again bearing a new amount of energy." It was for this conserved "atom" that Lewis, in December 1926, coined a new name:

> It would seem inappropriate to speak of one of these hypothetical entities as a particle of light, a corpuscle of light, a light quantum, or a light quant, if we are to assume that it spends only a minute fraction of its existence as a carrier of radiant energy, while the rest of the time it remains as an important structural element within the atom. It would also cause confusion to call it merely a quantum, for later it will be necessary to distinguish between the number of these entities present in the atom and the so-called quantum number. I therefore take the liberty of proposing for this hypothetical new atom, which is not light but plays an essential part in every process of radiation, the name *photon*.[210]

Clearly, Lewis's "photon" is not the modern photon. Indeed, by modern standards the properties Lewis attached to his photon were most unusual, especially the one which he regarded as most important: that in "any isolated system the total number of photons is constant." Since Lewis's "conservation of photons" is in direct conflict with a key assumption in Bose's derivation of Planck's law, it seems that Lewis was either unaware of this important earlier work, or he attached no significance to it. A similar fate befell Lewis's photon concept, for while his *term* was promptly incorporated into physics, particularly through its free use at the fifth Solvay Conference in Brussels ten months later, his photon *concept* promptly disappeared. His theoretical ideas on the nature of radiation were scarcely mentioned at the fifth Solvay conference.[211]

In contrast, Lewis's original theory of radiation, or at least the key element in it, did not die—in spite of Einstein's criticism. Two decades later, Wheeler and Feynman saw that the emitting and absorbing atoms should be coupled in a *probabilistic* sense, and they used this as their point of departure for their theory of direct interparticle interaction. In their judgment: "Lewis went nearly as far as it is possible to go without explicitly recognizing the importance of other absorbing matter in the system, a point touched upon by Tetrode, and shown . . . [by us] to be essential for the existence of the normal radiative mechanism."[212]

Similarly, Lewis's profound conviction that time is symmetric—the fundamental point of departure for all his work on radiation theory—took on entirely new significance after Lee and Yang's discovery of the nonconservation of parity in 1956, a decade after Lewis's death. This discovery led to a thorough re-examination of symmetry principles in physics; and the question of whether or not time is symmetric, which Lewis raised and pursued so persistently, generally in the face of skepticism and disinterest, is currently of major concern to physicists.[213]

12. Compton's Ideas on the Nature of Radiation

It is natural and appropriate to conclude our preliminary survey of post-Compton effect constructive theories of radiation by asking what Compton himself believed after 1923 regarding the ultimate nature of radiation. This question cannot be answered precisely, since Compton never formulated a quantitative theory; rather, one can only describe his attitude qualitatively, as it appears in his technical and popular writings.

Compton described the major lesson he learned from his and Simon's experiment of mid-1925 as follows:

> Since other experiments have shown that these scattered x-rays can be diffracted by crystals, and are thus subject to the usual laws of interference, there is no reason to suppose that other forms of radiant energy possess an essentially different structure. It thus becomes highly probable that all electromagnetic radiation is constituted of discrete quanta proceeding in definite directions. It is not impossible to express this result in terms of waves if we suppose that a wave train possessing a single quantum of energy can produce an effect only in a certain predetermined direction.[214]

Compton's ideas on the nature of the relationship between waves and quanta became clearer in a 1925 article in the *Scientific American*.[215] In the first place, he felt "that most physicists look forward to a final solution of the problem of the nature of light in some combination of the wave and quantum theories." Just what this "combination" might turn out to be was "very obscure," but Compton offered the following assessment of the current situation:

An attempt at the reconciliation of these two viewpoints, which seems to be growing in favor, is to suppose that electric waves are not ether waves, but successive sheets of radiation quanta, somewhat as one sometimes sees sheets of drops in rainstorms. Where according to the usual theory the electric field is strong, there will be a concentration of quanta, and there quanta will be moving forward just as the wave does. Also there will be some quanta directed upward and some down, just as part of the electric field in the wave is directed upward and part down. From a 1,000-watt broadcasting station of 300 meters wavelength, it can be shown that, at a distance of 1,000 kilometers, there would be more than 400 such quanta in each cubic centimeter of the wave. Thus, the wave would be to all appearances continuous, though actually made up of discrete particles, just as water seems continuous though made up of molecules.[216]

This marriage of particle and wave, as Compton put it two years later (1927), "confronted the dilemma of having before us a convincing evidence that radiation consists of waves, and at the same time that it consists of corpuscles."[217] By this time, of course, Bohr had enunciated his principle of complementarity, first at the Volta Conference (Como, September 16, 1927) and then at the Solvay Conference (Brussels, October 24–29, 1927)—with Compton in both audiences. That Compton was fully aware of Bohr's principle was shown a short time later when he wrote: "We need not think of these two views as necessarily alternative. It may well be that the two conceptions are complementary. Perhaps the corpuscle is related to the wave in somewhat the same manner that the molecule is related to matter in bulk; or there may be a guiding wave which directs the corpuscles which carry the energy."[218]

Therefore, while fully aware of Bohr's principle of complementarity, Compton did not embrace its meaning in the full sense intended by Bohr. He was still hoping and searching for a reconciliation of some type. In fact, there was a distinct asymmetry in the functions which Compton assigned to waves and to quanta. He emphasized, first, that it was "unescapable that radiation consists of directed quanta of energy, *i.e.,* of photons, and that energy and momentum are conserved when these photons interact with electrons or atoms." He then added:

Let me say again that this result does not mean that there is no truth in the concept of waves of radiation. The concluson is rather that the energy is not transmitted by such waves. The power of the wave concept in problems of interference, refraction, etc., is too well known to require emphasis. Whether the waves serve to guide the photons, or whether there is some other relation between photons and waves is another and a difficult question.[219]

Several times during the next few years Compton reiterated this distinction. At the end of 1928, agreeing with Einstein that "light remains as perhaps the darkest of our physical problems,"[220] Compton conjectured that probably "as far as we can go" is to observe that if "light consists of waves, it should act as waves do; and if it consists of corpuscles it should act as do corpuscles."[221] One "suggested solution," which was derived from wave mechanics, was to accept the "mathematical proof that the dynamics of a particle may be expressed in terms of the propagation of a group of waves," but even here the asymmetry between waves and quanta could not be avoided: "It is a corollary that the energy of the radiation lies in the photons, and not in the waves. For we mean by energy the ability to do work, and we find that when radiation does anything it acts in particles." One therefore had to "seize both horns of the dilemma": "In reply to our questions, what is light? the answer seems to come, waves and particles, light is both."[222]

In March 1929, again in a *Scientific American* article, Compton, whose deep-seated and lasting commitment to the electromagnetic theory is by now evident, summarized the recent developments in radiation theory, to which he himself had contributed so fundamentally, as follows:

> For centuries it has been thought that the corpuscular and wave conceptions are contradictory; but when we are confronted with apparently convincing evidence that light consists of particles, the two conceptions must in some way be reconcilable. The theoretical physicists are hard at work on a reconciliation of the two theories. One suggestion is that the energy of radiation is carried by the particles and that the waves serve merely to guide the particles. According to the second view, the particles of radiation exist in any true sense only when the radiation is acting on atoms or electrons, and that between such events the radiation moves as waves. These ideas, however, are difficult to state in any satisfactory form.
>
> Perhaps the best picture that one can give of the relation between waves and particles is the analogy of sheets of rain which one sometimes sees in a thunderstorm. We may liken the waves to the sheet of rain that one sees sweeping down the street or across the fields. The radiation particles or photons would correspond to the rain drops of which the sheet is composed.[223]

He closed on an undisguised theological note: "Having carried the analysis of the universe as far as we are able, there thus remain the proton, the electron, and the photon—these three. And, one is tempted to add, the greatest of these is the photon, for it is the life of the atom."[224]

D. Bohr's Principle of Complementarity (1927)

1. Its Enunciation and Consequences for Radiation Theory

The preceding attempts to reconcile waves and quanta—and future research may well disclose still more—adequately demonstrate that the hope of a great many physicists in the years immediately following the discovery of the Compton effect was to be able to formulate a satisfactory constructive theory of radiation. "The rivalry of classical wave theory and modernist quantum theory has of late years dominated the physics of radiation," wrote E. N. da C. Andrade in July 1927, "and in spite of all our efforts we can scarcely say that a really satisfactory solution is yet in sight."[225] The entire endeavor was profoundly frustrating, but the goal was clear. As it was put by Planck, who had "no doubt that finally a satisfactory solution will be found," the goal was to find a theory in which the wave and particle regions "will no longer appear fundamentally different from one another but will represent only the opposite ends of a single region which includes them both."[226]

All of the physicists who shared Planck's hope, however, were outside the mainstream of physics—if the mainstream is defined to be the ideas put forth by the "body of true believers," as Pauli characterized them,[227] who worked primarily in Copenhagen and Göttingen. For it was against the background of the diverse attempts previously described that Bohr consolidated his thinking on complementarity in early 1927, and first publicly presented his views in Como in the fall.[228] The purpose of his considerations, he told his audience, was to make some "general remarks regarding the principles underlying the description of atomic phenomena, which I hope may help to harmonise the different views, apparently so divergent, concerning this subject." His salient point was that: "The very nature of the quantum theory . . . forces us to regard the space–time co-ordination and the claim of causality, the union of which characterises the classical theories, as complementary but exclusive features of the description, symbolising the idealisation of observation and definition respectively."[229]

As regards light, its propagation in space and time is adequately expressed by the electromagnetic theory. Especially the interference phenomena *in vacuo* and the optical properties of material media are completely governed by the wave theory superposition principle. Nevertheless, the conservation of energy and momentum during the interaction between radiation and matter, as evident in the photoelectric and Compton effect, finds its adequate expression just in the light quantum idea put forward by Einstein. As is well known, the doubts regarding the validity of the superposition principle on one hand and of the conservation laws on the other, which were suggested by this apparent contradiction, have been definitely disproved through direct experiments. . . . The two views of the nature of light are rather to be considered as different attempts at an interpretation of experimental evidence in which the limitation of the classical concepts is expressed in complementary ways.[230]

Bohr added that an "important contribution to the problem of a consistent application of these methods has been made lately by Heisenberg," in stressing "the peculiar reciprocal uncertainty which affects all measurements of atomic properties."[231]

In May of the following year, J. C. Slater, recognizing that "the connections between quanta and light waves . . . no doubt are familiar to many persons," explicitly and quantitatively illustrated how Heisenberg's uncertainty principle precludes the possibility of carrying out an experiment to simultaneously detect the wave and particle natures of radiation or matter.[232] He summarized the relationship between waves and quanta that had emerged from the work of Schrödinger, Heisenberg, Born, and others by observing that to

> . . . solve the problem of the motion of quanta . . . , one solves the optical wave equation, and the intensity of the resulting wave at any point measures the probability of the existence of a quantum at that point. This of course is a connection between waves and quanta frequently proposed before the development of the wave mechanics. Previously, it had to be admitted that this theory was incomplete, in that it did not precisely define the paths of the individual quanta; but now it is seen that this indefiniteness is just what one would expect from Heisenberg's principle of indeterminateness.[233]

In those experiments, such as the Compton effect, "where a definite localization of quanta seems to be observed, this localization is to be described, to the accuracy with which it can be observed, by setting up wave packets."[234] One error, however, was "particularly to be avoided. This is the supposition that our wave packet is a dynamical 'model' of a light quantum, or that it can give any information about the 'dimensions' of a quantum. Such questions are entirely foreign to the theory; the dimensions of the wave packet . . . are derived entirely from the initial conditions of the problem, as regulated by experimental circumstances."[235]

2. *The Meaning of Complementarity as Conveyed Through Metaphors and Similes*

The Copenhagen interpretation of quantum mechanics, as embodied in Heisenberg's uncertainty principle and Bohr's principle of complementarity, with their probabilistic aspects,[236] was soon accepted by a number of physicists and physical chemists, among them E. U. Condon,[237] A. Sommerfeld,[238] Irving Langmuir,[239] and G. P. Thomson[240]—and ultimately by the majority of physicists. For these physicists, "the dualistic nature of radiation," in E. C. Kemble's words, "is a brute fact to be described rather than explained or exorcised."[241] Indeed, it is rather striking that in order to actually describe this "brute fact," in order to convey its primary meaning both to each other and to laymen, physicists who accepted it tended to take refuge in metaphors and similes—a variety of which were invented very early. Thus, even before Bohr spoke of complementarity, C. D. Ellis compared the wave and particle theories to "the plan and elevation of an engineering drawing. The wave-theory and the quantum-theory may be just the plan and elevation of the true theory of light, and the marked contradictions between them may be due to the language we use to describe the experimental results, just as the differences between a plan and elevation reflect our attempt to describe in two dimensions a three-dimensional object."[242] C. J. Davisson, in May 1928, summarized the particle behavior of light and the wave behavior of electrons in a humorous vein, suggesting that the confusion between wave and particle might be comparable to confusing the behavior of a rabbit and a cat.[243] And Paul R. Heyl in 1929, quoting Sir Oliver Lodge, wrote that the wave and particle concepts "are like a shark and a tiger, each supreme in its own element and helpless in that of the other."[244]

Some of the most long-lived metaphors and similes of all were described in a 1927 lecture by H. S. Allen.[245] Allen noted that the wave and particle theories constitute two separate edifices which have been built up over the years, and then continued:

> Up to the present no one has bridged the gulf between these two buildings. Many attempts have been made to build a bridge, but the keystone of the arch has not been fitted. Physicists are obliged to live sometimes in one building, sometimes in the other. We use either hypothesis according to the nature of the problem that we have under consideration, or, as Sir William Bragg expressed it in his presidential address to the British Association at Glasgow:
>
> "On Mondays, Wednesdays, and Fridays we adopt the one hypothesis, on Tuesdays, Thursdays, and Saturdays the other. We know that we cannot be seeing clearly and fully in either case, but are perfectly content to work and wait for the complete understanding."[246]

In line with this, "we are assured by Bohr that we have to deal not with *contradictory* but with *complementary* pictures." "Perhaps," Allen suggested, "Bohr's latest work may be regarded as an attempt to dig an underground passage between the two [buildings], but the tunnel is dark and gloomy, and the atmosphere scarcely fit for human respiration." We might, however, wish to adopt Alice's solution in *Through the Looking-Glass,* that is, that Tweedledee and Tweedledum "live in the same house." But we had better, concluded Allen, "abandon the simile of the house and try another analogy. We may liken the 'complementary' theory of Bohr to a see-saw on which Tom Particle and Mary Wave are so evenly balanced that a touch will send one end of the plank up or down. If we attempt to fix one end to mother earth, the other is suspended in mid-air."[247]

In 1929 E. C. Kemble asked: "What is the upshot of the battle between the wave theory of light and the corpuscular theory? The answer is, 'A deadlock.' " Kemble then went on to explain what he meant in words that partly would, and partly would not, be accepted by most physicists today:

> Of course, the existence of two apparently conflicting sets of characteristics for radiation has been a commonplace for many years and to many physicists the adoption of a dualistic point of view as the starting point for a fresh attack on the fundamental problems of physics will seem an evasion of the fundamental question, *"Why* does light act in some respects like an assemblage of corpuscles and in other respects like a spreading wave phenomenon?" We assert, however, that in the last analysis the function of theoretical physics is to describe rather than to explain. Science seeks to interpret the infinitely complex world of direct experience as the outcome of fundamentally simple laws. The reduction of complexity to simplicity is the goal, and when it is attained, we prove that order underlies chaos and leave the question, "Why" still essentially untouched. Hence, discarding this question as ultimately unanswerable, we may address ourselves to the task of describing what we observe in the most compact manner possible. If the behavior of radiation can be at least approximately described by means of the dualistic point of view, its temporary adoption will be a step in advance. No claim of ultimate validity is made for the theory, however.[248]

It is Kemble's last sentences that would be disavowed today by most physicists, who generally accept the dualistic structure of radiation as a fundamental fact of Nature. This conviction was, of course, greatly strengthened in 1927–1928, when Dirac, Jordan and Klein, and Jordan and Wigner developed the method of second quantization,[249] which proved that the particle and wave pictures for radiation and matter are formally equivalent, and hence indicated that these pictures "are merely two different aspects of one and the same physical reality." Nevertheless, some physicists could never accept the wave-particle duality as the end of the quest, and thereby abandon hope for some overriding constructive theory of radiation. As everyone knows, foremost among these physicists was Albert Einstein, who relentlessly resisted complementarity throughout his life. Even in 1951, four years before his death, Einstein wrote to his old friend Michele Besso that: "All the fifty years of conscious brooding have brought me no closer to the answer to the question, 'What are light quanta?' " He added, however: "Of course today every rascal thinks he knows the answer, but he is deluding himself."[250]

E. Arthur Holly Compton: Nobel Laureate in Physics 1927

Most of the events described in this final chapter would not have occurred in the way they did, or at the time they did, if the Compton effect had not been discovered when it was. This discovery, therefore, truly represented a turning point in physics. Its far-reaching significance was formally recognized in December 1927 when Arthur Compton received the Nobel Prize for Physics. As was reported in *Nature:* "The technique of these experiments was so difficult that it was some time before the results were confirmed by other workers, and some controversy arose as to the genuineness of the effect. In the end, however, the Compton effect was finally established, and it stands to-day as the firmest individual piece of evidence in favour of the hypothesis of localised light quanta."[251] Appropriately, Compton shared the Prize with C. T. R. Wilson, whose cloud chamber experiments first provided direct visual evidence for Compton's recoil electrons.[252]

By a curious twist of fate, which exemplifies the twists and turns in the history of physics, and the differences in the perceptions of physicists, just as Bohr had argued against Einstein's light quantum hypothesis in his Nobel lecture in December 1922,[253] shortly after Compton had made his discovery, so Compton argued against a truly dualistic theory of light in his Nobel lecture in December 1927,[254] shortly after Bohr had enunciated his principle of complementarity. Such differences permeate physics, and are very much alive today.

F. References

[1] W. Heisenberg, "Die Entwicklung der Quantentheorie 1918–1928," *Naturwiss.* 17 (1929):491.

[2] K. M. G. Siegbahn in 1927 recognized this point also when he introduced Compton on the occasion of the latter's Nobel award. He remarked that the Compton effect was "now so important that, in the future, no atomic theory can be accepted that does not explain it and lead to the laws established by its discoverer." And Compton himself noted in his Nobel Lecture: "It is these changes in the laws of optics when extended to the realm of X-rays that have been in large measure responsible for the recent revision of our ideas regarding the nature of the atom and of radiation." See *Nobel Lectures: Physics,* Vol. 2 (Amsterdam: Elsevier, 1964), pp. 171 and 189.

[3] A. H. Compton, "The Scattering of X-Rays," *J. Franklin Inst.* 198 (1924):57; lecture delivered before Section B of the AAAS on December 28, 1923.

[4] A. Einstein, "Das Komptonsche Experiment," *Berliner Tageblatt* (April 20, 1924), 1. Beiblatt.

[5] A. Sommerfeld, *Atombau und Spektrallinien.* 4 Aufl. (Braunschweig: Friedrich Vieweg, 1924), pp. 56–59 and 758–763.

[6] The "Centennial of the Undulatory Theory of Light" was observed in Pasadena, California, at least, in 1926 when P. S. Epstein delivered an address of that title to the Astronomy and Physics Club of Pasadena. See *Science* 63 (1926):387–393.

[7] R. A. Millikan, "The Last Fifteen Years of Physics," *Proc. Am. Phil. Soc.* 65 (1926):76.

[8] H. A. Lorentz, "The Radiation of Light," *Nature* 113 (1924):608. Lorentz still held many of the same views at the time of his death. See "How Can Atoms Radiate?" *J. Franklin Inst.* 205 (1928):449–471.

[9] W. Pauli, "Über das thermische Gleichgewicht zwischen Strahlung und freien Elektronen," *Z. Phys.* 17 (1923):272–286. For further discussion and references to the literature see Martin J. Klein, "Einstein and the Wave-Particle Duality," *Nat. Phil.* Vol. 3 (New York: Blaisdell, 1964), pp. 22–24; and "The First Phase of the Bohr-Einstein Dialogue," in Russell McCormmach (ed.), *Historical Studies in the Physical Sciences* Vol. 2 (Philadelphia: University of Pennsylvania Press, 1970), p. 16.

[10] Compton, "Scattering," (reference 3) p. 69.

[11] A. Einstein and P. Ehrenfest, "Zur Quantentheorie des Strahlungsgleichgewichts," *Z. Phys.* 19 (1923):301–306.

[12] It was not, however, accepted without debate. See G. Breit, "Free electrons in black body radiation," (Abstract) *Phys. Rev.* 23 (1924):772, and the discussion of Ornstein and Burger's work in Chapter 7, section C.2.

[13] L. de Broglie, "Rayonnement noir et de lumière," *J. Phys. Radium* 3 (1922):422–428.

[14] See Max Jammer, *The Conceptual Development of Quantum Mechanics* (New York: McGraw-Hill, 1966), pp. 239–24.

[15] L. de Broglie, "A Tentative Theory of Light Quanta," *Phil. Mag.* 47 (1924):446. See also "Waves and Quanta," *Nature* 112 (1923):540. The former paper represents a summary statement of de Broglie's ideas on light quanta and matter waves which he published in three short notes published in the *Compt. Rend.* in the fall of 1923. See Jammer, pp. 243–245.

[16] Klein, "Wave-Particle Duality," (reference 9) pp. 26–46.

[17] Jammer, *Conceptual Development*, pp. 246–280.

[18] V. V. Raman and Paul Forman, "Why was it Schrödinger who Developed de Broglie's Ideas?" in Russell McCormmach (ed.), *Historical Studies in the Physical Sciences*, Vol. 1 (Philadelphia: University of Pennsylvania Press, 1969), pp. 291–314.

[19] W. Gordon, "Der Comptoneffekt nach der Schrödingerschen Theorie," *Z. Phys.* 40 (1926):117–133.

[20] E. Schrödinger, "Über den Comptoneffekt," *Ann. Physik.* 82 (1927):257–264; translated and reprinted in *Collected Papers on Wave Mechanics* (London: Blackie & Son, 1928), pp. 124–129. See also G. Wentzel, "Zur Theorie des Comptoneffekts," *Z. Phys.* 43 (1927):1–8 and 779–787. An carly attempt at a wave theoretical treatment was made by Carl Eckart in "The Wave Theory of the Compton Effect," *Phys. Rev.* 24 (1924):591–595.

[21] For a discussion of these theories and references to the literature see, for example, Compton's *X-Rays and Electrons* (New York: Van Nostrand, 1926), pp. 296–305, and his paper "Some Experimental Difficulties with the Electromagnetic Theory of Radiation," *J. Franklin Inst.* 205 (1928):172.

[22] O. Klein and Y. Nishina, "Über die Streuung von Strahlung durch freie Elektronen nach der neuen relativistischen Quantendynamik von Dirac," *Z. Phys.* 52 (1928):853–868.

[23] For discussions of these theories, as well as the state of the experimental situation and references to the literature, see the review articles by G. Wentzel, "Die Theorien des Compton-Effektes. I," *Phys. Z.* 26 (1925):436–454, and H. Kallmann and H. Mark, "Der Comptonsche Streuprozess," in *Ergebnisse der Exakten Naturwissenschaften,* Bd. 5 (Berlin: Springer, 1926), pp. 267–325. See also Y. H. Woo, "The Intensity of the Scattering of X-Rays by Recoiling Electrons," *Phys. Rev.* 25 (1925):444–451, and G. Breit, "A Correspondence Principle in the Compton Effect," *Phys. Rev.* 27 (1926):362–372.

[24] Wentzel, "Theorien," p. 438.

[25] Kallmann and Mark, "Comptonsche Streuprozess," p. 294.

[26] *Ibid.,* p. 299. See also Compton, "Experimental Difficulties," (reference 21) pp. 169–170.

[27] Professor Slater recently gave an account of his work in his paper "The Development of Quantum Mechanics in the Period 1924–1926," Report No. 297 (July 28, 1972), Quantum Theory Project for Research in Atomic, Molecular and Solid-State Chemistry and Physics, University of Florida, Gainesville. See also Jammer, *Conceptual Development,* (reference 14) pp. 182–195, and Klein, "First Phase," (reference 9) pp. 23–39. Other discussions of Slater's work and of the Bohr-Kramers-Slater paper were given, for example, by L. Rosenfeld in his "Men and Ideas in the History of Atomic Theory," *Arch. Hist. Exact Sci.* 7 (1971):69–90, especially pp. 75–81, and by Klaus M. Meyer-Abich in his *Korrespondenz, Individualität und Komplementarität* (Wiesbaden: Franz Steiner, 1965), especially pp. 102–133.

[28] N. Bohr, H. A. Kramers, and J. C. Slater, "The Quantum Theory of Radiation," *Phil. Mag.* 47 (1924):785–802. A German version appeared in the *Z. Phys.* 24 (1924):69–87.

[29] Heisenberg, "Entwicklung," (reference 1) p. 492. Heisenberg later made the more extravagant claim that the Bohr-Kramers-Slater paper represented the "first serious attempt to resolve the paradoxes of radiation in rational physics," but there are difficulties with this contention. See Klein, "First Phase," pp. 37–38.

[30] Darwin (Chapter 6, reference 15) held this view at least as early as 1919. In 1923 he reiterated it in his paper "The Wave Theory and the Quantum Theory," *Nature* 111 (1923):771–773.

[31] J. C. Slater, "Radiation and Atoms," *Nature* 113 (1924):307.

[32] *Ibid.,* pp. 307–308.

[33] Slater to Kramers, December 8, 1923, on deposit in the Archive for History of Quantum Physics (AHQP) in Philadelphia, Berkeley, and Copenhagen.

[34] Slater to Bohr, March 22, 1923, AHQP. Slater enclosed Lyman's and Kemble's letters of recommendations in this letter to Bohr.

[35] Slater, "Radiation and Atoms," p. 308.

[36] *Ibid.*

[37] Slater to van der Waerden, November 4, 1964. Quoted in B. L. van der Waerden (ed.), *Sources of Quantum Mechanics* (New York: Dover, 1968), p. 13.

[38] *Ibid.*

[39] J. C. Slater, "The Nature of Radiation," *Nature* 116 (1925):278.

[40] Bohr to Slater, January 10, 1925, AHQP.

[41] Darwin to Bohr, July 20, 1919, AHQP. This letter and its aftermath was discussed by Martin J. Klein in his "First Phase," (reference 9) pp. 20–22. In "The Wave Theory and the Quantum Theory," *Nature* 111 (1923):771, for example, Darwin wrote: " . . . I think there is no doubt that many physicists consider a breach of the law of conservation as a serious objection to any theory. If we are to believe at all in the wave theory it is much more reasonable to maintain the exact opposite."

[42] N. Bohr, "On the Application of the Quantum Theory to Atomic Structure, Part I: The Fundamental Postulates of the Quantum Theory," *Proc. Cambridge Phil. Soc.* (Suppl., 1924):42; also quoted in Klein, "First Phase," (reference 9) p. 22. Bohr of course took Einstein's 1917 ideas to indicate that one had to be content with "considerations of probability" (reference 28, p. 788), a point of view that Einstein was unwilling to adopt.

[43] Bohr to Michelson, February 7, 1924, AHQP.

[44] The only equation in the entire paper (p. 788) was $h\nu = E_1 - E_2$.

[45] Bohr, Kramers, and Slater, (reference 28) pp. 799–800.

[46] A. H. Compton, "Experimental Difficulties," (reference 21) p. 171.

[47] A. H. Compton, "Scattering," (reference 3) p. 71.

[48] A. Sommerfeld, "Grundlagen der Quantentheorie und des Bohrschen Atommodelles," *Naturwiss.* 12 (1924):1049.

[49] Pauli to Bohr, February 21, 1924, AHQP.

[50] Pauli to Bohr, October 2, 1924, AHQP.

[51] Klein, "Wave-Particle Duality," (reference 9) pp. 25–26, and "First Phase," (reference 9) pp. 32–33; Einstein's letter to Ehrenfest is quoted in the latter reference on page 33. Klein also notes that in April 1924, Einstein also expressed his qualms to Hedwig Born, Max Born's wife.

[52] Pauli to Bohr, October 2, 1924, AHQP.

[53] J. D. van der Waals, Jr., "Remarques relatives à des questions du domaine de la théorie des quanta," *Arch. Néerl.* 8 (1925):300; Klein, "First Phase," p. 29.

[54] C. D. Ellis, "The Light-Quantum Theory," *Nature* 117 (1926):896.

[55] E. C. Stoner, "The Structure of Radiation," *Proc. Cambridge Phil. Soc.* 22 (1925):577–591.

[56] *Ibid.*, pp. 581–582. Stoner's detailed arguments for directed quanta may be found on p. 580.

[57] Heisenberg to Bohr, January 8, 1925, AHQP.

[58] Fowler's lecture, which was delivered in May 1925 and which is on deposit in the AHQP, clearly reveals his acceptance of Bohr's statistical conservation ideas.

[59] R. Becker, Über Absorption und Dispersion in Bohrs Quantentheorie," *Z. Phys.* 27 (1924):173–188; Klein, "First Phase," p. 29.

[60] J. C. Slater, "A Quantum Theory of Optical Phenomena," *Phys. Rev.* 25 (1925):402, footnote 9.

[61] P. Jordan, "Zur Theorie der Quantenstrahlung," *Z. Phys.* 30 (1924):297–319; Klein, "First Phase," (reference 9) p. 30.

[62] A. Einstein, "Bermerkung zu P. Jordans Abhandlung 'Zur Theorie der Quantenstrahlung'," *Z. Phys.* 31 (1925):784–785.

[63] Schrödinger to Bohr, May 24, 1924, AHQP.

[64] Schrödinger, "Bohrs neue Strahlungshypothese und der Energiesatz," *Naturwiss.* 12 (1924):720–724.

[65] *Ibid.*, p. 724. Kramers began drafting a long paper in response to Schrödinger's qualms which is on deposit in the AHQP. His point of view regarding the point made in the next paragraph was expressed in a letter to Sommerfeld on September 6, 1924 (AHQP): "That . . . we renounce the strict preservation of the energy principle shows our striving for a description of the processes without making any assumptions for which observations provide no basis." For Schrödinger's discussion of the "Vedantic Vision," see his *My View of the World* (Cambridge: Cambridge University Press, 1964), pp. 18–22.

[66] C. D. Ellis, "Theory," (reference 54) p. 896.

[67] W. Bothe and H. Geiger, "Ein Weg zur experimentellen Nachprüfung der Theorie von Bohr, Kramers und Slater," *Z. Phys.* 25 (1924):44. Bothe's next paper on the subject dealt with "Die Polarisation der gestreuten Röntgenstrahlen," and was published in the *Z. Phys.* 31 (1925):24–25.

[68] W. Bothe and H. Geiger, "Experimentelles zur Theorie von Bohr, Kramers und Slater," *Naturwiss.* 13 (1925):440–441.

[69] W. Bothe and H. Geiger, "Über das Wesen des Comptoneffekts: ein experimenteller Beitrag zur Theorie der Strahlung," *Z. Phys.* 32 (1925):639–663.

[70] Born to Bohr, January 1, 1925, AHQP.

[71] Bohr to Born, January 18, 1925, AHQP.

[72] Bothe and Geiger, "Wesen," (reference 69) pp. 662–663.

[73] Geiger to Bohr, April 17, 1925, AHQP.

[74] Bohr to Geiger, April 21, 1925, AHQP.

[75] Bohr to Fowler, April 21, 1925, AHQP. Heisenberg still did not know Bohr's attitude toward the Bothe-Geiger experiment by May 16—see his letter to Bohr of that date deposited in the AHQP.

[76] A. H. Compton and A. W. Simon, "Directed Quanta of Scattered X-Rays," *Phys. Rev.* 26 (1925):290, footnote 6. See also Compton's *X-Rays and Electrons,* (reference 21) p. 282, footnote 1. Compton regarded this experiment of his and Simon's as a "crucial test" between his quantum theory of scattering and the Bohr-Kramers-Slater theory.

[77] A. H. Compton and A. W. Simon, "Measurements of β-rays Associated with Scattered X-Rays," *Phys. Rev.* 25 (1925):107 (Abstract) and 306–313. For a discussion of related work, for example, by L. Meitner, see Kallmann and Mark's review (reference 23).

[78] *Ibid.*, p. 313.

[79] Compton and Simon, "Directed Quanta," pp. 289–299. Most of Compton's and Simon's photographic plates are still preserved in the Washington University Archives.

[80] *Ibid.*, pp. 298–299. For some later work of Compton's in India, see "An Attempt to find a Unidirectional effect of x-ray Photons" (with K. N. Mathur and H. R. Sarna), Phys. Rev. 31 (1928):159; *Indian J. Phys.* 3 (1929):463–466.

[81] In "Ein Versuch zur Strahlungsstatistik," *Naturwiss.* 14 (1926):321, and in "Über die Kopplung zwischen elementaren Strahlungsvorgängen," *Z. Phys.* 37 (1926):547–567, Bothe carried out essentially the inverse of the Bothe-Geiger experiment by proving that *no* coincidences were produced in point counters placed on either side of a thin copper foil emitting fluorescent K radiation. See also "Lichtquanten und Lichtwellen," *Naturwiss.* 14 (1926):1280–1281. In "Lichtquanten und Interferenz," *Z. Phys.* 41 (1927): 333–344, Bothe concluded that Breit's (1923) and Jordan's (1924) doubts about Einstein's 1917 conclusion that quanta possess momentum were not valid (p. 341, footnote 1); see also "Zur Statistik der Hohlraumstrahlung," *Z. Phys.* 41 (1927):345–351. Other work was carried out, for example, by R. D. Bennett, "An Attempt to test the Quantum Theory of X-Ray Scattering," *Proc. Natl. Acad. Sci.* 11 (1925):601–602; A. J. Dempster and H. F. Batho, "Light Quanta and Interference," *Phys. Rev.* 30 (1927):644–648; and D. M. Dennison, "A Proposed Experiment on the Nature of Light," *Proc. Natl. Acad. Sci.* 14 (1928):580–581 (see also the letter of Kramers to Dennison, June 25, [1928 ?], AHQP). For references to the work of Compton himself, Jauncey, F. Bubb, P. A. Ross, and other physicists, see Compton's *X-Rays and Electrons* (reference 21) and other reference works.

[82] See for example A. Landé, "Neue Wege der Quantentheorie," *Naturwiss.* 14 (1926):458.

[83] H. A. Kramers, "The Law of Dispersion and Bohr's Theory of Spectra," *Nature* 113 (1924):673–676.

[84] M. Born, "Über Quantenmechanik," *Z. Phys.* 26 (1924):379–395.

[85] H. A. Kramers, "The Quantum Theory of Dispersion," *Nature* 114 (1924):310–311. See also G. Breit, "The Quantum Theory of Dispersion," *Nature* 114 (1924):310. Additional insight into Kramers' work of this period may be gained from his letter of September 6, 1924, to Sommerfeld and from Sommerfeld's response of October 1 (AHQP).

[86] J. H. Van Vleck, "The Absorption of Radiation by Multiply Periodic Orbits, and its Relation to the Correspondence Principle and the Rayleigh-Jeans Law. Part I: Some Extensions of the Correspondence Principle." *Phys. Rev.* 24 (1924):330–346; "Part II: Calculation of Absorption by Multiply Periodic Orbits," *Phys. Rev.* 24 (1924):347–365. Professor Van Vleck pointed out the great importance of research in midwestern universities at this time in his "Reminiscences of the First Decade of Quantum Mechanics," *Internat. J. Quantum Chem.* 5 (1971):3–20. His recollections are often at variance with the impression one gains from Stanley Coben's article "The Scientific Establishment and the Transmission of Quantum Mechanics to the United States, 1919–32," *Am. Hist. Rev.* 76 (1971):442–466.

[87] J. C. Slater, "Quantum Theory of Optical Phenomena," (reference 60) p. 401.

[88] Slater to Bohr, January 6, 1925, AHQP.

[89] Bohr to Slater, January 10, 1925, AHQP. Slater told E. C. Kemble of the unfavorable reaction from Copenhagen in two letters, written on June 26 and August 5, 1924, AHQP.

[90] H. A. Kramers and W. Heisenberg, "Über die Streuung von Strahlen durch Atome," *Z. Phys.* 31 (1925):681–708. Kramers' full commitment to the Bohr-Kramers-Slater program at this time is thoroughly illustrated in an unpublished typescript of February 1925 entitled, "On the Characteristics of Atoms in a Radiation Field," on deposit in the AHQP.

[91] W. Heisenberg, "Quantenmechanik," *Naturwiss.* 14 (1926):993. A similar statement by J. H. Van Vleck is quoted in Klein, "First Phase," (reference 9) p. 13.

[92] Einstein to Ehrenfest, August 18, 1925, quoted in Klein, "First Phase," p. 35.

[93] *Ibid.*, p. 36.

[94] J. C. Slater, "Nature of Radiation," (reference 39) p. 278.

[95] *Ibid.;* for Swann's paper see "The Trend of Thought in Physics," *Science* 61 (1925):425–435.

[96] Pauli to Kramers, July 27, 1925, AHQP.

[97] G. Mie, "Bremsstrahlung und Comptonsche Streustrahlung." *Z. Phys.* 33 (1925): 33–41; "Zur Theorie der Bremsstrahlung und der Comptonschen Streustrahlung," *Phys. Z.* 26 (1925):665–668. In the discussion following the latter paper, Smekal, Sommerfeld, and Bothe immediately criticized Mie's attempts to revise and extend the Bohr-Kramers-Slater program. Earlier, as indicated in a letter of December 9, 1924 (AHQP) which Mie wrote to Bohr, Mie was much more cautious about accepting the Bohr-Kramers-Slater theory. He continued to work on radiation problems for a number of years. In 1925, A. Landé also reexamined the nature of atomic transitions in his "Zur Quantentheorie der Strahlung," *Z. Phys.* 35 (1925):317–322, and concluded (p. 322, footnote 2), among other things, that an earlier theory of G. Wentzel, "Zur Quantenoptik," *Z. Phys.* 24 (1924):193–199, was in error. For other comments on Wentzel's and Bohr's theories, see R. de L. Kronig, "Zur Einseitigkeit der Quantenstrahlung," *Z. Phys.* 27 (1924):383–386. One can also remark that even in 1926 F. Zwicky concluded that the virtual oscillators had a physically real existence. See his "Quantum Theory and the Behavior of Slow Electrons in Gases," *Proc. Natl. Acad. Sci.* 12 (1926):461–466.

[98] Born to Bohr, April 24, 1925, AHQP. Bohr's reply to Born on May 1, 1925 (AHQP), was not encouraging, expressing views very similar to those which he had written to Geiger on April 21 (reference 74). Kramers was more reluctant to give up the Bohr-Kramers-Slater program, as is evident from a letter he wrote to Born dated May 13, 1925 (AHQP).

[99] Schrödinger to Sommerfeld, July 21, 1925. AHQP.

[100] E. Schrödinger, "Quantisierung als Eigenwertproblem," *Ann. Physik* 79 (1926): 361–376, received January 27, 1926. His remaining three papers, which were published in *Ann. Physik* 79 (1926):489–527, and *ibid.* 80, pp. 437–490, and 81, pp. 109–139, were received on February 23, May 10, and June 21, respectively. For an English summary, see "An Undulatory Theory of the Mechanics of Atoms and Molecules," *Phys. Rev.* 28 (1926):1049–1070.

[101] Dirac's account of how Schrödinger arrived at his discovery is given in Jammer, *Conceptual Development,* (reference 14) pp. 257–258.

[102] W. Heisenberg, "Über quantentheoretische Umdeutung kinematischer und mechanischer Beziehungen," *Z. Phys.* 33 (1925):879–893.

[103] M. Born and P. Jordan, "Zur Quantenmechanik," *Z. Phys.* 34 (1925):858–888.

[104] M. Born, W. Heisenberg, and P. Jordan, "Zur Quantenmechanik II," *Z. Phys.* 35 (1926):557–615. A week earlier (November 7, 1926), P. A. M. Dirac's paper, "The Fundamental Equations of Quantum Mechanics," *Proc. Roy. Soc. (London)* 109 [A] (1926):642–653, had been received for publication.

[105] Slater to Bohr, May 27, 1926, AHQP.

[106] Bohr to Slater, January 28, 1926, AHQP.

[107] See reference 1.

[108] G. Beck, "Comptoneffect und Quantenmechanik," *Z. Phys.* 38 (1926):144–148.

[109] P. A. M. Dirac, "Relativity Quantum Mechanics with Application to Compton Scattering," *Proc. Roy. Soc. (London)* 111 (1926):405–423; see also "The Compton Effect in Wave Mechanics," *Proc. Cambridge Phil. Soc.* 23 (1926):500–507. For a discussion of Dirac's fundamental paper on quantum electrodynamics, "The Quantum Theory of the Emission and Absorption of Radiation," *Proc. Roy. Soc. (London)* 114 (1927): 243–265, see, for example, Joan Bromberg, "Dirac's Quantum Electrodynamics and the Wave-Particle Equivalence," forthcoming in the Summer 1972 proceedings of the International School of Physics, "Enrico Fermi." See also Jagdish Mehra, " 'The Golden Age of Theoretical Physics': P. A. M. Dirac's Scientific Work from 1924 to 1933," in A. Salam and E. P. Wigner (eds.), *Aspects of Quantum Theory* (Cambridge: Cambridge University Press, 1972). E. Fermi discussed Dirac's Theory in his "Quantum Theory of Radiation," *Rev. Mod. Phys.* 4 (1932):87–132, especially pp. 109–112.

[110] Bohr published his reaction in his "Atomic Theory and Mechanics," *Suppl. Nature* (December 5, 1925):845–852. For a German version, see "Atomtheorie und Mechanik," *Naturwiss.* 14 (1926):1–9. In the former, see especially p. 848.

[111] For further information on Einstein's endeavors, see Klein's papers in reference 9, as well as Russell McCormmach, "Einstein, Lorentz, and the Electron Theory," in Russell McCormmach (ed.), *Historical Studies in the Physical Sciences, Vol. 2* (Philadelphia: University of Pennsylvania Press, 1970), pp. 41–87.

[112] L. S. Ornstein and H. C. Burger, "Die Dimension der Einsteinschen Lichtquanten," *Z. Phys.* 20 (1924):345–350; "Zur Dynamik des Stosses zwischen einem Lichtquant und einem Elektron," *Z. Phys.* 20 (1924):351–357.

[113] *Ibid.*, p. 345. The exact value cited for the constant of proportionality varied from about 1/16 (Ornstein and Burger) to $3/4\pi$ (E. Marx, reference 117) to $1/\pi$ (R. J. Piersol, reference 116).

[114] *Ibid.*, pp. 356–357.

[115] W. Pauli, "Bemerkungen zu den Arbeiten 'Die Dimension der Einsteinschen Lichtquanten' and 'Zur Dynamik des Stosses zwischen einem Lichtquant und einem Elektron' von L. S. Ornstein und H. C. Burger," *Z. Phys.* 22 (1924):261–265.

[116] L. S. Ornstein and H. C. Burger, "Zusammenwirken von Lichtquanten und Plancksches Gesetz," *Z. Phys.* 21 (1924):358–365; "Lichtbrechung und Zerstreuung nach der Lichtquantentheorie," *Z. Phys.* 30 (1924):253–257. See also R. J. Piersol, "Cross-sectional area of an Einstein quantum" (Abstract) *Phys. Rev.* 27 (1926):509.

[117] Erich Marx, "Die Dimension der Einsteinschen Lichtquanten," *Z. Phys.* 27 (1924):248–253.

[118] K. Schaposchnikow, "Newtonsche Mechanik und Lichtquanten," *Z. Phys.* 30 (1924):228–230; V. S. Vrkljan, "Bemerkung zu der Arbeit von K. Schaposchnikow: 'Newtonsche Mechanik und Lichtquanten'," *Ibid.*, 35 (1925):495–498; K. Schaposchnikow, "Über Zusammenstösse von Lichtquanten," *Ibid.*, 33 (1925):706–709; "Zur Mechanik der Lichtquanten," *Ibid.* 36 (1926):73–80; "Ein neues Prinzip in der Dynamik der Lichtquanten," *Ibid.* 41 (1927):352–354; "Grundlagen einer Elektronen-und-Lichtquanten-Dynamik," *Ibid.* 41 (1927):927–930; V. S. Vrkljan, "Bemerkung zu der Arbeit von K. Schaposchnikow: 'Ein neues Prinzip in der Dynamik der Lichtquanten'," *Ibid.*, 43 (1927):516–518.

[119] See for example G. Beck, "Über einige Folgerungen aus dem Satz von der Analogie zwischen Lichtquant und Elektron," *Z. Phys.* 43 (1927):658–674; A. E. Ruark and H. C. Urey, "The Impulse Moment of the Light Quantum," *Proc. Natl. Acad. Sci.* 13 (1927):763–770; P. Jordan "Über die Polarisation der Lichtquanten," *Z. Phys.* 44 (1927):292–300; W. Anderson, "Über die Struktur der Lichtquanten," *Ibid.* 58 (1927):841–857.

[120] G. P. Thomson, "Test of a Theory of Radiation," *Proc. Roy. Soc. (London)* [A] 104 (1923):115–120.

[121] E. O. Lawrence and J. W. Beams, "On the Nature of Light," *Proc. Natl. Acad. Sci.* 13 (1927):207–212.

[122] *Ibid.*, p. 210.

[123] *Ibid.*, p. 212.

[124] G. Breit, "The Length of Light Quanta," *Nature* 119 (1927): 280–281.

[125] See reference 81.

[126] Ellis, "Light-Quantum Theory," (reference 54) p. 895.

[127] M. Planck, "The Physical Reality of Light-Quanta," *J. Franklin Inst.* 204 (1927):18; the German version appeared in *Naturwiss.* 15 (1927):529–531 (entire paper).

[128] H. A. Lorentz, "Radiation," (reference 8) p. 611. Kallmann and Mark also concluded their long review article (reference 23) with the remark: "How one can combine such different characteristics in a single picture . . . one can still not clearly explain on the basis of the present theory." (p. 323)

[129] Compton, "Scattering," (reference 3) p. 66.

[130] See Chapter 6, section F.2.

[131] P. S. Epstein and P. Ehrenfest, "The Quantum Theory of the Fraunhofer Diffraction," *Proc. Natl. Acad. Sci.* 10 (1924):133–139.

[132] *Ibid.*, pp. 136–137.

[133] R. A. Millikan, "Atomism in Modern Physics," *J. Chem. Soc.* 125 (1924):1417.

[134] P. Ehrenfest and P. S. Epstein, "Remarks on the Quantum Theory of Diffraction," *Proc. Natl. Acad. Sci.* 13 (1927):407.

[135] *Ibid.*, p. 408, footnote 1. See also G. E. M. Jauncey and A. H. Compton, "Coherence of the Reflected X-Rays from Crystals," *Nature* 120 (1927):549.

[136] W. Duane, "The General Radiation," *Science* 66 (1927):637–640.

[137] P. Jordan, "Bemerkung über einen Zusammenhang zwischen Duanes Quantentheorie der Interferenz und den de Brogliesсhen Wellen," *Z. Phys.* 37 (1926):376–382. Jordan cites Elsasser's earlier paper published in *Naturwiss.* 13 (1925):711. See also N. R. Campbell, "Philosophical Foundations of Quantum Theory," *Nature* 119 (1927):779 for a criterion of some of Jordan's ideas. For more recent comments on Duane's quantum condition, see Alfred Landé, "Quantum Fact and Fiction," *Am. J. Phys.* 33 (1965):123–127.

[138] J. H. Van Vleck, "Quantum Principles and Line Spectra," *Bull. Natl. Res. Counc.*, No. 54 (1926), p. 287. This is also quoted by Klein, in "First Phase," (reference 9) p. 35, where O. D. Chwolson's similar sentiments are also quoted.

[139] Edwin B. Wilson, "Some Recent Speculations on the Nature of Light," *Science* 65 (1927):271.

[140] Darwin, "Wave Theory and Quantum Theory," (reference 30) p. 771. See also Stoner, "Structure," (reference 55) p. 577, who held similar views.

[141] See references 13–15 for complete citations to de Broglie's papers discussed in this paragraph.

[142] De Broglie, "Tentative Theory," (reference 15) p. 447.

[143] *Ibid.*, pp. 449–453.

[144] *Ibid.*, p. 452.

[145] *Ibid.*, p. 454.

[146] W. Anderson, "A Consequence of the Theory of M. Louis de Broglie," *Phil. Mag.* 47 (1924):872; G. E. M. Jauncey, "De Broglie's Theory of the Quantum and the Doppler Principle," *Nature* 114 (1924):51. De Broglie's reply is appended to the latter paper, occupying pp. 51–52.

[147] L. de Broglie, "Researches sur la théorie des quanta," *Ann. Phys. (Paris)* 3 (1925):22–128.

[148] L. de Broglie, "Sur la possibilité de relier les phénomènes d'intérference et de diffraction à la théorie des quanta de lumière," *Compt. Rend.* 183 (1926):447–448; "Interference and Corpuscular Light," *Nature* 118 (1926):441–442.

[149] W. Bothe, "Lichtquanten und Lichtwellen," *Naturwiss.* 14 (1926):1280–1281.

[150] E. B. Wilson, "Speculations," (reference 139) p. 270.

[151] For complete references to Bateman's papers, see F. D. Murnaghan's obituary notice, "Harry Bateman 1882–1946," *Natl. Acad. Sci. Biog. Mem.* 25 (1949):241–249, and A. Erdélyi's obituary notice in *Roy. Soc. Obit. Not. (London)* 5 (1945–1948):591–618.

[152] H. Bateman, "On the Theory of Light-Quanta," *Phil. Mag.* 46 (1923):977–991.

[153] *Ibid.*, p. 980.

[154] *Ibid.*, p. 982.

[155] *Ibid.*, pp. 985–986.

[156] *Ibid.*, p. 989.

[157] H. S. Allen, "Faraday's 'Magnetic Lines' as Quanta—Part II," *Phil. Mag.* 48 (1924):429–445.

[158] H. S. Allen, "Light and Electrons," *Nature* 112 (1923):279.

[159] H. A. Senftleben, "Über eine Formulierung der elektromagnetischen Gesetze, welche eine Eingliederung der Quantentheorie gestatten könnte," *Z. Phys.* 31 (1925):627–636.

[160] F. Russell Bichowsky, "An electromagnetic theory of quanta," (Abstract) *Phys. Rev.* 25 (1925):244.

[161] L. V. King, "Gyromagnetic electrons and a classical theory of atomic structure and radiation," (Abstract) *Phys. Rev.* 27 (1926):804; *Nature* 118 (1926):354.

[162] G. Wataghin, "Versuch einer korpuskularen Theorie der Interferenz und Beugung," *Z. Phys.* 51 (1928):593–604.

[163] J. Stark, "Folgerungen aus der atomistischen Konstitution der Lichtenergie," *Ann. Physik.* 86 (1928):1037–1040; "Über den elementaren Vorgang der Emission und Absorption des Lichtes," *Ann. Physik.* 87 (1928):909–926. For other suggestions or speculations, see F. J. v. Wiśniewski, "Über eine mögliche Interpretation des elektromagnetischen Feldes des Lichtes," *Z. Phys.* 56 (1929):713–716; G. I. Pokrowski, "Welcher Natur ist die Ruhemasse der Lichtquanten?" *Z. Phys.* 57 (1929):566–569.

[164] J. H. Jeans, *Report on Radiation and the Quantum-Theory,* 2d ed. (London: Fleetway Press, 1924). Reviewed in *Nature* 113 (1924):701–702.

[165] J. H. Jeans, "Electric Forces and Quanta," *Suppl. Nature March* 7, 1952): 361–368. In general, Einstein would have had difficulty accepting the ideas Jeans ascribed to him in this lecture.

[166] J. H. Jeans, *Atomicity and Quanta* (Cambridge: Cambridge University Press, 1926). Reviewed in *Suppl. Nature* (June 12, 1926):46, and *Phys. Rev.* 28 (1926):1048.

[167] Jeans, *Atomicity,* pp. 57–58.

[168] *Ibid.,* pp. 60–62.

[169] Stoner, "Structure," (reference 55), p. 579.

[170] Review in *Phys. Rev.* (reference 166).

[171] Review in *Suppl. Nature* (reference 166).

[172] Stoner, "Structure," (reference 55) pp. 577–594.

[173] *Ibid.,* pp. 582–585.

[174] See Chapter 2, section B.1.

[175] Stoner, "Structure," pp. 585–586. See also W. S. Franklin, "The Significance and Scope of the Idea of Frequency in Physics," (Abstract) *Phys. Rev.* 29 (1927):362; *Science* 65 (1927):221–223.

[176] *Ibid.,* p. 587. Later weak light interference experiments were performed under the direction of A. J. Dempster at the University of Chicago (see reference 81).

[177] See reference 58.

[178] J. J. Thomson, "A Suggestion as to the Structure of Light," *Phil. Mag.* 48 (1924):737–747; the quotation is on p. 737.

[179] *Ibid.*, p. 740.

[180] *Ibid.*, p. 741.

[181] Wilson, "Speculations," (reference 139) p. 268.

[182] *Ibid.*

[183] E. T. Whittaker, "On the Adjustment of Sir J. J. Thomson's Theory of Light to the Classical Electromagnetic Theory," *Proc. Roy. Soc. (Edinburgh)* 46 (1925–1926):116–125; the quotation is on p. 117.

[184] *Ibid.*, pp. 118–123.

[185] Wilson, "Speculations" p. 270.

[186] J. J. Thomson, "The Structure of Light," *Phil. Mag.* 50 (1925):1181–1196.

[187] N. P. Kasterin, "J. J. Thomson's Model of a Light-Quantum," *Phil. Mag.* 2 (1926):1208–1212.

[188] Swann, "Trend of Thought" (reference 95).

[189] *Ibid.*, p. 433.

[190] *Ibid.*, footnote 6. Ellis, "Light Quantum Theory," (reference 54) p. 897, characterized these theories, in general, as theories which required the quanta to move along the "tramlines" defined by the Poynting vector.

[191] See section C.2, preceding.

[192] W. F. G. Swann, "Contemporary Theories of Light," *J. Opt. Soc. Am.* 20 (1930):484–523.

[193] *Ibid.*, p. 500.

[194] G. N. Lewis, "A New Principle of Equilibrium," *Proc. Natl. Acad. Sci.* 11 (1925):182–183.

[195] G. N. Lewis, "The Distribution of Energy in Thermal Radiation and the Law of Entire Equilibrium," *Proc. Natl. Acad. Sci.* 11 (1925):423.

[196] Tolman to Lewis, May 14, 1925; June 13, 1925; June 29, 1925; all in University of California, Berkeley, Archives. R. C. Tolman, "The Principle of Microscopic Reversibility," *Proc. Natl. Acad. Sci.* 11 (1925):436–439.

[197] Fowler to Lewis, June 6, [1925]; August 30, [1925]; in University of California, Berkeley, Archives.

[198] Lewis to Einstein, July 27, 1926; in University of California, Berkeley, Archives.

[199] Einstein to Lewis, August, 22, 1926, *Ibid.*

[200] G. N. Lewis, *The Anatomy of Science* (New Haven: Yale University Press, 1926).

[201] Joel H. Hildebrand, "Gilbert Newton Lewis," *Natl. Acad. Sci. Biog. Mem.* 31 (1958):223.

[202] G. N. Lewis and R. C. Tolman, "The Principle of Relativity, and Non-Newtonian Mechanics," *Proc. Am. Acad. Arts Sci.* 44 (1909):711ff.; *Phil. Mag.* 18 (1909):510ff.

[203] G. N. Lewis, "The Nature of Light," *Proc. Natl. Acad. Sci.* 12 (1926):24.

[204] Lewis, *Anatomy*, p. 132.

[205] G. N. Lewis, "Light Waves and Light Corpuscles," *Nature* 117 (1926):237.

[206] Lewis, "Nature of Light," p. 28.

[207] Wilson, "Speculations," (reference 139) p. 266. In addition, Wilson discussed some ideas of Gibbs; also ideas of Cox and Hubbard.

[208] Tolman to Lewis, June 3, 1926, University of California, Berkeley, Archives; R. C. Tolman and S. Smith, "On the Nature of Light," *Proc. Natl. Acad. Sci.* 12 (1926):343–347. Einstein to Lewis (reference 199).

[209] Lewis to Einstein, November 3, 1926, University of California, Berkeley, Archives.

[210] G. N. Lewis, "The Conservation of Photons," *Nature* 118 (1926):874. Compton echoed Lewis' words in his "Experimental Difficulties," (reference 21) pp. 156–157.

[211] See *Electrons et photons—Rapports et discussions du cinquième conseil de physique tenu à Bruxelles du 24 au 29 Octobre 1927 sous les auspices de l'Institut International de Physique Solvay* (Paris: Gauthier-Villars, 1928), pp. 139–140.

[212] J. A. Wheeler and R. P. Feynman, "Interaction with the Absorber as the Mechanism of Radiation," *Rev. Mod. Phys.* 17 (1945):159, footnote 10. See also "Classical Electrodynamics in Terms of Direct Interparticle Action," *Rev. Mod. Phys.* 21 (1949):425–433.

[213] For some of the issues involved, see, for example, T. Gold, (ed.), *The Nature of Time* (Ithaca, N. Y.: Cornell University Press, 1967).

[214] Compton and Simon, "Directed Quanta," (reference 76) p. 299.

[215] A. H. Compton, "Light Waves or Light Bullets?" *Sci. Am.* 133 (1925):246–247.

[216] *Ibid.*, p. 247.

[217] A. H. Compton, "X-rays as a Branch of Optics," in *Nobel Lectures,* (reference 2) p. 189.

[218] Compton, "Experimental Difficulties," (reference 21) p. 156.

[219] *Ibid.*, pp. 177–178.

[220] A. H. Compton, "What is Light?" *Sigma Xi Quarterly* 17 (1929):14.

[221] *Ibid.*, p. 18.

[222] *Ibid.*, p. 33. Compton expressed essentially the same views in his paper "The Corpuscular Properties of Light," *Rev. Mod. Phys.* 1 (1929):74–89; *Naturwiss.* 17 (1929):507–515.

[223] A. H. Compton, "What Things Are Made of—II," *Sci. Am.* 140 (1929):235–236.

[224] *Ibid.*, p. 236.

[225] See E. N. da C. Andrade's joint review of Compton's *X-Rays and Electrons* (New York: Van Nostrand, 1926) and K. K. Darrow's *Introduction to Contemporary Physics* (London: Macmillan, 1927), *Nature* 120 (1927):143.

[226] Planck, "Reality of Light-Quanta," (reference 127) p. 18. To avoid thinking of "the energy of light-quant as concentrated at separate points in space," Planck suggested (p. 18) that "there goes out from each light-quantum a kind of action-at-a-distance, and indeed not only at a distant place but also at a distant time, for according to the Theory of Relativity we cannot distinguish in this connection between space and time."

[227] See reference 96.

[228] N. Bohr, "The Quantum Postulate and the Recent Development of Atomic Theory," *Suppl. Nature* 121 (1928):580–590; see *Naturwiss.* 16 (1928):245–257, for a German version. Bohr's principle of complementarity was discussed in Jammer, *Conceptual Development*, pp. 345–361, and very thoroughly by Meyer-Abich (reference 27). There is also an extensive philosophical literature on it, which I have not attempted to analyze or cite here.

[229] *Ibid.*, p. 580.

[230] *Ibid.*, pp. 580–581.

[231] For Heisenberg's paper, see "Über den anschaulichen Inhalt der quantentheoretischen Kinematik und Mechanik," *Z. Phys.* 43 (1927):172–198. See also Jammer, *Conceptual Development* (reference 14), pp. 323–345.

[232] J. C. Slater, "Light Quanta and Wave Mechanics," *Phys. Rev.* 31 (1928):895–899.

[233] *Ibid.*, p. 896.

[234] *Ibid.*, p. 895.

[235] *Ibid.*, p. 898. See also E. H. Kennard, "Note on Heisenberg's Indetermination Principle," *Phys. Rev.* 31 (1928):344–348. Heisenberg himself discussed all of these questions in detail in his University of Chicago lectures in early 1929. See *The Physical Principles of Quantum Theory,* translated by C. Eckart and F. C. Hoyt (Chicago: University of Chicago Press, 1930).

[236] For a discussion of the contributions of M. Born and others, see Jammer, *Conceptual Development,* pp. 281–293.

[237] E. U. Condon, "Recent Developments in Quantum Mechanics," *Science* 68 (1928):193–195.

[238] A. Sommerfeld, "Einige grundsätzliche Bemerkungen zur Wellenmechanik," *Phys. Z.* 30 (1929):866–871, especially p. 870.

[239] Irving Langmuir, "Modern Concepts in Physics and their Relation to Chemistry," *Science* 70 (1929):385–396.

[240] G. P. Thomson, "Waves and Particles," *Science* 70 (1929):541–546. Thomson asserted (p. 544) that: "This dual aspect of things as waves and particles must be very fundamental in the world. There is little doubt that protons would show it also, though experimental proof has so far not been possible. There is even strong, though indirect, evidence that a completed atom has a wave as a whole as well as component waves for its individual electrons. One reason for regarding the duality as really fundamental is that it holds for such different things as electrons and quanta. For in spite of this one point of resemblance they are essentially different."

[241] E. C. Kemble, "The General Principles of Quantum Mechanics. Part I," *Rev. Mod. Phys.* 1 (1929):158. It may be worth mentioning that F. J. Selby in an unorthodox move suggested in his note on "The Quantum Postulate and Atomic Theory," *Nature* 121 (1928):828, that "Bohr's complementarity is rather a trinity, of which the third member is the 'conscious' observing mind. . . ."

[242] Ellis, "Light-Quantum Theory," (reference 54) p. 897. In an insightful pre-Bohr statement, Ellis remarked that "it is an important practical achievement that a complete

description can be given by means of two theories, providing the appropriate theory is used for each phenomenon." That our physical concepts may be strongly linguistically dependent has been recently emphasized by Yehuda Elkana in a series of lectures at the Summer 1972 International School of Physics, "Enrico Fermi" in Varenna.

[243] C. J. Davisson, "Are Electrons Waves?" *J. Franklin Inst.* 205 (1928):599.

[244] Paul R. Heyl, "The History and Present Status of the Physicist's Concept of Light," *J. Opt. Soc. Am.* 18 (1929):189.

[245] H. S. Allen, "The Quantum Theory," *Suppl. Nature* 122 (1928):887–895.

[246] *Ibid.*, p. 887.

[247] *Ibid.,* pp. 893–894.

[248] Kemble, "General Principles," (reference 241) pp. 158–160.

[249] For Dirac's paper and Joan Bromberg's and Jagdish Mehra's discussions of it, see reference 109. See also P. Jordan and O. Klein, "Zum Mehrkörperproblem der Quantentheorie," *Z. Phys.* 45 (1927):751–765; P. Jordan and E. Wigner, "Über das Paulische Äquivalenzverbot," *Ibid.* 47 (1928):631–651. This work is also briefly discussed in Jammer, *Conceptual Development,* p. 365, where the quotation appears.

[250] Quoted in Klein, "First Phase," (reference 9) pp. 38–39.

[251] *Nature* 120 (1927):737.

[252] See reference 2.

[253] See Chapter 6, section B.3.

[254] See reference 217.

APPENDIX I

Compton Scattering, Thomson Scattering, and Related Phenomena

A. Compton Scattering and Thomson Scattering

The basic features of Compton scattering are sketched in Chapter 6, as well as in any number of physics textbooks.[1] The salient facts are as follows: (a) The incident photon (of energy $h\nu_0 = hc/\lambda_0$), as a consequence of colliding elastically with a "free" electron at rest (energy and momentum being conserved in the process), is scattered through an angle θ and experiences a change in wavelength given by

$$\Delta\lambda = \frac{h}{mc}(1 - \cos\theta), \tag{6.18}$$

where m is the rest mass of the electron. (b) The recoil electron is scattered through an angle ϕ and possesses kinetic energy

$$T = \frac{2\,\alpha^2\,mc^2}{1 + 2\alpha + (1 + \alpha)^2 \tan\phi}, \tag{I.1}$$

where $\alpha = h\nu_0/mc^2$. (c) The scattering angles ϕ and θ are related by the following expression:

$$\tan\phi = -\frac{1}{(1 + \alpha)\tan(\theta/2)}, \tag{6.26}$$

which shows that whereas the incident photon may be scattered through any angle whatsoever ($0 \leqq \theta \leqq \pi$), the recoil electron is confined to the forward direction ($0 \leqq \phi \leqq \pi/2$).

A more detailed discussion of Compton scattering is greatly facilitated by an article by Ann T. Nelms,[2] who analyzed the problem in depth and plotted a variety of pertinent graphs. One, for example, shows how the recoil electron acquires an ever-increasing fraction of the total energy as the energy of the incident photon increases.[3] For our purposes, however, her discussion of the relationship of Comp-

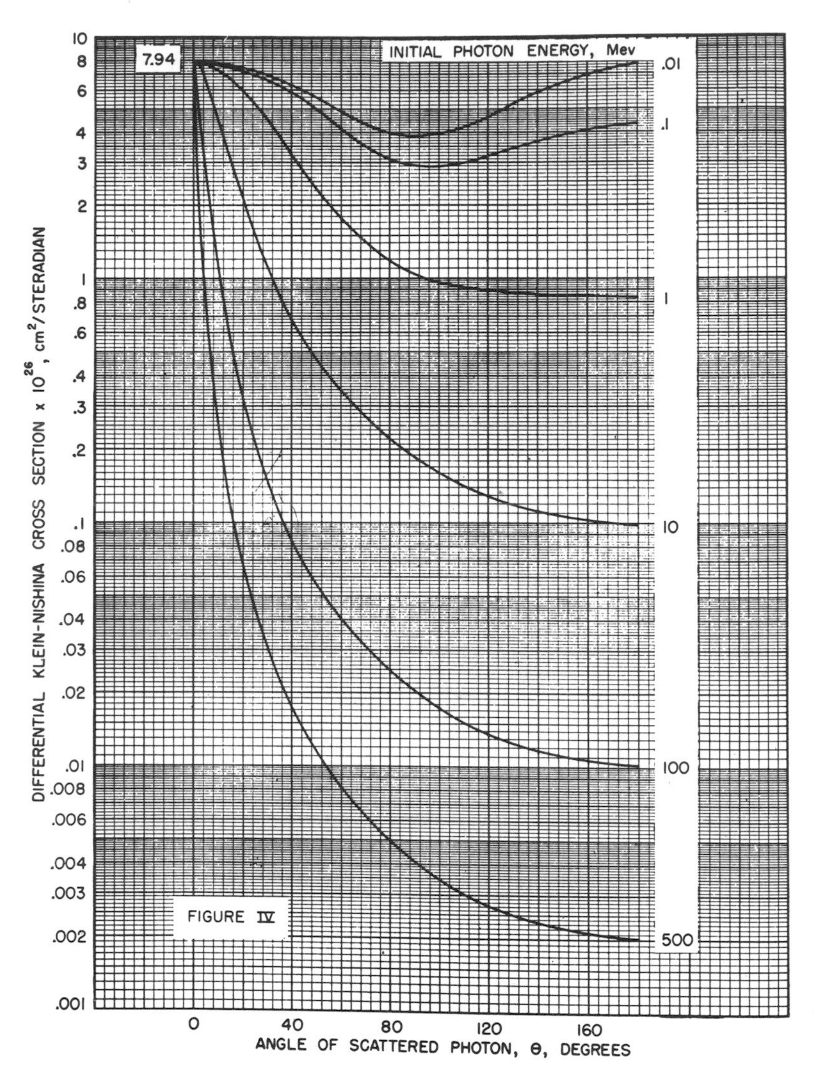

FIG. 1. Nelms' plots (her Figure IV) of $d\sigma/d\Omega_\theta$ *vs* θ for incident photon energies between 0.01 and 500 Mev.

ton scattering to Thomson scattering is most significant. This relationship may be seen most readily by considering the so-called differential scattering cross section $d\sigma/d\Omega$, which is a measure of the *intensity* of the particles (either photons or electrons) expected to be scattered into the solid angle $d\Omega = 2\pi \sin\theta d\theta$.[4] To a very good approximation ($\lesssim 1\%$ error), Klein and Nishina's 1929 analysis suffices.[5] Their result for the scattered photons, as quoted by Nelms,[6] is

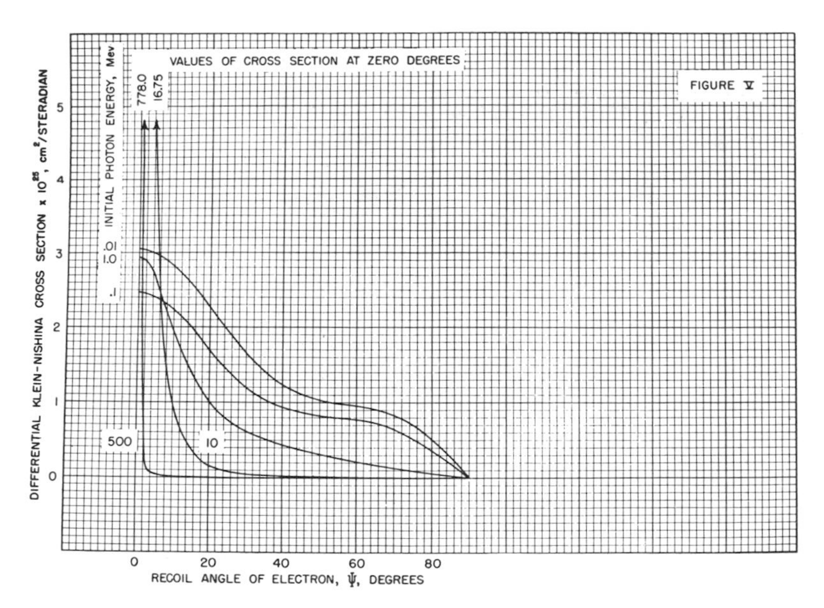

FIG. 2. Nelms' plots (her Figure V) of $d\sigma/d\Omega_\Psi$ *vs* Ψ for incident photon energies between 0.01 and 500 Mev.

$$\frac{d\sigma}{d\Omega_\theta} = \frac{e^4}{2m^2c^4}\left\{\frac{(1+\cos^2\theta)}{[1+\alpha(1-\cos\theta)]^2}\right\}\left\{1+\frac{\alpha^2(1-\cos\theta)^2}{[1+\cos^2\theta][1+\alpha(1-\cos\theta)]}\right\} \tag{I.2}$$

which is plotted for various incident photon energies in Fig. 1.[7] From these plots one can see very clearly how the asymmetrical Compton scattering goes over to the symmetrical Thomson scattering as the energy of the incident photons decreases—in the Compton scattering limit the scattered photon appears predominantly in the forward direction, while in the Thomson scattering limit as many photons are back-scattered as forward-scattered.

For the recoil electrons, which can *only* be forward-scattered, one finds[8] that

$$\frac{d\sigma}{d\Omega_\phi} = \frac{d\sigma}{d\Omega_\theta}\frac{(1+\alpha)^2(1-\cos\theta)^2}{\cos^3\phi}, \tag{I.3}$$

which is plotted by Nelms for various incident photon energies in Fig. 2[9] (note that Nelms uses the symbol Ψ instead of ϕ for the recoil electron scattering angle). From these plots one can see that as the incident photon energy increases, the recoil electrons are scattered more and more predominantly through smaller and smaller scattering angles in the forward direction.

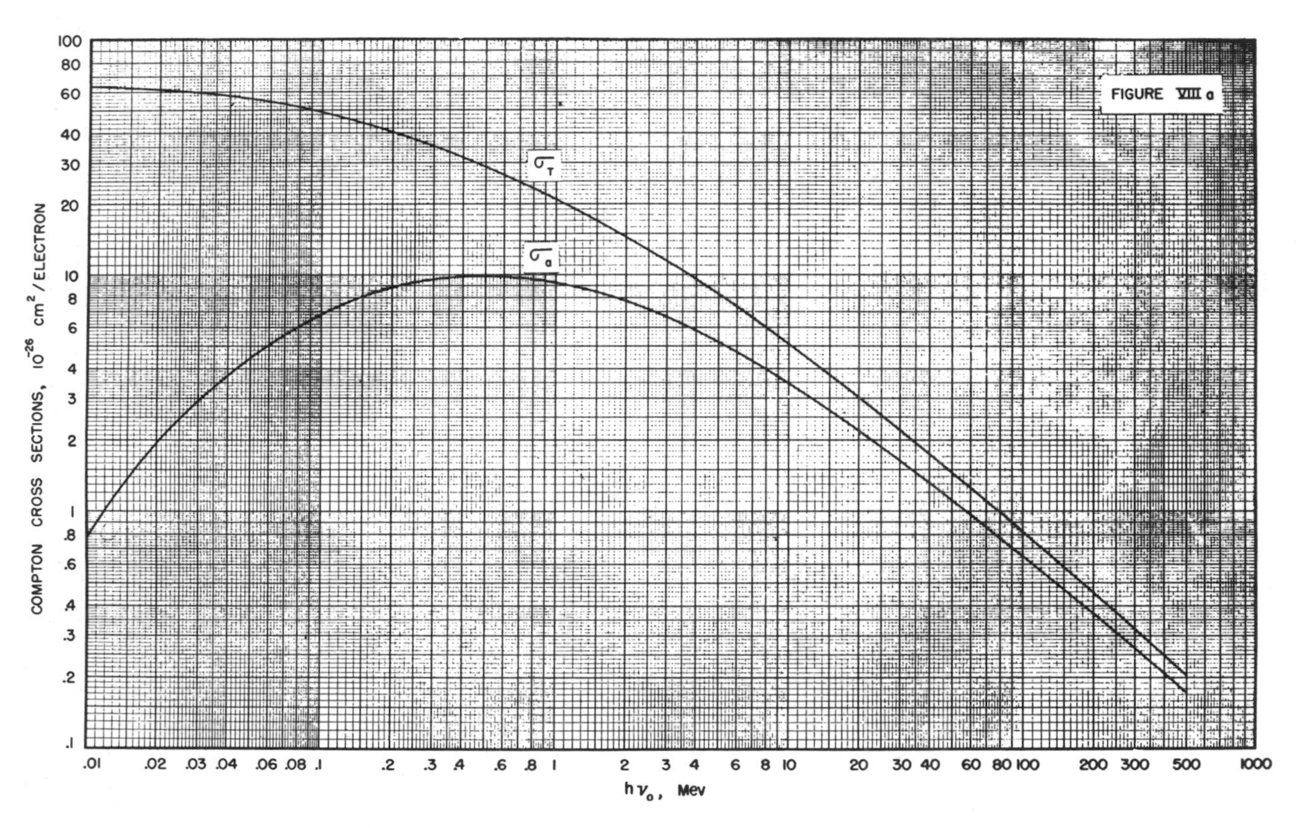

FIG. 3. Nelms' plots (her Figure VIIIa) of σ_T and σ_a *vs* incident photon energy $h\nu_o$.

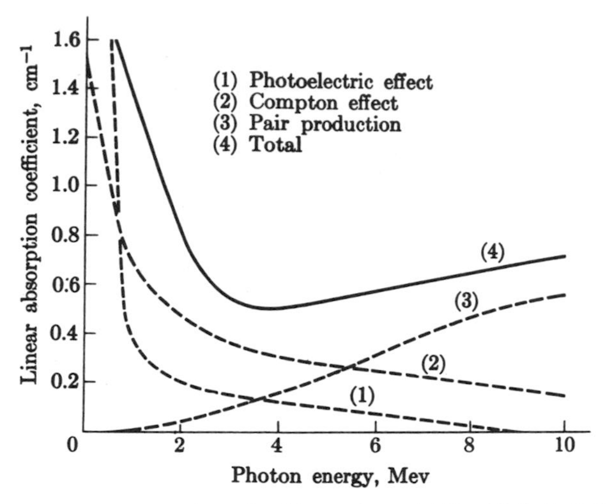

Reproduced by permission of Addison-Wesley Publishing Company, Inc.

FIG. 4. Plots of linear-absorption coefficients, for three major absorption processes, *vs* incident photon energy.

The *total* scattering cross section σ_T is obtained by simply integrating over the differential scattering cross section. For the scattered photons, one would therefore calculate

$$\sigma_T = 2\pi \int_0^\pi (d\sigma/d\Omega_\theta) \sin\theta d\theta.$$

The result is displayed graphically in Fig. 3,[10] which shows that the *total* energy absorbed through Compton scattering decreases dramatically at higher incident photon energies. The curve labeled σ_a represents the energy variation of the "true absorption cross section" $\sigma_a = f\sigma_T$, where f is the fraction of the incident energy, previously mentioned, which is actually transferred to the absorbing electron.

In general, of course, one must always realize that Compton scattering is only one of three major absorption processes that occur—the other two arise (a) from the ordinary photoelectric effect, and (b) from pair production (above 1.02 Mev). A rough idea of the relative importance of these three processes may be obtained from the plots in Fig. 4,[11] which show how the respective linear-absorption coefficients vary with incident energy. One can see that although Compton scattering is the principal absorption mechanism for incident energies between about 1 Mev and 5 Mev (0.01 Å $\lesssim \lambda_0 \lesssim 0.002$ Å), it is masked by the photoelectric effect on the low energy side and by pair production on the high energy side.

A basic assumption underlying the entire discussion above is that the struck electron is essentially "free," i.e., that it is bound so loosely to its parent atom that no energy is effectively expended in tearing it loose. This assumption has the physical consequence that no parameter characteristic of the *atom* enters into the expressions for the change in wavelength and for the differential scattering cross sections. Only m and e enter, both of which refer to the *electron.* Moreover, in Compton scattering, the change in wavelength is also found to be completely independent of the wavelength of the incident radiation; in fact, it depends *solely* on the scattering angle θ.

In Thomson scattering, the electron is also assumed to be "free"—but only in the sense that it is "free" to oscillate about a fixed position in space, in response to the force exerted on it by the electric vector of the incident radiation. Once again, this assumption has the physical consequence that no parameter characteristic of its parent atom enters into the expression for the Thomson differential scattering cross section, which is given by

$$\left(\frac{d\sigma}{d\Omega_\theta}\right)_0 = \frac{e^4}{m^2c^4}\,\frac{(1+\cos^2\theta)}{2}. \tag{I.4}$$

Moreover, since the frequencies of the incident radiation, the oscillating electron, and the scattered radiation are all precisely identical, no parameter characteristic of the *incident radiation* appears in equation (I.4) (nor of course in the corresponding expression for the Thomson total cross section).

This is in complete contrast to expression (I.2) for the Compton differential scattering cross section, where the energy (or frequency) of the incident radiation enters through the parameter $\alpha = h\nu_0/mc^2$. This contrast, in fact, enables us to understand the bridge between Compton scattering and Thomson scattering from another perspective. Thus, in Compton scattering the "freedom" of the electron is so great that it may be knocked clear out of the atom with effectively zero expenditure of energy. Consequently, Compton scattering predominates when the incident photon energy is *high,* that is, when α is large. At the other extreme is Thomson scattering, where the electron, while "free" to oscillate, nevertheless remains attached to its parent atom. Consequently, Thomson scattering predominates when the incident photon energy is *low,* that is, when $\alpha \to 0$. Indeed, one may easily see by comparing expressions (I.2) and (I.4) that

$$\lim_{\alpha\to 0}\left(\frac{d\sigma}{d\Omega_\theta}\right) = \left(\frac{d\sigma}{d\Omega_\theta}\right)_0 \text{ and hence also that } \lim_{\alpha\to 0}\sigma_T = \sigma_0.$$

Two other factors, in addition to the energy of the incident radiation, determine whether or not the struck electron is effectively "free": (1) The atomic weight (or atomic number Z) of its parent atom, and (2) the "directness" of the collision. In general, electrons in light atoms are bound less tightly than those in heavy atoms, and an electron struck "head on" will appear to be less tightly bound than one struck glancingly. The result is that the Klein-Nishina formula, equation (I.2), must be modulated by a certain factor S, which was first calculated independently by W. Heisenberg[12] and L. Bewilogua[13] in 1931. Nelms shows, for example, in the case of 300 kev photons incident on electrons in lead atoms ($Z = 82$), that this factor predicts a huge decrease in the differential scattering cross section at the smaller scattering angles.[14]

B. Absorption and Emission of X-Rays; Radioactivity

In addition to those bearing directly on the Compton effect, concepts relating more generally to the absorption and emission of X-rays, and to radioactivity, frequently appear in this book. Hence, these concepts should also be discussed briefly at this point.

If one examines experimentally the absorption of hard X-rays, one finds that the variation of the mass-absorption coefficient μ/ρ with incident wavelength λ generally obeys the equation $\mu/\rho = aZ^4\lambda^3 + b$, where a and b are constants. At certain points, however—at the so-called "absorption edges"—the curve shows sharp discontinuities. As the incident wavelength decreases (or as the incident energy increases) certain points are reached when the incident photons have enough energy to expel deep-lying electrons from progressively lower and lower energy levels in the atom, and it is at these points that the K, L_I, L_{II}, L_{III}, . . . absorption edges appear.[15] Complete tables showing the location of these absorption edges for all elements may be found, for example, in G. L. Clark's *Encyclopedia of X-Rays and Gamma Rays.*[16]

X-ray *emission* is generally produced by bombarding a substance with electrons. If the energy of the electrons is high enough, the substance will emit in general two types of X-rays: (1) "Bremsstrahlung" resulting from the rapid deceleration of the electrons, and (2) those X-rays resulting from the liberation of an electron from a low-lying atomic energy level. The former X-ray spectrum is continuous and has a high-frequency cut-off which is determined by the energy of the incident electrons; the latter spectrum, which is superimposed on the former, is a discontinuous line spectrum characteristic of the bombarded substance.[17] The principal X-ray emission lines of all elements may also be found, for example, in Clark's *Encyclopedia.*[18]

A major reason for discussing radioactivity is that many of the early experimentalists used radioactive substances, especially radium, as sources of beta and gamma radiation. Sometimes these sources were shielded and sometimes they were not, and since alpha radiation is emitted besides beta and gamma radiation, it is essential to understand at least the basic facts regarding the relative penetrating powers of these radiations.

Rough estimates of the penetrating powers of α- and β-particles may be obtained, for example, from formulae and graphs in Robert B. Leighton's *Principles of Modern Physics.*[19] A rule of thumb is that 5 Mev α-particles will penetrate about 3 cm of air, while 2 Mev β-particles will penetrate about 1.5 mm of lead. It is therefore clear, for instance, that a radium source surrounded by a lead shield a few millimeters thick will be a source of γ-rays only, since γ-rays can pass through even a few centimeters of lead, while the α- and β-particles will have been completely absorbed by the shield. These α- and β-radiation energies are typical; more detailed values for various elements may be obtained from readily available sources.[20] Similarly, one may consult, for example, C. E. Crouthamel's *Applied Gamma-Ray Spectrometry*[21] to find detailed information on the energy spectra of the γ-rays emitted by radioactive elements.

C. References

[1] For example, J. A. Richards, F. W. Sears, M. R. Wehr, and M. W. Zemansky, *Modern University Physics* (Reading, Mass.: Addison-Wesley, 1960), pp. 816–820.

[2] "Graphs of the Compton Energy-Angle Relationship and the Klein-Nishina Formula," U. S. *Natl. Bur. Std. Cir.* 542.

[3] *Ibid.*, p. 89.

[4] For a full discussion of this concept, see Herbert Goldstein, *Classical Mechanics* (Reading, Mass.: Addison-Wesley, 1950), pp. 81ff.

[5] Nelms, p. 2. For Klein and Nishina's paper, see Chapter 6, reference 57.

[6] Nelms, p. 5.

[7] *Ibid.*, p. 38.

[8] *Ibid.*, p. 6.

[9] *Ibid.*, p. 52.

[10] *Ibid.*, p. 88.

[11] Richards, *et. al.*, p. 821.

[12] Über die inkohärente Streuung von Röntgenstrahlung," *Phys. Z.* 32 (1931):737–740.

[13] Über die inkohärente Streuung von Röntgenstrahlung," *Phys. Z.* 32 (1931):740–744.

[14] Nelms, p. 3.

[15] For typical plots, see for example, Robert B. Leighton, *Principles of Modern Physics* (New York: McGraw-Hill, 1959), p. 425.

[16] (New York: Van Nostrand Reinhold, 1963), pp. 1124–1125.

[17] For a typical composite spectrum, see Leighton, p. 407.

[18] Clark, pp. 1126–28, 1130.

[19] See p. 502.

[20] See for example, Richards, *et al.*, pp. 868–870, where the old designations RaA, RaC, etc., are also conveniently identified by their modern terminology. True β-ray spectra (not internal conversion spectra) are of course continuous, while γ-ray and α-ray spectra are discrete. Maximum energies of β-rays (the largest is 3.18 MeV for RaC) from natural β-emitters are given, for example, in W. E. Burcham, *Nuclear Physics: An Introduction* (New York: McGraw-Hill, 1963), p. 54.

[21] (Oxford: Pergamon, 1960), pp. 411–414.

APPENDIX II

Abbreviations of Sources Cited

Am. Hist. Rev.—American Historical Review

Am. J. Phys.—American Journal of Physics

Ann. Physik—Annalen der Physik

Ann. Phys. (Paris)—Annales de Physique

Ann. Rept. Inst.—Annual Reports of the Smithsonian Institution

Arch. Ges. Med.—Archiv für Geschichte der Medizin

Arch. Hist. Exact Sci.—Archive for History of Exact Sciences

Arch. Néerl.—Archives Néerlandaises des sciences exactes et naturelles

Bell Lab. Record—Bell Laboratories Record

Bell System Tech. J.—Bell System Technical Journal

Ber. Akad. Wiss. (Wien)—Berichte der Akademie der Wissenschaften (Wein)

Ber. Deut. Physik. Ges.—Berichte der deutschen physikalilischen Gesellschaft

Ber. K. Bayer. Akad. (München)—Berichte der königlich-Bayerischen Akademie der Wissenschaften (Müchen)

Biog. Mem. Fell. Roy. Soc.—Biographical Memoirs of Fellows of the Royal Society of London

Brit. J. Hist. Sci.—British Journal for the History of Science

Brit. J. Phil. Sci.—British Journal for the Philosophy of Science

Bull. Natl. Res. Counc.—Bulletin of the National Research Council

Centaurus—Centaurus

Compt. Rend.—Comptes rendus hebdomadaires des séances de l'académie des sciences

Dict. Sci. Biog.—Dictionary of Scientific Biography

En. Brit.—Encyclopedia Britannica

Eng.—Engineering

Experientia—Experientia

Harvard Gaz.—Harvard University Gazette

Indian J. Phys.—Indian Journal of Physics

Internat. J. Quantum Chem.—International Journal of Quantum Chemistry

Isis—Isis

J. Am. Chem. Soc.—Journal of the American Chemical Society

J. Chem. Soc.—Journal of the Chemical Society (London)

J. Phys. Radium—Journal de physique et le radium

J. Franklin Inst.—Journal of the Franklin Institute

J. Opt. Am.—Journal of the Optical Society of America

J. Röntgen Soc.—Journal of the Röntgen Society

J. Wash. Acad. Sci.—Journal of the Washington University Academy of Sciences

Radium—Le Radium, la radioactivité, les radiations, l'ionization

Mem. Manchester Soc.—Memoirs and Proceedings of the Manchester Literary and Philosophical Society

Natl. Acad. Sci. Biog. Mem.—National Academy of Sciences Biographical Memoirs

Nat. Phil.—The Natural Philosopher

Nature—Nature

Naturwiss.—Naturwissenschaften

Nuovo cim.—Nuovo Cimento

Osiris—Osiris

Phil. Mag.—Philosophical Magazine

Phil. Trans. Roy. Soc. London—Philosophical Transactions of the Royal Society of London

Physica—Physica

Phys. Rev.—The Physical Review

Phys. Today—Physics Today

Phys. Z.—Physikalische Zeitschrift

Proc. Am. Acad. Arts Sci.—Proceedings of the American Academy of Arts and Sciences

Proc. Am. Phil. Soc.—Proceedings of the American Philosophical Society

Proc. Cambridge Phil. Soc.—Proceedings of the Cambridge Philosophical Society

Proc. Konink. Akad. Wetensch.—Koninklijke Nederlandse Akademie van Wetenschappen (Amsterdam), Proceedings

Proc. Natl. Acad. Sci. U.S.—Proceedings of the National Academy of Sciences

Proc. Phys. Soc. (London)—Proceedings of the Physical Society (London)

Proc. Roy. Inst.—Proceedings of the Royal Institution

Proc. Roy. Soc. (London)—Proceedings of the Royal Society (London)

Proc. Roy. Soc. (Edinburgh)—Proceedings of the Royal Society (Edinburgh)

Repts. BAAS—Reports of the British Association for the Advancement of Science

Rev. Mod. Phys.—Reviews of Modern Physics

Roy. Soc. Obit. Not. (London)—Obituary Notices of Fellows of the Royal Society (London)

Schriften Physik.-ökon. Ges. (Königsberg)—Schriften der physikalisch-ökonomischen Gesellschaft zu Königsberg in Preussen

Sci. Am.—Scientific American

Sci. Am. Suppl.—Scientific American Supplement

Science—Science

Sitzber. Akad. (München)—Sitzungsberichte der mathematicsch-physikalischen Klasse der königlich-bayerischen Akademie der Wissenschaften zu München

Sitzber. Phys.-ökon Ges. (Königsberg)—Sitzungsberichte der physikalisch-ökonomischen Gesellschaft zu Königsberg

Sitzber. Würz. Ges.—Sitzungsberichte der Würzberger physikmedizinschen Gesellschaft

Sitzber. Preuss. Akad. Wiss.—Sitzungsberichte der preussischen Akademie der Wissenschaft

Smith. Misc. Coll.—Smithsonian Miscellaneous Collections

Suppl. Nature—Supplement to Nature

Trans. Am. Illum. Eng. Soc.—Transactions of the American Illuminating Engineering Society

Trans. Roy. Soc. Can.—Transactions of the Royal Society of Canada

Trans. Roy. Soc. S. Aust.—Transactions of the Royal Society of South Australia (Adelaide)

U. S. Natl. Bur. Std. Circ.—United States National Bureau of Standards Circular

Wash. U. Mag.—Washington University Magazine

Wash. U. Stud.—Washington University Studies

Wied. Ann.—Wiedemanns Annalen der Physik
Z. Phys.—Zeitschrift für Physik

Acknowledgment: I have used long quotations from many of the journals listed here, and I would like to thank the appropriate editors for permission to reproduce these quotations in my book.

ILLUSTRATION ACKNOWLEDGMENTS

We should like to express our sincere appreciation for permission to reproduce a number of the illustrations in this volume. Especial thanks are due:

Mrs. Arthur H. Compton, for figures 9 and 14 in chapter 4; figure 1 in chapter 5; and figure 10 in chapter 6.

Professor A. O. C. Nier, for figure 1 in chapter 3.

Professor R. E. Norberg, for figure 5 in chapter 5 and figure 6 in chapter 6.

Dr. Ing. H. Sommerfeld, for figure 10 in chapter 6.

Addison-Wesley Publishing Company, Inc., for figure 4 in appendix I.

Proceedings of the Cambridge Philosophical Society, for figure 3 in chapter 1.

Cavendish Laboratory, Cambridge University, for figure 8 in chapter 4.

Journal of the Franklin Institute, for figure 12 in chapter 4.

Proceedings of the National Academy of Sciences, for figures 11, 12, 13, 14, 15, 16, 17, 18, 19, 20, and 21 in chapter 6.

Bulletin of the National Research Council, for figures 3, 4, and 7 in chapter 5.

Journal of the Optical Society of America, for figure 10 in chapter 4.

The Philosophical Magazine, for figures 1, 2, 4, and 5 in chapter 1 and figures 1, 2, 3, 4, and 13 in chapter 4.

The Physical Review, for figure 2 in chapter 2; figures 2, 3, 7, 8, 9, 10, 11, and 12 in chapter 3; figure 6 in chapter 5; and figures 1, 2, 3, 4, 5, 7, and 8 in chapter 6.

Physikalische Zeitschrift, for figures 6 and 7 in chapter 1.

Proceedings of the Royal Society of London, for figure 9 in chapter 6.

Washington University Studies, for figures 5, 6, and 7 in chapter 4.

Journal of the Washington University Academy of Sciences, for figures 4, 5, and 6 in chapter 3.

United States National Bureau of Standards Circular, for figures 1, 2, and 3 in appendix I.

Name Index